Global Water Ethics

Scholarly interest in water ethics is increasing, motivated by the urgency of climate change, water scarcity, privatization and conflicts over water resources. Water ethics can provide both conceptual perspectives and practical methodologies for identifying outcomes which are environmentally sustainable and socially just. This book assesses the implications of ongoing research in framing a new discipline of water ethics in practice.

Contributions consider the difficult ethical and epistemological questions of water ethics in a global context, as well as offering local, empirical perspectives. Case study chapters focus on a range of countries including Canada, China, Germany, India, South Africa and the USA. The respective insights are brought together in the final section concerning the practical project of a universal water ethics charter, alongside theoretical questions about the legitimacy of a global water ethics.

Overall the book provides a stimulating examination of water ethics in theory and practice, relevant to academics and professionals in the fields of water resource management and governance, environmental ethics, geography, law and political science.

Rafael Ziegler is Head of Research at GETIDOS (Getting things done sustainably), based in Greifswald, Germany. He has worked as a lecturer at McGill University, ECLA and FU Berlin, and as a Deputy Professor of environmental ethics at the University of Greifswald.

David Groenfeldt is Director of the Water-Culture Institute in Santa Fe, New Mexico, and an Adjunct Associate Professor of Anthropology at the University of New Mexico, Albuquerque, USA. He is the author of *Water Ethics: A Values Approach to Solving the Water Crisis* (Routledge, 2013).

Earthscan Studies in Water Resource Management

Rules, Norms and NGO Advocacy Strategies
Hydropower Development on the Mekong River
Yumiko Yasuda

The Mekong: A Socio-legal Approach to River Basin Development
Ben Boer, Philip Hirsch, Fleur Johns, Ben Saul and Natalia Scurrah

Trans-jurisdictional Water Law and Governance
Edited by Janice Gray, Cameron Holley and Rosemary Rayfuse

Water and Rural Communities
Local Politics, Meaning and Place
Lia Bryant with Jodie George

Land and Hydropolitics in the Nile River Basin
Challenges and New Investments
Edited by Emil Sandström, Anders Jägerskog and Terje Oestigaard

Legal Frameworks for Transparency in Water Utilities Regulation
A Comparative Perspective
Mohamad Mova Al'Afghani

Water Regimes
Beyond the Public and Private Sector Debate
Edited by Dominique Lorrain and Franck Poupeau

Negotiating for Water Resources
Bridging Transboundary River Basins
Andrea Haefner

Environmental Water Markets and Regulation
A Comparative Legal Approach
Katherine Owens

The Politics of Fresh Water
Access, Conflict and Identity
Edited by Catherine M. Ashcraft and Tamar Mayer

Global Water Ethics
Towards a Global Ethics Charter
Edited by Rafael Ziegler and David Groenfeldt

Water, Knowledge and the Environment in Asia
Epistemologies, Practices and Locales
Edited by Ravi Baghel, Lea Stepan and Joseph K.W. Hill

For more information and to view forthcoming titles in this series, please visit the Routledge website: www.routledge.com/books/series/ECWRM/

Global Water Ethics

Towards a Global Ethics Charter

Edited by Rafael Ziegler and David Groenfeldt

Routledge
Taylor & Francis Group
LONDON AND NEW YORK
earthscan
from Routledge

First published 2017 by Routledge

2 Park Square, Milton Park, Abingdon, Oxfordshire OX14 4RN
52 Vanderbilt Avenue, New York, NY 10017

Routledge is an imprint of the Taylor & Francis Group, an informa business

First issued in paperback 2019

British Library Cataloguing-in-Publication Data
A catalogue record for this book is available from the British Library

Library of Congress Cataloging in Publication Data
Names: Ziegler, Rafael, editor. | Groenfeldt, David, editor.
Title: Global water ethics : towards a global ethics charter / edited by Rafael Ziegler and David Groenfeldt.
Description: Abingdon, Oxon ; New York, NY : Routledge, [2017] | Series: Earthscan studies in water resource management | Includes bibliographical references and index.
Identifiers: LCCN 2016042619| ISBN 9781138204294 (hbk) | ISBN 9781315469690 (ebk)
Subjects: LCSH: Water supply—Management—Moral and ethical aspects. | Water quality management—Moral and ethical aspects. | Water resources development—Moral and ethical aspects. | Water use—Moral and ethical aspects. | Environmental ethics.
Classification: LCC TD353 .G57 2017 | DDC 178—dc23
LC record available at https://lccn.loc.gov/2016042619

ISBN: 978-1-138-20429-4 (hbk)
ISBN: 978-0-367-22681-7 (pbk)

Typeset in Times New Roman
by Swales & Willis Ltd, Exeter, Devon, UK

Contents

Illustrations

Figures

Tables

Box

Preface

Our present epoch, the Anthropocene, had its beginnings in the industrial revolution and its conquest of nature. Decisions about rivers were made by engineers and political leaders. Napoleon's conquests of German kingdoms, ecclesiastical territories and free towns permitted engineer Johann Gottfried Tulla to 'straighten' the 'wild' Rhine in the nineteenth century and turn it into one of Europe's biggest 'water streets'. It is probably safe to assume that no professional ethicists were consulted. A century later, one of the most famous photographs of post-World War II Germany shows West German Minister of the Environment Klaus Töpfer jumping into the Rhine, demonstrating with his own body that after a long period of decline, the water quality of the Rhine had improved. The minister wanted to make a point about how we (should) think about rivers. Indeed, in the early twenty-first century the European River Swimming day 'Big Jump' has become an international institution celebrating river protection – meanwhile worshippers continue bathing in the Ganges and other holy rivers, however polluted they have become. While river quality and water management might improve in some places, economic globalization and industrialization is putting serious pressure on other places. Even within Europe, the status of the Rhine and other rivers improves, while the energy thirst of economic development provokes a hydro-power boom in the Balkan countries and Turkey that puts some of the last 'wild' rivers of Europe at threat of extinction. Thus, the short history of the Anthropocene includes multiple, seemingly contradictory histories and ways of relating to water.

And yet, there are interconnections. There is a growing awareness across the globe about the need for caution in the way we intervene in natural water ecosystems. Can water ethics deepen and explicate this awareness? Is there a need for 'global water ethics'? As we will discuss in the introductory chapter, there are economic, political, environmental and scientific reasons for taking 'global water ethics' seriously. However, this point is not to be confused with the claim that there is somehow a global guideline for water use. To be sure, there are a great many guidance systems available already, catering to the values of their proponents. Economists favour cost–benefit analysis; social activists emphasize human rights and water justice; environmentalists would like to see more protected areas; and scientists recently argued that we need to stay within planetary boundaries, including for freshwater. Everyone, it seems, has an agenda. How can we, and how *should* we, assess the merits of one agenda over another?

The philosophical subfield of ethics deals with such quandaries, and the particular theme of environmental ethics has grappled with the relationship between humans and what is labelled as the 'natural' world. This book explores that relationship through an overview of the philosophical literature relevant to water ethics (Part I), ethical questions of water use in a globalized world (Part II) and varieties of water ethics charters (Part III). So while this book does not offer a global guideline for water use, it does provide philosophical tools and ideas for thinking about the issues of water in a globalized world, and it suggests some normative guidelines and methods for water charting.

It is fitting that this book originates from a philosophical conference on the theme of environmental ethics. In July 2015, the 12th conference of the International Society for Environmental Ethics (ISEE) meeting in Kiel, Germany included a special set of three sessions on the topic of water ethics, organized by the editors of this book. Setting aside multiple sessions on a seemingly marginal topic such as 'water ethics' was unheard of in the history of the ISEE conferences. We would like to thank the conference organizers, notably Konrad Ott and his team, for their willingness to provide the opportunity for an extended exploration of water ethics within the context of the conference.

The three sessions at ISEE 2015 confirmed that water ethics has become a compelling topic for an increasing number of scholars from environmental ethics and interdisciplinary researchers from anthropology, hydrology, geography, law, management studies, philosophy and political science to name just the main disciplines contributing to this book. This interest in water ethics flows both from academic curiosity and from practical problems. The overall theme of the ISEE conference was 'Environmental Ethics between Action and Reflection', which nicely captures the dynamic role we are proposing for the emerging sub-field on water ethics. We aim to be firmly grounded in foundational questions and reflections about water and at the same time applying these philosophical perspectives to practical issues, questions and dilemmas about water use and access, and our relationship to water ecosystems.

Both water ecosystems and human social systems are at the same time local and global, and everything in between. Turmoil in Syria can affect Europe, Asia and the Americas; local water supplies in Peru are used to grow asparagus and produce copper, both traded globally. The global water crisis is simultaneously local and global and the ethics of addressing water governance also needs to be understood at a range of scales. This book has as a central theme the notion of Water Charters and water charting as a way to implement water ethics. Water charters constitute a special kind of boundary object and boundary practices that constantly lead to new questions, while at the same time the practical goals of a town, a people, a profession or a community are constitutive for the charting effort. Thus, we hope that this volume will provide a fresh perspective on water ethics – 'between action and reflection'.

Most of the chapters of this book were first presented at the ISEE 2015 and the joint discussion of water ethics at this lively conference. They were further developed via two rounds of comments from the editors in fall 2015 and spring 2016. In addition, Dorothea Härlin and Adrian Sym joined the book effort later on, providing, respectively, the case study of Berlin's Water Charter, and a presentation of the business-oriented Water Stewardship Standard.

We would like to thank all contributors for actively participating in this process of book creation, and Mara Nothers for compiling the index. We hope this book can contribute to an improved understanding of our relation to water and of the ways we live with water. The shortcomings of current water governance render water ethics and water management a multi-generational task par excellence: for the young to learn about water as a cultural heritage (and the many ways in which aspects of this heritage have been marginalized), for their elders to learn about water as something that we borrow from the future – and for both to think about water as something that we share with the rest of life on the 'blue' planet.

Through reflecting on the 'boundary' between water's past and future, the task of editing this book has provided us as editors many opportunities to discuss water ethics in Skype conversations across the Atlantic, often in the early morning or late evening. We would like to thank our families for their tolerance for these odd hours, and all the support in between!

Rafael Ziegler and David Groenfeldt
Berlin and New Mexico, July 2016

Contributors

Maite Martínez Aldaya
Universidad Pública de Navarra
Departamento de Proyectos e Ingeniería Rural
Ed. Los Olivos 2ª planta
Campus de Arrosadia
31006 Pamplona, Navarra, Spain

Professor Petra Döll
Institute of Physical Geography
Goethe University Frankfurt
Altenhöferallee 1
60348 Frankfurt am Main, Germany

Professor Eran Feitelson
Department of Geography
The Hebrew University of Jerusalem
Mount Scopus
Jerusalem 91905, Israel

Professor Dieter Gerten
Potsdam Institute for Climate Impact Research (PIK)
Research Domain I: Earth System Analysis
Telegrafenberg A62
D-14473 Potsdam, Germany

Dr David Groenfeldt
Director, Water-Culture Institute
1021 Camino Santander
Santa Fe, NM 87505, USA

Dorothea Härlin
Graefestr.61A
10967 Berlin, Germany

M.A. Eleanor Hayman
Chair for Geography and Geographical
Remote Sensing
LMU München
Luisenstraße 37
80333 München, Germany

Xiang Huang
Department of Environmental
Science and Engineering
Zhongnan University of Economics and Law
503 Wenyong Building
No.182 Nanhu Avenue
East Lake High-Tech Development Zone
Wuhan 430073, China

Dr Angela Kallhoff
Universitätsprofessorin
Universtität Wien
NIG, 3.Stock, A0306
Universitätsstr. 7
1010 Wien, Austria

Dr Martin Kowarsch
Mercator Research Institute on Global
Commons and Climate Change (MCC)
Torgauer Str. 12–15
10829 Berlin, Germany

Professor M. Ramón Llamas
Director Water Observatory Botin Foundation
Department of Geodynamics
Complutense University
28040 Madrid, Spain

Jie Liu
Institute of Water Sciences
Peking University
1004 Wangkezhen Building
No. 5 Yiheyuan Road
Haidian District
Beijing 100871, China

Pedro Martínez-Santos
Profesor Titular de Universidad
Departamento de Geodinámica
Facultad de Ciencias Geológicas
Universidad Complutense de Madrid
28040 Ciudad Universitaria
Madrid, Spain

Dr Simon Meisch
Head of Junior Research Group 'Ethics of Science in the Research for Sustainable Development'
International Centre for Ethics in the Sciences and Humanities (IZEW)
Eberhard Karls University Tübingen
Wilhelmstr. 19
72074 Tübingen, Germany

Dr Daniel Mishori
The Porter School of Environmental Studies &
The Department of Geography and the Human Environment
Tel Aviv University
PO Box 39040
Tel Aviv 6997801, Israel

Lucy Rodina
Institute for Resources, Environment and Sustainability
The University of British Columbia
Vancouver Campus
AERL Building
429-2202 Main Mall
Vancouver, BC, Canada V6T 1Z4

Susan Lea Smith
Professor of Law
Director, Sustainable Environmental, Energy and Resources Law
Willamette University
College of Law
245 Winter St. SE
Salem, OR 97301, USA

Adrian Sym
Alliance for Water Stewardship (AWS)
Edinburgh Centre for Carbon Innovation
High School Yards, Infirmary Street
Edinburgh EH1 1LZ, Scotland, UK

Dr Rafael Ziegler
Head of Social-Ecological Research Platform GETIDOS
Ernst-Moritz-Arndt University Greifswald
Soldmannstraße 15
17489 Greifswald, Germany

1 Introduction

Global water ethics: towards a water ethics charter

Rafael Ziegler and David Groenfeldt

The Earth as seen from space shows a finite and fragile entity in a dark universe. It shows the Earth as our home planet. The 'blue marble', as the iconic Apollo 17 image of the Earth from 1972 has come to be known, promised an important turning point for humanity; and NGOs, governments and business were quick to draw on the power of this image (Sachs 1999). One might imagine that the image of the fragile blue marble floating in space would foster a consciousness of humans having a shared destiny and responsibility on a finite planet, and lead to a consciousness of responsibility for this blue planet and its waters.

This has not happened. Along with forests, soils and swamps, waters have been relentlessly exploited. While the satellite perspective of the blue marble has fostered a lively discussion of human impacts on the planet, and the recognition of these impacts as constituting a new geological era called the Anthropocene, the popular discussion about this new era has given scant attention to the implications for social justice and environmental sustainability. Rather, the Anthropocene is defined via quantitative thresholds, where people have come to dominate nature rather than the other way around. The playful and humble metaphor of the blue marble has been replaced by the image of a 'safe operating space' for humans (Steffen et al. 2015).

Water ethics embraces both the marble and the operating space, in the sense that the authors of the following chapters are concerned about people and the space where human societies and economies can thrive, but also recognize a larger responsibility to the planet's species and ecosystems (the marble). The challenge we face in this human-dominated space of the Anthropocene is to recognize a larger sphere of ethical responsibility, particularly because we humans have so much power. From the discussion of climate change, we know that the world is far more closely inter-connected than our laws recognize. Contamination that affects the entire planet is inadequately regulated by national laws. Thus it is hardly a surprise that the Anthropocene discussions are mostly triggered by the 'atmospheric Anthropocene' and climate change (Preston 2015). A growing policy discourse has developed about alternative forms of global governance responses to climate change, ranging from global caps on emissions to climate engineering solutions.

What about water? The challenge of protecting freshwater is even more difficult than protecting the atmosphere, because freshwater sources are local. For Americans, the drying up of a third of China's rivers has no direct impact, whereas the whole world is affected even by localized CO_2 emissions. The atmosphere is very effective at spreading the CO_2 around the planet, albeit with disproportionate effects at the poles. By contrast, freshwater management belongs more to a 'terrestrial Anthropocene'. Drying up a river, or restoring it, happens at a specific place, in a river basin, but there do not seem to be evident planetary implications of these processes.[1] Should stewards of land and water therefore resist the image and the narrative of the Anthropocene?

In spite of a rapidly growing literature exploring diverse perspectives, methods and theories about water governance, the topic of water ethics has been largely missing from the discourse at all levels – locally, nationally or globally. Our purpose in compiling this book is to help fill this gap. We seek to show that there is an ethical dimension to water governance, and that attention to the ethical use, management and governance of water is fundamental to the transformation we need for attaining the goal of sustainable water governance. Through an exploration of water ethics and the inter-connections of multiple values, the domain of water governance will become broader, more interesting and more effective.

Global water ethics: some clarifications

What is global water ethics?[2] In this section, we disentangle the question into its constituent three words, and discuss what we mean by 'ethics', and more specifically 'water ethics' and, finally, 'global water ethics'.

Ethics refers to the study of practical reasoning and its concepts of 'good', 'right', 'virtue', etc. Descriptive ethics accounts for the specific conceptions and use of practical reasoning in a place or context: the ethics of the Athenians and Spartans, the ethics of water engineers and of anti-dam activists, etc. Normative ethics analyses and contributes to the justifications provided for a specific ethic, view of the good life, of justice, etc. In practice, descriptive and normative ethics are interlinked. Normative ethics typically draws on ethical intuitions and ideas of rationality of a place or tradition at least as a starting point of discussion.

Water ethics is the study of practical reasoning with a focus on human practices and water. To this end, water ethics focuses on domains of human action such as drinking water and hygiene, irrigation and agriculture, environmental and nature conservation; it analyses the way concepts of the good and the right inform these practices, not just taking them as a 'given' but also questioning them, asking for the justification and possibly change of some practices.[3] Environmental ethics adds to water ethics a specific concern: should practical considerations take water to be there as something for humans only or should other living beings also be considered? Do rivers and lakes, and water itself, have intrinsic value and not just instrumental value for others? While an anthropocentric focus prevails in many discussions of water ethics, the major positions in environmental ethics all have implications for water ethics.[4] So as to overcome a shallow anthropocentrism, it is important to differentiate the different uses and users of water. Eran Feitelson proposes to speak of 'waters' rather than 'water' so as to better differentiate human water needs and wants, ecosystem needs, community needs, etc. (Feitelson in this volume).

Global water ethics finally suggests a specific, emerging focus in water ethics that traditionally, and for many goods reasons, has focused on places, particular watersheds and river basins, practices of agriculture, spiritual and recreational human practices around rivers, lakes and springs, etc. Yet, 'in the world of today, people in Japan indirectly affect the hydrological system and people in Europe indirectly impact the regional water system in Brazil', as Arjen Hoekstra and Ashok Chapagain observe in their book on the globalization of water (Hoekstra & Chapagain 2008, p. 1). There are at least two aspects to this use of globalization: a focus on interdependent processes and effects across the global, and a historical diagnosis that something new and different is happening 'in the world of today'. In a phrase coined by Tony Allan, the watershed is not necessarily the problemshed (Allan 2001). Globalization scholars define globalization as 'a process (or set of processes) which embodies a transformation in the spatial organization of social relations and transactions – assessed in terms of their extensity, intensity, velocity and impact – generating transcontinental flows or interregional flows and networks of activity, interaction, and the exercise of power' (Held et al. 1999). While trade and economic globalization are a main aspect of these processes, political, social, cultural and environmental aspects co-shape globalization.

On this basis, we can further clarify the focus of global water ethics. Global water ethics is not a variety of the idea that globalization is 'good'. Rather, it opens a space for discussion of the global aspects of human water use, their ethical aspects and the governance implications this might have. This point is apparent already in the different outlooks adopted in the globalization discussion more generally. Globalization scholars (Held et al. 1999) identify *hyperglobalists*, who typically with a focus on economic globalization perceive the opportunity of a new global age that will benefit all, for example, due to efficient trade in virtual water. By contrast, they also identify *sceptics* who acknowledge the presence of globalization trends, but evaluate them negatively due to increasing inequality, absent global democracy and a threat to the diversity of highly adapted lifestyles around the world. According to the sceptics, states should be strengthened. With a view to water, this position is likely to call for a focus on the governance of local watersheds and the protection of national standards in water legislation. Finally, *transformationalists* seek a balance between the need to find appropriate institutions for global challenges while acknowledging and appropriately reconfiguring local institutions.[5]

Global water ethics is not an exclusively economic question. While especially trade is no doubt a main driver also of water challenges, further differentiations are needed so as not to render invisible cultural, social, political and environmental aspects of globalization that sometimes are affected by economic globalization, that sometimes shape economic globalization themselves, and that sometimes might be simply relatively independent. There are different ways of doing this. David Groenfeldt (2013) distinguishes environmental, social, cultural and economic values in water ethics. If we think that justice is likely the major normative concern for global water ethics, then also more specific distinctions might be appropriate. For example, Angela Kallhoff distinguishes distributive, ecological, cultural and procedural justice (Kallhoff 2014), and yet others draw on the concepts of dignity and the internally plural notion of capabilities to discuss the complex normative aspects of water management (Ziegler et al. 2014; Meisch in this volume). These distinctions are closely linked to the 'waters' mentioned above. Either way, a key point is not to render invisible the different aspects of global water ethics.

Global water ethics is not to be identified with a specific stance in the ethical debate of universalist versus particularist, communitarian or place-based ethics. There are a number of practical initiatives aimed at formalizing a set of universal ethical principles regarding water, including the Global Water Ethics Charter (Groenfeldt in this volume), the Water Stewardship Standard (Sym in this volume) and the Water Justice Framework (Smith in this volume). There are universalist, theoretical perspectives (Meisch; Ziegler et al. in this volume). But there are also particularist (Levi and Mishori in this volume) and pragmatic perspectives (Kowarsch in this volume). Whereas universalists might find it quite 'natural' to approach global issues, they face the de facto plurality of views on justice and of the good in relation to water around the world. Whereas particularists might find it more 'natural' to focus on the local watershed or river basin and the variety of approaches in different places, they face the challenge of articulating ethical responses to globalization trends and their effects.

The need for global water ethics

With these qualifications in mind, we can ask why a *global* water ethic is needed. We isolate for analytical purposes some key reasons, while recognizing that they are interrelated in practice.

A first reason is the fact of economic globalization, and specifically the water dimension of trade. According to water footprint accounting, the global annual average footprint (1996–2005) was 9087 gm^3/year (Hoekstra & Mekonnen 2012): about 92% of the water footprint are due to agricultural activities; about one-fifth of the global water footprint results from production for export; national water footprints differ considerably with consumers in the United States, for example, having a footprint in this period of almost triple the size of consumers in India and China. Such estimates indicate that national or regional consumer patterns are not just life-style questions but directly affect production and consumption elsewhere (and so do the supply-chains in production that enable this consumption). They call for an ethical evaluation of the challenge of water scarcity (Postel 2013) that the footprint estimates as such cannot deliver (see the chapter by Aldaya et al. in this volume).

The estimates indicate a further argument for global water ethics: the extent and the dynamic of trade in virtual water suggest the insufficiency of purely watershed approaches to sustainable water management. Assume a country or a region adopts a sustainable water management in its region (rainwater harvesting, groundwater recharge, good environmental flow of rivers, no water pollution, etc.). Yet agricultural or industrial production moves to other regions of the world, where water management is unsustainable. Products are imported back and the overall result is prima facie hardly sustainable from a global perspective.

The phenomenon of large-scale land acquisition further underscores this point and points to a political economy dynamic beyond trade. According to the global land matrix data base, as of 2014 there are 986 known *land deals* covering a territory of 373,673 km^2, more than the entire territory of Germany. The land matrix project cautiously defines a land deal as a change in ownership or control of an area larger than 2 km^{2,} since 2002, and with the primary purpose of agricultural, forestry or industrial production in countries with low or medium income, including a potential change of land use from decentralized local production and use to large-scale commercial use. Such land deals are pursued by investors (investment funds, food industry and states). The change in land use for export of products such as food and biofuels is frequently water intensive and has provoked the discussion of 'water grabbing'. Mehta et al. (2012, p. 197) define 'water grabbing' as a situation 'where powerful actors are able to take control of, or reallocate to their own benefits, water resources already used by local communities or feeding aquatic ecosystems on which their livelihoods are based'. Thus, the political economy of globalization provides a first set of reasons for a global water ethic with trade as a key dynamic, but not to be identified with a narrow focus on efficient markets. Political power and the outlooks and interests of states are just as important.

In this light, it is not surprising that there are also multiple 'default' systems of 'global water ethics' being promoted as an implicit feature of water governance initiatives. Chief among these is integrated water resources management (IWRM), the dominant global paradigm for defining good water governance (Aldana et al. in this volume). Thus, a second political reason for global water ethics is due to the need to make explicit the ethics of such default paradigms (Brown & Schmidt 2010). Are these the ethics that we – and the 'we' needs to be defined – wish to advance through our governance of water? Can some consensus be discovered in these default systems (Groenfeldt 2013)? Do some ethical themes remain marginalized or even excluded in the discussion (Conca 2006)?

A third reason for a global focus on water ethics is climate change. Via emissions, resulting temperature increases and changes in the climate, consumption in one place affects water and its management elsewhere as illustrated by the discussion of adaptation to climate change, which in many ways is a water management discussion: floods and droughts are the widely discussed extremes of a worldwide need to adapt water management to a changing climate. A global water–climate ethics question accordingly concerns the responsibility for mitigating the effects, and for dealing with the consequences of climate change. In addition, climate change policy provides new reasons for 'old' water management approaches. Watershed development and rainwater harvesting in arid regions receive a renewed importance, though in the light of an external pressure that watershed development itself cannot control. Much more controversially, large dams are legitimated as renewable, climate-friendly hydropower.

The last point suggests a fourth reason: aquatic ecosystems are among the most pressured ecosystems worldwide with specifically high risks to species and their extinction (Vörösmarty et al. 2010; Cuttleod et al. 2011). The extinction of species as an ethical issue in itself as well as a loss of options for future generations points to the relevance of discussions of global water ethics.

Finally, a fifth reason for a global water ethic is the growing awareness of human agency at a global scale and our ability to measure our ever-increasing planetary footprint. The image of the 'blue planet' and other visualizations of the Earth speak to the emergence of a variety of methods of global data gathering, mapping and image-making that over the last decades have allowed an increasingly detailed astronaut's perspective (Sachs 1999) on planet Earth. This has changed the perception of the Earth, including increasingly detailed knowledge about the limits of water resources on a finite planet (Schmidt & Peppard 2014, referring to the study of Schiklomanov et al.). In combination with the ICT revolution and the massive increase in data and data availability, there arises a 'sense-making' need: what are we to make of all these facts about the blue planet and its complex interactions? What are their implications for human practices on a shared planet? Probably, the most well-known recent articulation of this sense-making need is the discussion of the Anthropocene as a putative new, geological epoch – and contested narrative of exit from the Holocene.

Charting moments: global water ethics in practice

This takes us back to our starting point: that 'the blue planet' has become the iconic image of the hope for a more sustainable development, for the realization that current ways in which economic globalization affects societies and the environment are unsustainable, and that there are multiple crises across the planet. For freshwater, the main challenge is to live, and live well, with limited freshwater supplies while supporting a growing human population and growing global economy. 'Water scarcity' is the economic term for this global challenge.

One type of creative response to a crisis is to reflect on ethics. The Socratic reflection on Athenian politics and the rich ethical reflections it spawned in the Platonic, Aristotelian and Stoic schools of thought is just one famous example. Societies in crisis can attempt to spell out ethical principles to guide behaviour, be it 'new principles' or re-discovering the tradition (Daly & Cobb 1994; Levi and Mishori in this volume). From the Bible's Ten Commandments, Buddhism's Four Noble Truths, the Magna Carta to the Earth Charter, history shows charters (in a wide sense) as creative responses to societal challenges.

The features of these charters are that they can be stated in clear language, the ethical principles are concise, and compliance is sought in the first and most important place by the voluntary agreement of individuals. Dialogue and education are therefore central for charters. Charters might be codified by law, enforced by powerful institutions and actor networks, but ultimately they call for confirmation via reflected agreement or further refinement based on the questioning of all involved. For this reason, *charting* is even more important than charters; it creates space for reflection on values, their relation to practices and to waters that broadens water governance and the choices and options it includes once properly considered. This point is illustrated well by the practice of counter-mapping and collaborative charting of waterways that Eleanor Hayman writes about in this volume. A further example are the reflections of the Berlin Water Table members about what principles ought to govern urban water management, as discussed by Härlin in this volume. In both these cases the local reflection very much includes situating oneself in a global context – not only because global development affects the local, but also because local actors seek to co-shape these global developments.

Accordingly, this book tracks 'charter moments' as a moment of reflection in global water governance and its multiple challenges.[6] Currently water ethics is even incorporated into a global 'water ethics charter', which intends to be universal in scope and relevant to people and cultures around the planet (see the chapter by Groenfeldt in this volume). The process of developing this charter pays attention to geographic, cultural, gender and age diversity as well as disciplinary perspectives. Groenfeldt emphasizes the process aspect of this global chartering project, and the complementarity of this effort with local water ethics charters that could eventually draw from a systematic global effort for their inspiration, while the global process continuously needs to be responsive to the local efforts, as well as more actor specific efforts (such as charters for businesses, for specific religions and so on). Here we document this process, and situate it vis-à-vis other charting efforts from civil society, business and religion.

The chapters

Chapter 2 by Susan Smith provides the historical background for this book. It begins with the most influential initiatives for global water ethics: those undertaken by the United Nations (UN). From the informal acknowledgement of the human right to water for almost a half century to the formal recognition of an international human right to water and sanitation in 2010, from the water-related Millennium Development Goals in 2000 to the adoption of the Sustainable Development Goals in 2015, the UN has sparked discussions on ethics and water.

However, these discussions typically draw on specific issues and political problems, whereas more general reflection on ethics and philosophy are either missing or very much in the background. The next chapters help fill this gap. In Chapter 3, Simon Meisch introduces important distinctions and considerations from meta-ethics and normative ethics. He identifies water ethics as an emerging field in application-oriented ethics, and proposes that the theory of sustainability and a focus on dignity and human capabilities should take a central place to address and deal with the many uncertainties of water management. In Chapter 4, Martin Kowarsch complements the philosophical background with considerations from the philosophy of science. He argues that the philosophy of pragmatism – as developed by John Dewey – provides a methodological approach to water problems that steers away from both abstract universalism and from ethical relativism. When thinking about ends, it is key, he argues, not to separate them from the consideration of means, and more specifically a broad range of practical implications of different means for achieving alternative policy objectives for water management. In a further chapter reflecting on the use of philosophical considerations for water ethics, Angela Kallhoff specifically focuses on water co-operation. The theory of water co-operation, she argues, rests normatively on a set of shared values and epistemologically on theories of joint agency. On this basis, we can understand how communities are able to develop plans for improved practices of water usage based on a shared goal understanding. In a further contribution to this part of the book, David Groenfeldt provides the link to decision-making and water governance. One way to think more generally about the role of ethics and philosophy in water management is precisely to note its role of making values explicit. While value-blind management might be an acceptable norm in circumstances of abundance, Groenfeldt argues that the need for reflection on co-operation precisely arises where distributive conflicts are the norm. In such circumstances, water ethics is important to understand priorities and then to develop appropriate strategies to bring our behaviour in line with our values.

Part II of the book turns to more specific normative concepts and positions in water ethics. Chapter 8, by Maite Aldaya, Pedro Martínez-Santos and M. Ramón Llamas, examines integrated water resources management (IWRM), and thus a central concept of water governance. Making ethical aspects of this concept explicit and linking them to specific instruments for the analysis of these values in relation to water uses, provides ways to understand the dynamics within the notoriously vague IWRM. The authors argue that the combination of explicit values and analytical instruments such as the water footprint are needed to achieve a more just and sustainable water management. While their chapter draws on water use in Spain to illustrate the argument, Jie Liu and Xiang Huang in Chapter 11 turn to competing water demands in the Heihe River Basin of northwest China. Water resource allocation, they argue, is the redistribution of the water value and this typically leads to conflicts of interest between various social groups as well as political agencies. Water ethics, in their view, provides 'soft pathways' for water management, if it helps to integrate systematically the various water users and uses in water allocation.

Central for water management is clarity about the hierarchy of water needs, or so Eran Feitelson argues in Chapter 9. We need to distinguish the needs of water for a decent level of living and for the ecosystem functioning, from more controversial water wants. The needs should enjoy moral priority and basic protection from the contestation of politics and the creative destruction of markets. Above these 'basic needs' of the hierarchy, come water use questions that depend on politics for discussion and regulation. For example, agriculture in communities where there is no readily available alternative source of livelihood, Feitelson argues, might enjoy a support not provided to large-scale commercial farming. As such, decisions are very place-based; they should be left to the political sphere according to this view. Finally, water as a luxury good and 'mere' factor of production can be left to markets; users should be required to pay the full social costs of such production and consumption. In a variation of such a hierarchy argument, Rafael Ziegler, Dieter Gerten and Petra Döll suggest in Chapter 7 that the discussion of planetary boundaries and a 'safe operating space' for freshwater use remains morally opaque, if the question of sufficient water for all and of just distribution are not addressed 'within' this space. For this, they explore from a biocentric perspective the planetary boundary framework. A biocentric perspective, they argue, implies a shift from a safe space for humanity to an environmental ceiling that also respects the water needs of other living beings. A *safe, just and sufficient* space respects the water needs of humans as moral agents, but replaces the idea of a maximum appropriation of freshwater by humans with a search for just and sufficient freshwater use.

In a further contribution on basic needs and the use of water(s) in an unjust and more than human world, Lucy Rodina reflects in Chapter 10 on the human right to water. Drawing from a case study of impoverished, peri-urban and informal settlements in Cape Town, South Africa, she questions a 'mere' basic needs approach to implementing the human right to water. Such a focus, she argues, sidelines justice concerns regarding both current and future generations. Yet another spotlight on water ethics from a 'marginalized' perspective is provided by Ricki Levi and Daniel Mishori in Chapter 12 on Anumap Mishra and the rediscovery of a water tradition in Rajasthan. There is no single water ethics, they suggest, but many emerging from different regions. They offer the example of an environmental virtue ethics rediscovered by Mishra in a very arid climate. Mishra's virtue ethics and the special virtues it identifies are distinct to this place, and yet might serve as an inspiration to other regions of the world, Levi and Mishori propose.

Finally, Part III of the book turns to the variety of water charters and water charting. Based on collaborative research with the Carcross/Tagish First Nation inland Tlingit and Tagish community consultants Colleen James/G̲ooch Tláa and Mark Wedge/Aan Gooshú in the Yukon Territory, Canada, and coastal Tlingit Elder David Katzeek/Kingeisti in Alaska, United States, Eleanor Hayman introduces the aquacentric ontology of the Tlingit and Tagish in Chapter 13. Better understanding such ontologies is a necessary step in decolonizing water management and its knowledge claims, she argues. It calls for a counter-mapping of Western ways of seeing water (and 'not listening') and a deep charting of alternative ontologies and of the principles of this culture as she shows in her collaborative chapter. Such deep charting, she suggests, offers an alternative blueprint for global water ethics (and we might add, an invitation to actively listen to the sounds of global water ethics).

However, also in the Western traditions the need for a rethinking of water management is increasingly appreciated. As Susan Smith reports in Chapter 14, the Ecumenical Water Network (EWN) and the member churches of the World

Council of Churches (WCC) have developed a water justice framework that is based on 10 principles of water justice. Likewise, David Groenfeldt introduces a secular approach to developing a global water ethics charter in Chapter 15. In a process that so far has included over 80 water experts worldwide, a global water ethics charter emerges. This charter collects principles of water management and organizes those in terms of general, environmental, social, economic, cultural and governance categories. The idea of this charter is to facilitate a comprehensive process that more local or regional charter processes can draw inspiration from and adapt for their purposes.

Two examples of such more specific charting initiatives are discussed in the subsequent chapters. Dorothea Härlin, in Chapter 16 on the Berlin Water Charter, makes the case for an ethical city water charter so as to provide guidance for the difficult task of urban water governance. While many cities have re-municipalized their water supply, she notes that a narrow profit focus can prevail even in public hands. A charter that spells out the political, social, economic, environmental and legal principles of the use of water provides a space for debate and a legal framework for the governance of water as a commons, and thus provides the city with an opportunity to move beyond a reductive focus on one or a few aspects only. The need to achieve a wider and inclusive perspective is also highlighted by Adrian Sym in Chapter 17 on water ethics and water stewardship. The AWS International Water Stewardship Standard offers a (voluntary) framework for water users to understand their water use and impacts, and to work together and more transparently for sustainable water management within their catchment. A shared understanding of water ethics can help to engage a more diverse range of stakeholders in water stewardship processes, Sym argues.

There are different ways to read this book and to benefit from the relations between the different chapters. Here we have highlighted some of these relations, while chapter authors suggest further links between their work and the various chapters of the book, and thus further ways of reading it.

Notes

1 This is not to say that there are no tipping points in relation to water and the Earth System. But the science of such complex local to global interrelations is still very much at the beginning (Steffen et al. 2015).
2 For further discussion of water ethics and philosophical ethics, see especially the chapters by Kallhoff, Kowarsch and Meisch in this volume.
3 See Ziegler and Kerschbaumer (2012) for this specification of water ethics. For a recent literature overview, see Groenfeldt (2013) and Schmidt and Peppard (2014). Two edited volumes on the topic are Llamas et al. (2009) and Brown and Schmidt (2010).
4 For a holist approach, see Gorke (2010); for an ecocentric approach drawing on Aldo Leopold, see Armstrong (2006); for a biocentric approach drawing on Paul Taylor, see the contribution by Ziegler, Gerten and Döll in this volume. Also important are hermeneutic and narrative approaches, frequently drawing on indigenous perspectives – see Hayman in this volume. For a virtue-theoretical approach, see Levi and Mishori in this volume.
5 For a further discussion of these distinctions and various ethical responses to globalization, see Crocker (2009, chapter 11). This tension between the global and the

local is also present in the global justice discussion of political philosophers. Early contributors to this discussion frequently emphasized a globalist or cosmopolitan idea that the standards of justice that apply at the domestic level should also apply globally; by contrast, the statists insist that domestic egalitarian principles of distributive justice should be distinguished from international relations and, for example, the duty to assist each other. Finally, a third group seeks to overcome this dichotomy; as Laura Valentini observes: 'The dilemma between globalizing domestic principles of justice, on the one hand, and denying that justice has any role to play outside state borders, on the other, is a false one. More plausible . . . is the view that different principles of justice apply to different domains of human action' (Valentini 2011, p. 3). If these discussions from globalization studies and global justice are an indicator, then for global water ethics too the balance and interrelation of global and local concerns are a key issue and challenge.

6 A related idea is that of a blue covenant (see Barlow 2008).

References

Allan, JA 2001, *The Middle East Water Question: Hydropolitics and the Global Economy*, I.B. Tauris, London.

Armstrong, AC 2006, 'Ethical issues in water use and sustainability', *Area*, vol. 38, no. 1, pp. 9–15.

Barlow, M 2008, *Blue Covenant. The Global Water Crisis and the Coming Battle for the Right to Water*, The New Press, New York.

Brown, PG & Schmidt, J 2010, *Water Ethics: Foundational Readings for Students and Professionals*, Island Press, Washington, DC.

Conca, K 2006, *Governing Water: Contentious Transnational Politics and Global Institution Building*, Global environmental accords, MIT Press, Cambridge, MA.

Crocker, D 2009, *Ethics of Global Development*, Cambridge University Press, Cambridge.

Cuttelod, A, Seddon, M & Neubert, E 2011, *European Red List of Non-Marine Molluscs*, Publications Office of the European Union, Luxembourg.

Daly, HE & Cobb, JB 1994, *For the Common Good*, Beacon Press, Boston, MA.

Gorke, M 2010, *Eigenwert der Natur*, Hirzel, Stuttgart.

Groenfeldt, D, 2013 *Water Ethics: A Values Approach to Solving the Water Crisis*, Routledge, London.

Held, D, McGrew, A, Goldblatt, D & Perraton, J 1999, *Global Transformations: Politics, Economics, and Culture*, Stanford University Press, Stanford, CA.

Hoekstra, A & Chapagain, AK 2008, *Globalization of Water: Sharing the Planet's Freshwater Resources*, Blackwell, Oxford.

Hoekstra, A & Mekonnen, MM 2012, 'The water footprint of humanity', *PNAS*, vol. 109, no. 9, pp. 3232–7.

Kallhoff, A 2014, 'Water justice: a multilayer term and its role in cooperation', *Analyse & Kritik: Zeitschrift für Sozialtheorie*, vol. 36, no. 2, pp. 367–82.

Llamas, MR, Martínez-Cortina, L & Mukherji, A 2009, *Water Ethics*, CRC Press, Boca Raton, FL.

Mehta, L, Veldwisch, GJ & Franco, J 2012, 'Introduction to the special issue: water grabbing? Focus on the (re) appropriation of finite water resources', *Water Alternatives*, vol. 5, no. 2, pp. 193–207.

Postel, S 2013, 'Sustaining freshwater and its dependents', in E Assadourian, T Prugh & L Starke (eds), *Is Sustainability Still Possible?*, State of the world, vol. 2013, Island Press, Washington, DC, pp. 51–62.

Preston, CJ 2015, 'The multiple Anthropocenes: toward fracturing a totalizing discourse', *Environmental Ethics*, vol. 37, no. 3, pp. 307–20.

Sachs, W 1999, *Planet Dialects: Explorations in Environment and Development*, Zed Books, London.

Schmidt, JJ & Peppard, CZ 2014, 'Water ethics on a human-dominated planet: rationality, context and values in global governance', *Wiley Interdisciplinary Reviews: Water*, vol. 1, no. 6, pp. 533–47.

Steffen, W, Richardson, K, Rockström, J, Cornell, SE, Fetzer, I, Bennett, EM, Biggs, R, Carpenter, SR, Vries, W de, de Wit, Cynthia A, Folke, C, Gerten, D, Heinke, J, Mace, GM, Persson, LM, Ramanathan, V, Reyers, B & Sörlin, S 2015, 'Sustainability. Planetary boundaries: guiding human development on a changing planet' (eng), *Science (New York)*, vol. 347, no. 6223, p. 1259855.

Valentini, L 2011, *Justice in a Globalized World: A Normative Framework*, Oxford University Press, Oxford.

Vörösmarty, CJ, McIntyre, PB, Gessner, MO, Dudgeon, D, Prusevich, A, Green, P & Glidden, S et al. 2010, 'Global threats to human water security and river biodiversity', *Nature*, vol. 467, no. 7315, pp. 555–61.

Ziegler, R & Kerschbaumer, L 2012, 'Wasserethik – eine Übersicht', *Greifswald Environmental Ethics Papers*, no. 3, www.botanik.uni-greifswald.de/130.html?&L=0.

Ziegler, R, Partzsch, L, Gebauer, J, Henkel, M, Lodemann, J & Mohaupt, F 2014, *Social Entrepreneurship in the Water Sector: Getting Things Done Sustainably*, Edward Elgar, Cheltenham.

2 The historical and intellectual context of global water ethics

Susan Lea Smith

Before the world embraced sustainable development at the Rio Conference on Environment and Development in 1992, one of the world's most distinguished hydrologists, Luna Leopold, argued that ethical principles must underlie water resources management. In his 1990 Abel Wolman Distinguished Lecture, Leopold urged that water resources be managed in accordance with two guiding ethical principles (Leopold 1990). The first is an ethos that protects the integrity of the whole of the hydrological continuum, including all elements of what we call ecosystems or watersheds, the rocks and soil, flora and fauna, air, water, and humans and the physical, chemical and biological forces that affect them. The second is a principle of equity, 'a dedication to fairness, a desire to consider various interests and treat all with some measure of equality' (Leopold 1990). In doing so, Luna Leopold articulated an ethic for managing water resources derived in large part from the land ethic for natural resources management advocated by his famed conservationist father, Aldo Leopold (1949).[1]

This chapter examines efforts over the past few decades to address water ethics, including water resource professionals, corporations and scholars from many disciplines. However, it begins with the most influential initiatives that have affected water ethics: those undertaken by the UN. Those initiatives have included informal acknowledgement of the human right to water for almost a half century, formal recognition of an international human right to water and sanitation during the last decade, articulation of the Millennium Development Goal for water and sanitation starting in 2000, adoption of the Sustainable Development Goal for water and sanitation in 2015, and the spark the UN has provided for many discussions on water ethics.

UN recognition of the human right to safe water and adequate sanitation

The most fundamental ethical statement about water to emerge is the UN recognition of safe water and adequate sanitation as a human right. The human right to water as understood by the UN includes sufficient, safe, acceptable, physically accessible and affordable water for domestic uses such as drinking, personal sanitation, washing of clothes, food preparation, personal and household hygiene (United Nations 2016).

Early developments

The idea of a human right to water no doubt dates back millennia, but its roots within the United Nations date back at least a half-century. When the International Covenant on Economic, Social, and Cultural Rights was signed in 1966, the human right to water was considered part of the Article 11 right to an adequate standard of living, but was not explicitly recognized (United Nations 1966).[2] Explicit acknowledgement or assertion of the right to water began to emerge during mid-1970s and that right is referenced in many UN soft law documents, including the action plan from the 1977 Mar del Plata UN Water Conference,[3] 1992 Rio Declaration and Agenda 21.

However, as the UN sought to integrate sustainability with economic development during the 1990s, the schism between the neoliberal, economic growth-oriented development community and the environmental and social justice-oriented sustainability community continued to grow. For example, although intended to advance sustainable development, the 1992 International Conference on Water and Environment in Dublin fell captive to a neoliberal counter-movement. That conference adopted four water resources management principles known as the Dublin Statement.[4]

The Dublin Statement acknowledged the human right to water, but subordinated it to notions of the economic value of water, use of water pricing for water conservation and protection, and recovery of service delivery costs. The fourth Dublin principle proclaimed: 'Water has an economic value in all its competing uses and should be recognized as an economic good' (World Meteorological Organization 1992, unpaginated). The Dublin statement went on to explain:

> Within this principle, it is vital to recognize first the basic right of all human beings to have access to clean water and sanitation at an affordable price. Past failure to recognize the economic value of water has led to wasteful and environmentally damaging uses of the resource. Managing water as an economic good is an important way of achieving efficient and equitable use, and of encouraging conservation and protection of water resources.
>
> (World Meteorological Organization 1992)

The fourth Dublin principle promoting the recognition of water as an economic good was and remains controversial. Formulated at a time when the World Bank was compelling developing countries to privatize water services as a condition of receiving development assistance, the principle was seen as a direct challenge to the idea of water as a public or social good (Walters 2013).

The neoliberal counter-movement in control of the Dublin Conference ultimately gave rise to even more powerful water justice movements opposing privatization of water services, commodification of water and absolute private rights to water, while promoting recognition of the human right to water and sanitation (Walters 2013). Over time, the fourth Dublin Principle became so unacceptable that the Global Water Partnership actually rephrased the principle in presentations on integrated water resources management: 'Water has an economic value in all its competing uses and should be recognized as an economic good *as well as social good*' (added words in italics; Global Water Partnership 2012).[5]

General Comment 15

The human right to water and sanitation was formally recognized by the UN Committee on Economic, Social and Cultural Rights (CESCR) General Comment 15 in 2002 (Committee on Economic Social and Cultural Rights 2003). A substantial amount of UN work preceded General Comment 15 (UN Office of the High Commissioner for Human Rights 2007b). That work included the 1998 working paper of the UN Economic and Social Council's Sub-Commission on Prevention of Discrimination and Protection of Minorities, which outlined the basis for 'the right of access of everyone to drinking water supply and sanitation services' (UN Sub-Commission on the Promotion and Protection of Human Rights 1998) and the 2000 General Assembly resolution on the right to development, which reaffirmed that 'in the full realization of the right to development, the rights to food and clean water are fundamental human rights and their promotion constitutes a moral imperative both for national Governments and for the international community' (UN General Assembly 1999, p. 12).

General Comment 15 stated:

> The human right to water entitles everyone to sufficient, safe, acceptable, physically accessible and affordable water for personal and domestic uses. An adequate amount of safe water is necessary to prevent death from dehydration, to reduce the risk of water-related disease and to provide for consumption, cooking, personal and domestic hygienic requirements.
>
> (Committee on Economic Social and Cultural Rights 2003, p. 2)

The CESCR General Comment recognized that human beings are entitled to water for other purposes, including food production, securing livelihoods and enjoying cultural practices (Committee on Economic Social and Cultural Rights 2003, p. 6). However, it gave priority in water allocation to water for personal and domestic uses, water required to prevent starvation and disease, and water to ensure that disadvantaged and marginalized farmers, including women farmers, have equitable access to water and water management systems, including sustainable rain harvesting and irrigation technology (Committee on Economic Social and Cultural Rights 2003, pp. 6–7).

Subsequent UN work further articulated and clarified the human right to water and sanitation, including the draft guidelines for the realization of the right to drinking water supply and sanitation adopted by the Sub-Commission for the Promotion and Protection of Human Rights and submitted to the Human Rights Council in 2006 (UN Sub-Commission on the Promotion and Protection of Human Rights 1998) and the Office of the High Commissioner for Human

Rights consultation on the right to water (UN Office of the High Commissioner for Human Rights 2007a). During this period, the highly influential work of the Independent Expert and Special Rapporteur on the Right to Water and Sanitation, Catarina de Albuquerque, began (2008–2014).

2010 UN resolutions

Many commentators overlook General Comment 15. They mark the more widely publicized UN General Assembly resolution in July 2010 (UN General Assembly 2010) and the UN Human Rights Council resolution in September 2010 (UN Human Rights Council 2010) as the point of formal recognition of the right. Yet even in the 2010 UN General Assembly, the universality of recognition of the human right to water and sanitation continued to be dogged by the privatization/commodification issue. The 2010 General Assembly resolution boldly recognized 'the right to safe and clean drinking water and sanitation as a human right that is essential for the full enjoyment of life and all human rights' (UN General Assembly 2010). However, 41 countries abstained from voting on the resolution. Their abstentions were purportedly on procedural grounds, but in reality were caused by the unresolved privatization/commodification issue and the concerns of the United States and Canada that the human right of water and sanitation might become international law to which they were bound, either by treaty or customary law.[6]

The unanimous adoption of the UN Human Rights Council resolution later that year was possible for two reasons. First, international negotiators had developed two strategic adjustments of the UN General Assembly resolution. One explicitly tied the human right to water and sanitation to rights created by the ICESCR (creating an out for the United States who is not bound by the ICESCR). Another affirmed that countries could opt to include private actors, such as water service corporations, in their efforts to realize the right, a compromise that side-stepped the privatization issue.[7] Second, the human right to water had become a politically salient international issue thanks to the efforts of legions of water justice activists around the globe. In particular, UN recognition of the human right to water and sanitation as well as progress on addressing other water justice issues owes much to the unflagging advocacy of environmental activist and author Maude Barlow over the past two decades. During this time, she increased public understanding of the dangers of corporate control of water and other justice dimensions of water issues, and rallied public support of water justice, by publishing acclaimed popular books on water, including *Blue Gold* (Barlow & Clarke 2002) and *Blue Covenant* (Barlow 2007).[8] And, during the critical period immediately before the adoption of the 2010 UN resolutions (2008–2009), Barlow served as Special Advisor on Water Issues to the UN Secretary General.

As a matter of international law, the human right to water and sanitation derives from the right to an adequate standard of living and is recognized as being inextricably related to the right to health, the right to life and the right to human dignity. The human right to water and sanitation is commonly defined in terms of 'sufficient, safe, accessible, culturally acceptable and affordable water and sanitation services for personal and domestic uses, and which are delivered in a participatory, accountable and non-discriminatory manner'[9] (GLAAS 2014). The most recent UN General Assembly resolution provided an extensive definition of the rights to water and sanitation:

> the human right to safe drinking water entitles everyone, without discrimination, to have access to sufficient, safe, acceptable, physically accessible and affordable water for personal and domestic use, and that the human right to sanitation entitles everyone, without discrimination, to have physical and affordable access to sanitation, in all spheres of life, that is safe, hygienic, secure, socially and culturally acceptable and that provides privacy and ensures dignity.
>
> (UN General Assembly 2015)

As with all economic, social and cultural human rights, the international community expects governments to ensure the progressive realization of this right so that everyone gains access to water and sanitation services over a reasonable timeframe, by adopting appropriate legislation, policies and programmes that are adequately resourced and monitored. Under the ICESCR, governments are also bound to avoid obstructing exercise of this right and must not allow others to obstruct exercise of this right.

Sustainability, Millennium Development Goals and Sustainable Development Goals

The United Nations efforts on sustainability, Millennium Development Goals and Sustainable Development Goals over the past few decades represent the United Nations' greatest contribution to the development of global ethical principles affecting water resource management. The UN sustainability initiatives have created awareness of human dependence on the Earth and its ecosystems. They have sought to protect the Earth for the benefit of both current and future generations. At the same time, they have created a global commitment to social justice and what amounts to a global preference for the poor and marginalized. They are particularly significant because they incorporate national monitoring and reporting mechanisms and the results have become the focus of development-oriented international organizations.

Sustainability

In 1983, the UN fostered creation of the World Commission on Environment and Development (WCED), commonly called the Brundtland Commission after its famous chair, Gro Harlem Brundtland, then Prime Minister of Norway. WCED published its report *Our Common Future*, which crafted what is now the most often cited definition of sustainable development: 'Sustainable development is development that meets the needs of the present without compromising the ability of future generations to meet their own needs' (WCED 1987). The work of the Brundtland Commission was instrumental in convening the Earth Summit in Rio de Janeiro, Brazil in 1992 and in creating Rio's related treaty and soft law achievements.

The ideas of *Our Common Future* strongly influenced the outcomes of the Rio Summit and subsequent international meetings on sustainable development. The 1992 Rio Declaration essentially engrafted ethical commitments to ecosystem integrity and social justice onto the dominant economic development paradigm.[10] The problem is that sustainability got 'mainstreamed' within the global development community without displacing the economic paradigm. This condemned social justice and ecological sustainability to a peripheral role in political decision-making. This sidelining of ecosystem integrity and social justice concerns occurred not only in developed countries that had embraced the neoliberal economic agenda, but also in the bulk of developing countries, which were held captive by not always benign international financial institutions such as the World Bank and IMF.

Throughout the remainder of the twentieth century, the battle between the neoliberal economic paradigm and the sustainable economic paradigm continued to rage (Harvey 2005). The neoliberal solution to poverty was reliance on the indirect 'trickle-down' effect of economic growth to improve the lot of the poor since the metaphorical rising seas of increasing gross national product should float all boats, including those of the poor. To achieve an influx of private capital into developing countries, the World Bank imposed structural adjustment programs on developing countries as the price of receiving loans or development aid. These programs typically included privatization of many public industries and services; reduction in government spending, employment and deficits; economic deregulation and trade liberalization. Those embracing sustainable development tended to pay more direct and explicit attention to poverty and argued that trickle-down is an ineffective device to address the plight of the poor.

Formulated in the 1990s with the encouragement of WCED, the Earth Charter attempted to create a coherent vision of sustainability during that era. The Earth Charter articulated 'fundamental ethical principles for building a just, sustainable and peaceful global society in the 21st century', by which it sought to inspire 'a new sense of global interdependence and shared responsibility for the well-being of the whole human family, the greater community of life, and future generations'.[11] The Earth Charter gave first place to four ethical principles: respect for the Earth and life in all its diversity; care for the community of life with understanding, compassion and love; democratic societies that are just, participatory, sustainable and peaceful; and environmental security for present and future generations. It then articulated specific precepts regarding ecological integrity, economic and social justice, and democracy, non-violence and peace as conditions necessary to fulfil those commitments.[12] Foremost among the economic and social justice precepts was the ethical, social and environmental imperative to eradicate poverty.

Millennium Development Goals

Ultimately, at least on paper, the sustainable development forces won the war. Adoption of the UN Millennium Development Goals (MDGs) in 2000 represented a commitment by all national governments and international financial institutions to abandon economic growth as the primary measure of development in developing countries and focus instead on human development,[13] with the primary objective of reducing poverty. Adoption of the MDGs coincided with the World Bank's formal abandonment in 1999 of conditioning loans and development aid to developing countries on adoption of structural adjustment programs (SAPs). The Bank abandoned SAPs due to widespread political pressure based on evidence that, rather than alleviating poverty, they exacerbated poverty and income inequality. At the same time, the Bank adopted a pro-poor stance consistent with the MDGs. The Bank revised its lending and aid policy to instead require countries securing loans and aid to adopt a 'poverty reduction strategy'. Unfortunately, in reality, poverty reduction strategies adopted by developing countries frequently resembled previously adopted SAPs. Nonetheless, both adoption of the MDGs and the abandonment of SAPs represented not only a global ethical commitment to improve the lot of the poor, but also put that commitment into action by backing it with real money.

The MDGs were not envisioned to be the ultimate human development goals or the goals that must be achieved to eradicate poverty. Instead they were conceived as interim steps to be achieved between 2000 and 2015 in each developing country. MDG 7 on environmental sustainability included a halfway sub-goal on water and sanitation: to cut the percentage of people who lacked access to improved water sources or improved sanitation in 1990 in half.[14] Though developing countries fell far short of the sanitation goal, the water MDG was realized in 2010 – in large part due to diligent efforts in two enormous countries that enjoyed robust economic growth during that period: China and India.

Sustainable Development Goals

Recognizing how sustainability had been marginalized rather than integrated into the development model, the Rio+20 UN Conference on Sustainable Development (2012) called for adoption of Sustainable Development Goals (SDGs) to supersede the MDGs. SDGs were intended as the international development agenda beyond 2015. Through a remarkably open and thorough participatory process, the UN formulated 17 SDGs to be achieved not just by developing countries, but by every nation.[15] Those goals were adopted in September 2015 at a UN Summit of heads of government of UN members.

The SDGs are a remarkable vision of how the world should be – but with the crucial commitment at the level of heads of government to create such a world, by planning, action and periodic review of implementation – a commitment accompanied by ancillary commitments, however tenuous and partial, to fund achievement of the goals. The water and sanitation SDG Goal 6 reaches beyond universal access to water and sanitation to include water quality improvement, water efficiency gains, sustainable use of groundwater and surface water, restoration and protection of aquatic ecosystems, and implementation of integrated water resources management. The vision is not exactly of paradise or the kingdom of heaven, but it comes pretty close.

International discussions of water ethics fostered by the UN

COMEST/IHP (1998–2004)

Beginning in 1998, a working group of the World Commission on the Ethics of Science and Technology (COMEST) as well as the UNESCO International Hydrology Programme (IHP) conducted extensive discussions about water ethics.[16] Rather than undertake a fresh inquiry into water ethics, COMEST/IHP recognized nine previously articulated ethical principles and sought ways to promote best ethical practices based on those principles (Liu et al. 2011). The principles identified by COMEST/IHP included human dignity, participation, solidarity, human equality, common good, stewardship, inclusiveness, transparency and access to information and empowerment (COMEST 2004, pp. 7–8). COMEST/IHP described those principles in this manner:

> *Human dignity*: there is no life without water and those to whom it is denied are denied life;
>
> *Participation*: all individuals, especially the poor, must be involved in water planning and management with gender and poverty issues recognized in fostering this process;
>
> *Empowerment*: the requirement to facilitate participation in planning and management means much more than to allow an opportunity for consultation, for best ethical practice enables stakeholders to influence management;
>
> *Transparency and universal access to information*: if data are not accessible in an understandable form, interested parties may disadvantage others;
>
> *Solidarity*: upstream and downstream interdependence within a watershed continually poses challenges for water management resulting in the need for an integrated water management approach;
>
> *Human equality*: all persons ought to be provided with the basic necessities of life on an equitable basis;
>
> *Common good*: water is a common good, and without proper water management human potential and dignity diminishes;
>
> *Stewardship*: protection and careful use of water resources is needed for intergenerational and intra-generational equity and promotes the sustainable use of life-enabling ecosystems; and
>
> *Inclusiveness*: water management policies must address the interests of all who live in a water catchment area. Minority interests must be protected

> as well as those of the poor and other disadvantaged sectors. In the past few years the concept of Integrated Water Management (IWRM) has come to the fore as the means to ensure equitable, economically sound and environmentally sustainable management of water resources.
>
> (COMEST 2004)

The COMEST/IHP principles can be usefully categorized as belonging to one of four overarching meta-principles: (1) allocation of water to satisfy the human right to water necessary to meet basic needs (human dignity, human equality, inclusiveness, common good); (2) human right to involvement in water decision-making (participation, transparency and information access and empowerment); (3) human obligations arising from the physical reality of interrelatedness of water use that necessitates sharing (solidarity); and (4) human obligations to manage water resources sustainably (stewardship).

Note that the ethical principles recognized by COMEST/IHP were wholly anthropocentric – even the stewardship principle contemplated that ecosystems should be managed in an environmentally sound, sustainable manner only in order to address intra-generational and intergenerational equity among human beings.

Botin Foundation workshops on water ethics

Another international discussion of water ethics closely followed the COMEST/IHP effort. The Marcelino Botin Foundation held its third water workshop in Spain during June 2007, bringing together experts from different cultural, geographic and religious backgrounds to share their ethical perspectives. Although the workshop was independently sponsored, two UN groups (UNESCO-IHP and the United Nations University) contributed greatly by identifying experts to be invited.

The Botin Foundation workshop was not designed to articulate a common global water ethic, but to 'discuss the role of ethics in the deep roots, values, and the potential commonalities of the global water policy, and share perspectives'[17] (Llamas et al. 2009). A subsequent Botin Foundation seminar in 2010 focused on how ethics, faith, religion and spirituality intersect with water management decisions.[18]

UNESCO East Asian working group on water ethics (2011)

In 2011, the Asia and Pacific working group published *Water Ethics and Water Resources Management*, an interesting and provocative report articulating 12 ethical principles for water resources management (Liu et al. 2011). Perhaps because this initiative preceded formal UN recognition of the human right to water and sanitation, the working group placed considerable reliance on the 2005 UNESCO Universal Declaration of Bioethics and Human Rights (UDBHR)[19] and linked that document to three of the principles it articulated: (1) human dignity and the right to water; (2) equity in availability and applicability of water and the right to water; and (3) managing water to meet ecosystem requirements and foster a healthy environment.

The UNESCO working group report indicates *the principle of human dignity and the right to water* is drawn from UDBHR Article 14 on health and the UDHR Article 3 on human dignity, the *principle of equitable availability and access* is drawn from UNDBHR Articles 2 and 10, and *the principle of managing water to meet ecosystem requirements* may be drawn from UNDBHR Article 17 (Liu et al. 2011). These first two principles form the basis for the working group asserting not only the human right to water necessary for human dignity, but equal entitlement of all humans to water. The third principle goes beyond anthropocentric justifications for providing water for ecosystems, articulating an eco-centric justification for providing water to ecosystems. Taken as a whole, these principles reached far beyond UNDBHR Article 17.

The working group then articulated a variety of other principles: *Vicinity, Frugality, Transaction, Multiple and Beneficial Use of Water, Quantity and Quality Measures, Compensation and User Pays, Polluter Pays, Participation, Equitable and Reasonable Utilization.* Like the first three principles, several of these principles address the basis for allocation of water among users. These allocative principles include vicinity, frugality, multiple and beneficial use, and equitable and reasonable utilization.

The *principle of vicinity* gives preference to those who live close to a water source or upstream on a river, but also imposes duties to avoid waste, pollution and destruction of natural flow. The preference given to local users appears to be limited to riparian communities, because it is such communities that can use and return the water largely unimpaired in terms of quantity and quality and thus allow multiple use downstream.

The *principle of frugality* means that people should not use water exceeding their actual needs, so that the remaining water can be used for communities lacking sufficient water or, in the case of groundwater, kept for future generations.

The working group suggested that because of the *principle of multiple and beneficial use*, that all beneficial uses of water other than those for human drinking purposes, should be considered equal. However, in the very next breath they actually prioritized water uses. The working group accorded water necessary for ecosystem protection top priority, followed by all local area uses (residential, agriculture, industry and commercial), finally followed by diversions of water to external communities in need of water. The working group distinguished 'general water' and 'special water', in accordance with the Asian Pacific and civil law distinction between water typically held in common and waters that may be privately exploited. The working group did not restrict use of special waters to priority uses, but instead recognized that special waters such as springs would be subject to commercial use.

The *principle of equitable and reasonable utilization* is another allocative principle, which is also a customary principle of public international law on shared waters. This principle was recognized and endorsed in both the UNECE Convention on the Protection and Use of Transboundary Watercourses and International Lakes (1992) and the UN Convention on the Law of the Non-Navigational Uses of International Watercourses (1997) (Wolf 1999).

The working group also included several principles typically associated with the economic approach to water, including the *principles of transaction, compensation, user pays and polluter pays.* The *principle of transaction* means that the saved and surplus water from the allocated amount can be traded as a commodity in the water market either through water banks, water exchanges or transfers where users have a private right to the use and ownership of water resources. The *transaction principle* allows transfer of water to other users in a manner that promotes economic efficiency.

The *principle of compensation* requires compensation of people whose living standards have been degraded because of water resources being transferred away from them. It also requires 'compensation' of nature but the working group does not identify how that might occur. Such compensation assures that a solution is Pareto-optimal.

The *user pays principle* requires paying a royalty or fee for using a natural resource when a common resource such as water is privatized. User fees also can be used as incentive measures to facilitate proper utilization of water resources. User fees or water prices may be set to recover both operational costs for the water service and the resource costs (full-cost recovery), but such a policy risks undermining the human right to water. At a minimum, water pricing should ensure equitable and affordable access to the safe drinking water.

The *polluter pays principle* not only covers the costs of environmental restoration, but also the costs of pollution prevention. The principle avoids investments in activities that are not socially beneficial because the costs of prevention and mitigation are greater than the benefits derived from the activities.

The working group also included two principles tied to water governance: the *principle of mandatory application of quantity and quality measures* and the *principle of public participation.* The latter assures that the interests of all groups, especially the poor and under-represented groups, are fully represented and taken into account. The former provides the information necessary for effective participation in water resources management.

Overall impact of the United Nations on water ethics

Various groups working under United Nations auspices over the past three decades have advanced global society's understanding of ecological sustainability, social and economic justice, democratic governance and the ability of economic development to be channelled towards human development. Sustainability now dominates the discussion of resource issues around the world, even in the Anglo-American neoliberal bastions. The United Nations and the legions who have worked under its auspices have set the table for a new ethical paradigm to emerge to replace the economic paradigm in resource management.

Statements of professional and corporate water ethics

In the past decade, there has been substantial development on the ethical front in the private sector. Water resource professionals have adopted professional ethical standards that address broader social concerns. Corporations have begun to include water ethics within their social responsibility/sustainability commitments. Groups of stakeholders including civil society as well as those who profit from water resources have developed certification standards on water. These efforts are described below.

Professional ethics

Water professionals have formulated professional codes of ethics that address ethical concepts to be employed in their work; for example, in 2006, American water resources engineers formally incorporated sustainable development as an ethical principle within their code of ethics.[20] For the most part, professional codes of ethics focus on providing honest, competent professional services. However, they occasionally provide professionals with some broader ethical guidance. For example, the 2006 American Society of Civil Engineers Code of Ethics as well as the American Academy of Water Resources Engineers Codes of Ethics require engineers to 'hold paramount *the safety, health and welfare of the public* and shall strive to comply with the *principles of sustainable development* in the performance of their professional duties'[21] (American Society of Civil Engineers 2006). The ASCE Board of Directors adopted the following definition of sustainable development in 2009: 'Sustainable Development is the process of applying natural, human, and economic resources to enhance the safety, welfare, and quality of life for all of the society while maintaining the availability of the remaining natural resources' (American Society of Civil Engineers 2016).

Similarly, there are professional organizations that promote ethical practices in the broader area of geosciences, including the International Association for Promoting Geoethics and the International Association for Geoethics.[22] The IAPG has proposed a Geoethical Promise similar to the Hippocratic Oath that binds medical practitioners, promising among other things to practise 'being fully aware of the involved social implications', doing their 'best for the protection of geosphere for the benefit of mankind', understanding their 'responsibilities towards society, future generations and the Earth for a sustainable development', putting the interest of society at large in first place while on the job, and being 'fully respectful of the Earth's processes' in their work[23] (International Association for Promoting Geoethics 2015).

Certification of responsible corporate water practices

Another recent effort has been to develop a certification process akin to private certification process of the Forest Stewardship Council. The Alliance for Water Stewardship (AWS), created by a core group of leading corporations, environmental and other water resources groups, formulated a Water Stewardship Standard in 2014 (see Sym in this volume for a detailed presentation). The standard was developed over a five-year period through the Global Water Roundtable, a multi-stakeholder consensus-based process that involved meetings, online review and pilot testing. The standard, as its name implies, is intended to provide a standard of practice for corporate water users and water resources managers alike.

AWS conceives of freshwater resources as a public good that must be effectively managed by all. It seeks to drive all water resources management towards water stewardship, which it defines as 'the use of water that is socially equitable, environmentally sustainable and economically beneficial, achieved through a stakeholder-inclusive process that involves site and catchment-based actions' (Alliance for Water Stewardship 2014). AWS further explains the knowledge and actions required to accomplish water stewardship:

> Good water stewards understand their own water use, catchment context and shared risk in terms of water governance, water balance, water quality and important water-related areas; and then engage in meaningful individual and collective actions that benefit people and nature.

Corporate ethical codes

Major transnational corporations are required by the US Securities and Exchange Commission and the New York Stock Exchange, and similar EU requirements, to have corporate codes covering governance and business ethics issues; however, typically corporate social responsibility is not addressed in these codes. Nonetheless, corporate social responsibility, including water and other sustainability issues, has become a concern for many major transnational corporations. The major transnational corporations that are intimately involved in water resources (such as water service providers, water infrastructure construction firms and beverage bottlers) periodically issue sustainability reports addressing water.[24] They typically profess to comply with the standards provided by one or more of the competing sustainability initiatives and reporting systems, such as the UN Global Compact,[25] the CEO Water Mandate, sustainability reporting under the Global Reporting Initiative's[26] G4 Sustainability Reporting Guidelines[27] and corporate environmental management under ISO 14001 (2015).[28]

The CEO Water Mandate contains specific commitments that corporations make concerning water reporting in the areas of direct operations, supply chain management, collective action, public policy, community engagement and transparency.[29] While quite specific, the commitments are scarcely demanding, because all contain the caveat that the commitment is 'where appropriate, over time'.

Water ethics scholarship

Water ethics scholarship has its roots in the interdisciplinary field of environmental ethics, expounded by authors from disciplines as diverse as history, anthropology, geography, economics, theology and philosophy. In the past decade or two, many scholars have begun to focus on the ethics of water as a discrete area of inquiry. Though legions have made important contributions to this scholarship, several recent book-length treatments of water ethics deserve note.

Chamberlain's Troubled Waters *(2008)*

Chamberlain identifies elements of a water ethic or ethos from utilitarian, natural law, Buddhist, ecofeminist and environmental ethics traditions. Chamberlain begins by acknowledging the value of a welfare economics approach to water issues: 'Some form of utilitarian cost/benefit analysis can be helpful in arriving at decisions about the global water crisis and the policies and approaches to be used in water management. However, the narrow calculus of economic, even monetary value must be enlarged to include the social and environmental costs of water extraction and use' (Chamberlain 2008, p. 142).

He notes that Catholic social thought regards *water as a common good* which requires 'the state through some form of regulation and/ or ownership be integrally involved in access, distribution, pricing, and sale of water' (p. 142). Chamberlain also identifies other elements emerging from this natural law approach, including: *participation* in decision-making; *subsidiarity* delegating governmental decisions to local and bioregional level; *stewardship* of the goods of creation as opposed to ownership, which creates a very strong emphasis against absolute private property rights; *preferential protection for the poor and marginalized*; *human dignity*; and *solidarity* with others in their struggles.

The Buddhist tradition fosters 'the central value of attunement to nature through seeking a harmony between the human and natural worlds' (p. 143). That value cultivates a communitarian approach, an ethic of discipline over water desires as opposed to water needs, an attitude of thriftiness that prevents waste, and a mindfulness that leads to action to prevent suffering. Buddhism thereby contributes 'a powerful and much-needed stream to the growing river of a water ethic' (p. 144).

Ecofeminists, through their *rejection of domination* and their emphasis on *relationality* and *embodiment*, provide an antidote to the Western utilitarian misconception of autonomy that a separate, rational self acts out of self-interest with other equally separate, rational selves. As Chamberlain (p. 145) explains,

> an equally important aspect is the feminist theme of *embodiment*, i.e., being at home in one's body and citing the body, particularly women's bodies, as valuable dimensions of ethical discourse. Being at home in one's body is a key to being at home in the natural world, argue several ecofeminists. Our relationships with nature constitute our very moral selves. Such *at-homeness* is at the heart of ecology itself, for the word derives from the Greek words *oikos*, home, and *logos*, study, "the study of the home." We are one with the Earth-home.

Finally, Chamberlain identifies the foundational ecological principle of *sustainability* as the grounding of environmental ethics: 'The question is whether current practices and policies ensure not only that there is sufficient quantity and quality of water for the Earth and all inhabitants but that there will continue to be sufficient quantities and qualities in the future' (p. 146). Chamberlain suggests that sustainability, at least as understood by Aldo Leopold and many environmental ethicists, extends 'our moral concerns and relationships to the regional Earth communities of which we are part' (p. 146). He attributes several additional principles to sustainability (pp. 146–7):

subsidiarity: making decisions at local and bioregional levels rather than global or national level.

accountability of government and private actors to the public

equality: every human being has a right to clean water for domestic and personal use

participation: all affected must be represented in decision-making

sufficiency: a commitment to meet the basic material needs of all life possible

equitable use and no harm: equitable sharing among all users and avoidance of harm to other users

democratic control of water as a community resource.

Brown and Schmidt's Water Ethics *(2010)*

An anthology of readings on water ethics by Peter Brown and Jeremy Schmidt (2010) provides a guided tour through several ethical stances towards water. Brown and Schmidt identify five groups of ethical approaches. The first is religious conceptions of ethical responsibilities towards water, ranging from self-interested human domination to bio-centric care for all creation. The second is utilitarianism, heavily accented with water as an economic good and markets as the means to allocative efficiency. The third treats water as a communal resource, exploring various approaches to community-based management of water and giving emphasis to democratic management processes. The fourth extends ethical consideration to non-humans as part of the community that we call ecosystems. The final approach contemplates the complexity of systems involved in water resource management, whether adherents conclude that the complexity demands sophisticated management or a humble concession that the systems are simply too complex to manage (pp. 12–14). This latter position is taken by Brown and Schmidt, who advocate an ethic of 'compassionate retreat' from the Western narrative of human superiority and adoption of an ethic of respect and reciprocity towards all life on Earth (pp. 265–86).

Groenfeldt's Water Ethics *(2013)*

David Groenfeldt (2013) argues that, in making water decisions, we should discuss the values and ethics implicated by various courses of action in addition to the technical and economic analyses that currently dominate water decision-making. He suggests that promoting awareness of the tacit values and ethics involved in those decisions will result in more honest and focused debate as well as more democratic, equitable and sustainable water decisions. His faith in the potency of ethical awareness derives in large part from Groenfeldt's belief that ecological sustainability, democratic water governance and the human right to water are now shared ethical values (p. 156). He finds evidence of those shared values in the Dublin Principles and the widespread adoption of integrated water resources management.

Groenfeldt's ethical analysis seeks articulation of the economic, social, environmental and cultural ethical values implicated in water decisions (pp. 4–6). The values he identifies reflect the three 'pillars' of sustainable development as well as the special consideration recently given to protecting the cultural values of indigenous peoples. Groenfeldt's book then uses these categories of ethical values to explore four domains of water management: ecosystem management, water use (agricultural, municipal and industrial), governance and indigenous water management. Groenfeldt's work foreshadows his current endeavour to create a Water Ethics Charter, concluding with a call (p. 173) to assemble the pieces of the emerging water ethic and encourage corporations, organizations, people and governments to adopt it (see also Groenfeldt in this volume).

Peppard's **Just Water** *(2014)*

Christiana Peppard's book applies normative ethics to contemporary water issues. After a brief introduction to current ethical thought, Peppard (2014) explores the commodification debate, transboundary water allocation, agricultural water use, climate change and hydraulic fracking. Peppard identifies three key ethical principles: justice, sustainability and equity to current and future generations (p. 186). Given her analysis elsewhere, these principles might be summarized as follows. Peppard understands justice as requiring *preferential treatment for the poor* in water resources management, decentralizing water resources decisions to allow the people to control water (the *subsidiarity* principle), and engaging in the struggles of the poor to address institutional injustice (the *solidarity* principle). She appears to associate sustainability with *ecological sustainability*, the ecological reality that ties together the fate of humans with the fate of nature. Peppard speaks of equity to current and future generations, otherwise known as *intergenerational equity*, which she perhaps conceives of as a non-negotiable duty owed by current generations to limit their abuse of water and ecosystems so that future generations may use them. Based on these principles, she affirms the human right to water and the status of water as a public good and makes a number of other policy suggestions.

Both Chamberlain and Peppard explore water ethics from a Catholic social thought perspective. The most notable distinction between the two scholars is Peppard's thoroughgoing anthropocentrism. Although Chamberlain acknowledged the anthropocentric nature of Catholic social thought at the time he wrote, he argued for expanding the community of concern to include nature primarily based on Buddhist, ecofeminist and environmental ethics. Peppard, a younger scholar, was less comfortable stretching the boundaries of community. Perhaps the subsequent Catholic teachings in *Laudato Si: On Care for Our Common Home*, the encyclical letter of Pope Francis (2015), will free Catholic scholars to understand the connection between humanity and the rest of creation a bit differently.[30]

Conclusion

This review of recent attempts to articulate global water ethics suggests an emerging consensus about some elements of water ethics: ecological sustainability, social justice or equity, efficient water use and democratic governance. But deep unresolved questions remain about the actual content of those elements. What does social justice demand of us: merely providing the insignificant amount of water required for domestic and personal uses, accomplishing the more difficult task of equitably sharing all water resources, or indeed giving the poor and marginalized preferential treatment in access to water? Are these ethical values co-equal values to be considered or balanced in water decisions – or should some be treated as inflexible constraints or otherwise given priority? Can water be managed more democratically and locally without giving rich and powerful special interests a greater ability to dominate decisions? Can we achieve efficient water use through economic incentives given the gross income inequalities in most economies? We still need to answer these and many other difficult questions – but we can at least celebrate the fact that we are now asking the right questions!

Notes

1 Aldo Leopold's 'A Land Ethic' was the concluding essay in *The Sand County Almanac*.
2 Explicit hard law recognition of the right to water occurred in specifically limited contexts such as the 1979 Convention for the Elimination of Discrimination Against Women (CEDAW) and the 1989 Convention on the Rights of the Child (CRC). The CEDAW sets forth the agenda to end discrimination against women and explicitly references both water and sanitation (www.un.org/womenwatch/daw/cedaw/, accessed 22 July 2016). The CRC explicitly mentions water as well as sanitation and hygiene education as part of the rights of children (www2.ohchr.org/english/law/crc.htm, accessed 22 July 2016).
3 Arguably, the 1977 Action Plan from the United Nations Water Conference was the first explicit UN recognition of the right to water in a UN soft law document. It recognized water as a right for the first time, declaring that 'All peoples, whatever their stage of development and social and economic conditions, have the right to have access to drinking water in quantities and of a quality equal to their basic needs' (UN Office of the High Commissioner for Human Rights 2007b).
4 www.un-documents.net/h2o-dub.htm, accessed 22 July 2016.
5 The Global Water Partnership, originally founded by the World Bank and the UN Development Programme, seeks to foster global adoption of integrated water resources management (IWRM) in accordance with the Dublin Principles. Since 2012, the GWP has rephrased the fourth principle as 'Water is a public good and has a social and

economic value in all its competing uses', www.gwp.org/en/The-Challenge/What-is-IWRM/IWRM-Principles/Social-and-economic-value-of-water/, accessed 22 July 2016. However, GWP commentary makes clear that the GWP considers economic value as key to IWRM and only acknowledges a narrow right to water.

6 The United States has not ratified or signed the International Covenant on Economic, Social, and Cultural Rights (ICESCR), so absent recognition of the human right to water and sanitation as customary international law, it is not obligated to progressively realize that right.

7 Clause 7 recognizes 'that States, in accordance with their laws, regulations and public policies, may opt to involve non-State actors in the provision of safe drinking water and sanitation services'.

8 Barlow's most recent popular book on water is *Blue Future: Protecting Water for People and the Planet Forever* (2013, House of Anansi Press, Toronto).

9 GLAAS quoted a popular definition of the right to water and sanitation published in the Human Right to Water and Sanitation Toolbox of WaterLex, www.waterlex.org/waterlex-toolkit/what-is-the-human-right-to-safe-drinking-water-and-sanitation/, accessed 22 July 2016.

10 Principle 5 of the Rio Declaration made poverty eradication an integral part of sustainability. It states: 'All States and all people shall cooperate in the essential task of eradicating poverty as an indispensable requirement for sustainable development, in order to decrease the disparities in standards of living and better meet the needs of the majority of the people of the world' (www.un.org/documents/ga/conf151/aconf15126-1annex1.htm, accessed 22 July 2016).

11 http://earthcharter.org/about-eci/faqs/, accessed 22 July 2016.

12 http://earthcharter.org/discover/the-earth-charter/, accessed 22 July 2016.

13 More information on the MDGs and progress in meeting the goals is available at www.un.org/millenniumgoals/, accessed 22 July 2016.

14 Sanitation was not initially addressed by MDG 7; sanitation was added to MDG 7 at the Rio+10 Johannesburg Summit on Sustainable Development in 2002.

15 https://sustainabledevelopment.un.org

16 COMEST ultimately published some 14 essays on water ethics (Priscoli et al. 2004) along with Lord Selborne's summary report *Best Ethical Practice in Water Use* (COMEST 2004).

17 In 2009, the Botin Foundation published a compilation of the workshop papers (Llamas et al. 2009).

18 The papers from this workshop were published in a 2012 special issue of *Water Policy* (Vol. 15, suppl. 1).

19 UNESCO Universal Declaration of Bioethics and Human Rights (2005). The AP Water Ethics Working Group suggested that the UDBHR provides a framework for environmental or bioethics that can be used for normative ethical reflection on water issues.

20 A helpful compendium of professional codes of ethics for various water resources-related professions can be found at the *Geoethics* site: http://serc.carleton.edu/geoethics/prof_soc.html, accessed 22 July 2016.

21 www.asce.org/code-of-ethics/, accessed 22 July 2016.

22 Additional information can be found at www.icog.es/iageth/index.php/20-2/ and www.geoethics.org/#!who/c22as, accessed 22 July 2016.

23 www.geoethics.org/#!geopromise/c21cy. This proposal was based on the published work of Matteucci et al. (2014), accessed 22 July 2016.

24 See, for example, *American Water Corporate Responsibility Report, 2013–2014*, www.amwater.com/files/American-Water-CR-2013-14.pdf, accessed 22 July 2016 and *Nestle in Society: Creating Shared Value*, 2014, www.nestle.com/asset-library/documents/library/documents/corporate_social_responsibility/nestle-csv-full-report-2014-en.pdf, accessed 22 July 2016.

25 The UN Global Compact consists of 10 principles governing human rights, labour, the environment and corruption to which corporations have committed, including:

Principle 1: Businesses should support and respect the protection of internationally proclaimed human rights;
Principle 7: Businesses should support a precautionary approach to environmental challenges;
Principle 8: Undertake initiatives to promote greater environmental responsibility; and
Principle 9: Encourage the development and diffusion of environmentally friendly technologies.

26 www.globalreporting.org/Pages/default.aspx, accessed 22 July 2016.
27 www.globalreporting.org/standards/g4/Pages/default.aspx, accessed 22 July 2016.
28 www.iso.org/iso/iso14000, accessed 22 July 2016.
29 http://ceowatermandate.org/files/CEO_Water_Mandate.pdf, accessed 22 July 2016.
30 http://w2.vatican.va/content/francesco/en/encyclicals/documents/papa-francesco_20150524_enciclica-laudato-si.html, accessed 22 July 2016.

References

Alliance for Water Stewardship 2014, *The AWS International Water Stewardship Standard*, accessed 6 July 2016, www.allianceforwaterstewardship.org/aws-standard-system.html#aws-standard

American Society of Civil Engineers 2006, *Code of Ethics*, accessed 6 July 2016, www.asce.org/uploadedFiles/About_ASCE/Ethics/Content_Pieces/CodeofEthics2006.pdf

American Society of Civil Engineers 2016, *Policy Statement 418 – The Role of The Civil Engineer in Sustainable Development*, Board of Direction, American Society of Civil Engineers, 9 July 2016, accessed 16 November 2016, www.asce.org/issues-and-advocacy/public-policy/policy-statement-418---the-role-of-the-civil-engineer-in-sustainable-development/

Barlow, M 2007, *Blue Covenant: The Global Water Crisis and the Fight for the Right to Water*, McClelland & Stewart, Toronto.

Barlow, M & Clarke, T 2002, *Blue Gold: The Battle Against Corporate Theft of the World's Water*, Stoddart, Toronto.

Brown, P & Schmidt, JJ (eds) 2010, *Water Ethics: Foundational Readings for Students and Professionals.* Island Press, Washington, DC.

Chamberlain, G 2008, *Troubled Waters: Religion, Ethics and the Global Water Crisis*, Rowman & Littlefield, Lanham, MD.

COMEST 2004, *Best Ethical Practice in Water Use*, UNESCO, Paris.

Committee on Economic Social and Cultural Rights 2003, 'General Comment 15, the right to water, twenty-ninth session, 2002', *Compilation of General Comments and General Recommendations Adopted by Human Rights Treaty Bodies*, United Nations Document HRI/GEN/1/Rev.6, New York.

Francis, Pope 2015, *Laudato Si: On Care for Our Common Home*, Our Sunday Visitor, Huntington, NJ.

GLAAS 2014, *Global Analysis and Assessment of Sanitation and Drinking Water*, UN Water and WHO, Paris.

Global Water Partnership 2012, *Social and Economic Value of Water*, Global Water Partnership, Stockholm, accessed 6 July 2016, www.gwp.org/en/The-Challenge/What-is-IWRM/IWRM-Principles/Social-and-economic-value-of-water/

Groenfeldt, D 2013, *Water Ethics: A Values Approach to Solving the Water Crisis*, Routledge, London.

Harvey, D 2005, *A Brief History of Neoliberalism*, Oxford University Press, Oxford.

International Association for Promoting Geoethics 2015, *The Geoethical Promise*, accessed 6 July 2016, www.geoethics.org/#!geopromise/c21cy

Leopold, A 1949, *A Sand County Almanac*, Oxford University Press, New York.

Leopold, L 1990, 'Ethos, equity, and the water resource: The 1990 Abel Wolman Distinguished Lecture', *Environment: Science and Policy for Sustainable Development*, vol. 32, no. 2, pp. 16–42.

Liu, J, Dorjderem, A, Fu, J, Lei, X, Liu, H, Macer, D, Qiao, Q, Sun, A, Tachiyama, K, Yu, L & Zheng, Y 2011, *Water Ethics and Water Resource Management*, UNESCO Ethics and Climate Change in Asia and the Pacific Project Working Group 14 Report, UNESCO, Bangkok.

Llamas, MR, Martínez-Cortina, L & Mukherj, A (eds) 2009, *Water Ethics*, CRC Press/ Balkema, Leiden, The Netherlands.

Matteucci R, Gosso G, Peppoloni S, Piacente S & Wasowski J 2014, 'The geoethical promise: a proposal', *Episodes*, vol. 37, no. 3, pp. 190–1.

Peppard, CZ 2014, *Just Water: Theology, Ethics, and the Global Water Crisis*, Orbis Books, Maryknoll, NY.

Priscoli, J, Dooge, J & Llamas, R 2004, *Water and Ethics Overview*, UNESCO, Paris.

UN General Assembly 1999, *Resolution 54/175 The right to development*, United Nations General Assembly, adopted 17 December 1999, published 15 February 2000, accessed 16 November 2016, www.un.org/en/ga/search/view_doc.asp?symbol=A/RES/54/175

UN General Assembly 2010, *United Nations General Assembly Resolution A/Res/54/175: The Right to Development*, accessed 6 June 2016, https://documents-dds-ny.un.org/doc/UNDOC/GEN/N09/479/35/PDF/N0947935.pdf?OpenElement

UN General Assembly 2015, *Resolution 70/169: The Human Rights to Safe Drinking Water and Sanitation*, accessed 7 July 2016, www.un.org/en/ga/search/view_doc.asp?symbol=A/RES/70/169

UN Human Rights Council 2010, *Resolution 15/9: Human Rights and Access to Safe Drinking Water and Sanitation*, United Nations Human Rights Council, adopted 30 September 2010, published 6 October 2010, accessed 16 November 2016, https://documents-dds-ny.un.org/doc/UNDOC/GEN/G10/166/33/PDF/G1016633.pdf?OpenElement

UN Office of the High Commissioner for Human Rights 2007a, *Fact Sheet 35, the Right to Water*, UN Office of the High Commissioner for Human Rights, Geneva.

UN Office of the High Commissioner for Human Rights 2007b, *Consultation on Human Rights and Access to Water and Sanitation: Guide for Discussion*, UN Office of the High Commissioner for Human Rights, Geneva.

UN Sub-Commission on the Promotion and Protection of Human Rights 1998, *The Realization of Economic, Social and Cultural Rights: The Right of Access of Everyone to Drinking Water Supply and Sanitation Service*, accessed 10 June 2016, www.refworld.org/docid/3b00efbf30.html

United Nations 1966, International Covenant on Economic, Social, and Cultural Rights, 16 December 1966, United Nations Treaty Handbook, accessed 16 November 2016, https://treaties.un.org/doc/Publication/MTDSG/Volume%20I/Chapter%20IV/IV-3.en.pdf United Nations 2016, *International Decade of Action 'Water for Life' 2005–2015*, accessed 5 July 2016, www.un.org/waterforlifedecade/human_right_to_water.shtml Walters, V 2013, *Water, Democracy and Neoliberalism in India*, Routledge, London.

WCED 1987, *Our Common Future*, Oxford University Press, New York.

Wolf, AT 1999, 'Criteria for equitable allocations: the heart of international water conflict', *Natural Resources Journal*, vol. 23, no. 1, pp. 3–32.

World Meterological Organization 1992, *The Dublin Statement and Report of the Conference: International Conference on Water and the Environment: Development Issues for the 21st Century: 26–31 January 1992, Dublin, Ireland*, Hydrology and Water Resources Department, World Meteorological Organization, Geneva.

Part I

Ethics and epistemology

3 What is water ethics and to what end do we study it?

Lessons for the Water Ethics Charter

Simon Meisch

1. Introduction

Water ethics is an application-oriented ethics[1] addressing water as a field of action that is increasingly regarded as problematic. Water ethics deals with human water practices. It reviews the normative and evaluative presuppositions and implications of these practices and aims to argue for justified, ethically reflected practices. As such, water ethics is both critical and constructive: It critically reflects existing water concepts and policies and argues for a different and more just water practice in contexts in which present practices have become questioned.

Water ethics is an intriguing subject. In the academic literature, it is regarded as an emerging field of application-oriented ethics (e.g. Kowarsch 2011; Groenfeldt 2013; Ziegler & Kerschbaumer 2012):

> The notion of a water ethics has emerged over the past 10 years only and is still relatively unknown in the field of applied ethics. The extent of its body of literature is rather small up to now. It consists of a heterogeneous set of approaches to the value of water in different respects (cultural, religious, ecological and economical) and associated with the debates on sustainable development and environmental justice.
>
> (Grunwald 2016, p. 12)

Grunwald's observation is correct. There is an emerging ethical discourse regarding itself as water ethical. Yet, how is it possible then in view of a multi-thousand-year-old tradition of reflecting and regulating water that a water ethical discourse is still nascent?

In the following, I will try to give an answer to this question by dealing with what water ethics is and to what end we study it. We will see how academic water ethical discourses can help to produce a Water Ethics Charter.[2] This charter is meant as a self-commitment tool for a more ethically reflected governance of water bodies. It is promoted by the Water Culture Institute (Santa Fe, NM, USA) and the Botin Foundation (Santander, Spain). As the Water Ethics Charter is still in a draft stage, it is not useful to comment on its text in detail since the final version might not contain phrases or paragraphs discussed here. Instead, I discuss relevant normative issues by taking Sandra Postel's essay 'The missing piece: a water ethic' (2010) as a proxy for ethical issues relevant for and addressed in the Water Ethics Charter. Both Postel and the Charter voice similar critiques of water governance and come to comparable conclusions. Meanwhile, the charter mentions further aspects, which will be discussed subsequently.

The chapter first introduces Postel's (2010) call for a new water ethic (section 2). Subsequently, it discusses water ethics as an application-oriented ethics and asks what this means for the water ethic suggested by Postel (section 3). By arguing for a global water ethics, we are between the poles of justifying universal ethical norms (such as the human right to water) and respecting particular moral contexts. We learn how we can use the concepts of sustainable development and value-reflexive governance (Meisch 2014; Meisch et al. 2012) in order to address this task (section 4). I will then suggest how the Charter might benefit from an exchange with academic water ethics (section 5). Concluding, I will return to the reasons why water ethical discourse is seen as emergent.[3]

2. A (new) water ethic(s)

Postel argues for a water ethic as a means to cope with water scarcity. She starts with the question of why modern water management has failed to provide effective solutions against water scarcity in spite of 'ever greater amounts of money and ever more sophisticated engineering' (Postel 2010, p. 221). Her answer is simple: an insatiable human demand for water meets ultimately finite water resources. Although there are attempts to use it more efficiently, much more can and needs to be done. For Postel, saving water does not suffice, because something else and more fundamental is missing. She later argues that this missing piece is a (new) water ethic (Postel 2010, p. 222).

Postel claims that the reification and commodification of water by humans are responsible for maldevelopments. On the one hand, water receives a value only as a resource socially acquired for farming, production or domestic use. On the other hand, its real relevance for all life on Earth (human and non-human) remains un- or undervalued. Modern societies have taken for granted that they can use water but without taking the responsibility to 'preserve and protect it' (Postel 2010, p. 222). Postel claims that prevailing utilitarian approaches to water (and market models as their related social institutions) cannot and might never grasp the real value of water. In order to value water correctly, a new water ethics is needed, that is, 'a guide to right conduct in the face of complex decisions about natural systems that we do not and cannot fully understand' (Postel 2010, p. 222). Central to this water ethic is the protection of freshwater ecosystems that, according to Postel, present approaches achieve only insufficiently. Therefore, she calls for 'a historic shift away from the strictly utilitarian approach to water management and towards an integrated, holistic approach that views people and water as interconnected parts of a greater whole' (ibid.). Instead of legitimizing the reification and commodification of water and the manipulation and control of water bodies, a water ethic should ask 'how we can best satisfy human needs while accommodating the ecological requirements of freshwater ecosystems' (Postel 2010, p. 222). For Postel, this is about dealing with human values that help to reconcile human and ecosystems' need for water. She goes on to argue for a right to water for rivers and all life within them. As water is essential for all life and ecosystems, all should get their fair share 'before some get more than enough' (Postel 2010, p. 223). Referring to legal discussions, she introduces two principles (public trust and the precautionary principle) that might help to achieve a fair distribution of water (Postel 2010, p. 223–4).

Summing up, Postel calls for a new water ethic that extends moral consideration to the natural environment. By rejecting so-called utilitarian approaches to water and related market-based models, she makes two claims. First, these approaches are not 'universal'. This might mean, on the one hand, that they neither include all relevant moral subjects and, on the other hand, that they do not have universal validity (Postel 2010, p. 224). Second, present approaches can be changed. Postel regards ethics as dynamic and evolving 'with social consciousness' (Postel 2010, p. 224). Change is possible but it requires 'leaders, movements, advocates, and activists that collectively pulled society onto higher moral ground' (Postel 2010, p. 224). These change agents make societies aware of the detrimental 'consequences of global environmental change' and thereby trigger a new water ethic (Postel 2010, p. 224).

Groenfeldt (2011, p. 109) adds that it is more appropriate to speak of ethics (in plural) than of ethic. People have different normative and evaluative views on water. The plurality of values with regard to water is the central issue distinguishing the Water Ethics Charter from Postel's text. The charter focuses more intensely on different social, cultural and spiritual values and especially those of indigenous peoples. Furthermore, it considers more principles than public trust and the precautionary principle. The Water Ethics Charter and Postel's text share a value approach and the belief that water ethics is about providing a normative frame for a different conduct with water. Ethics thus understood has to provide means to come from different values to arguments for actions. As we will see later, this is no easy task (section 4.2).

3. Water ethics

3.1 Preliminary notes on terminology

The academic literature distinguishes between ethics and morals. Ethics is the philosophical discipline that critically and systematically analyses and reflects morals. Accordingly, morals are the subject area of ethics. Morals are individually or collectively held sets of values and norms guiding action. Moral judgements evaluate whether means or aims of action are morally good. Ethical reflection asks if moral claims are justified and reflects the validity of moral statements (Düwell 2008, p. 34; Ricken 2013, pp. 16–18). (Normative) ethics assesses moral statements with regard to means–end rationality, to questions of the normatively

Right or to questions of the evaluatively Good. In reality, boundaries between ethics and morals are fluent. On the one hand, individually or collectively held sets of values and norms have implicit assumptions about why they are valid. On the other hand, doing ethics can become an action and thus a moral practice as well (Dietrich 2007, pp. 35–6; Ricken 2013, p. 17).

Ethics as the philosophical reflection on morals is usually divided into three sub-disciplines: meta-ethics, normative ethics and descriptive ethics (Düwell et al. 2011, pp. 2–3; Düwell 2008, pp. 35–6). *Meta-ethics* is the epistemology of ethics: it reflects the necessary preconditions for ethical argumentation, that is, the language and logic of ethical discourses, the methods of ethical arguments and the potential of ethical theories (Pieper & Thurnherr 1998, p. 10; Düwell et al. 2011, p. 2). *Normative ethics* makes claims about what is morally correct and gives reasons for these claims. It reconstructs normative or evaluative claims, but also and more importantly argues what are morally correct claims and gives reasons why they are morally correct. For instance, it gives reasons for a human right to water and specifies what it means in particular contexts. Water ethics does not content itself by referring to political or legal authorities or to a consensus within a global NGO community but aims to give reasons for arguments (O'Neill 2009). Normative claims can be distinguished with regard to their scope. *Individual ethics* deals with the evaluative and normative foundations of the actions of individuals; *social ethics* asks for the just social order and moral foundations of social coexistence. Both perspectives are connected. While the former reflects an individual's options, it presupposes the latter, which sets its possibility conditions (Mieth 2011). For instance, in the first case, we would argue for legitimate claims an individual has to water and in the second, we would look for fair institutions securing these claims. *Descriptive ethics* describes and reconstructs an existing moral system or moral systems. Today, this is mainly done by different social or cultural sciences. One would ask which moral claims, attitudes and institutions exist in a certain country at a certain time. Descriptive ethics reflects on morals with an empirical-descriptive or an historical interest. However, it can neither normatively justify and criticize morals, nor can it normatively reconstruct moral systems. Both are the subject of normative ethics. This presupposes that one cannot extrapolate from the existence of a given set of moral convictions to its ethical justification. As can be seen with slavery and sexual or cultural discriminations, there might be accepted social practices that turn out to be deeply immoral (O'Neill 2009, pp. 221–3). Against the background of the aforementioned, we can conclude that water ethics particularly deals with questions of meta-ethics and normative ethics with regard to water practices.

Application-oriented ethics investigates and specifies ethical reflection in specific fields of action. Accordingly, it aims to argue for the morally right practice in individual and social contexts in which normative certainties fade away and reassurance is needed (Böhme 1997; Stoecker et al. 2011). As will be shown later, application-oriented ethics requires empirical knowledge about what is the case in a particular field of action as well as the ethical competence to assess and reflect relevant norms and values and to come to a moral judgement (section 3.7).

The outline of application-oriented ethics is thus always inter- as well as trans-disciplinary. It also has an emancipatory side because it intends to enable humans in their own fields of action to identify and process moral issues and to create a morally responsible practice themselves. With this, they do not need to delegate their concerns to moral authorities such as academic philosophy or theology, the churches or positivist legal opinion alone (Mieth 1993). Ethical issues are to be processed where they emerge and together with those persons for whom they emerge and who are involved. Application-oriented ethics makes existing normative implications explicit and accessible to a rational discourse in order to achieve a legitimate practice. It is thus not true that it moralizes social contexts that before were value-free. Water ethics understood as application-oriented connects to the motivation and intention of Postel and the Water Ethics Charter. Both observe that present water governance has become questionable because it aims to apply a one-size-fits-all-approach – utilitarianism, in their reading – to water. Instead, they want to solve water-related issues with the people to whom they emerge. Below, we will address (mainly meta-ethical) questions to which these water ethical approaches need to find an answer.

3.2 Morals or ethics?

Let's return to Postel's text. She calls for a different water practice and proposes an ethical approach that she believes can better justify this different practice. Her normative argument might be reconstructed as follows. All living creatures and ecosystems need water; therefore, they all have the same legitimate right to that water. So far, humanity has failed to understand and acknowledge this right. As the present moral practice (based on utilitarianism[4] and the social institution of the market) cannot protect the legitimate right of all living beings and ecosystems, a different normative and evaluative approach (ethical holism) is necessary that better protects the needs of all water users.

However, I think as water ethicists we cannot stop here, that is, criticizing an existing moral practice and calling for another one. O'Neill (2009, p. 220) reminds us of the high self-expectations of application-oriented ethics:

> Attempts to provide ethical instruction or edification, or to shape readers' character, outlook or cognitive capacities, are likely to be seen as old-fashioned, even naïve, and as making normative assumptions for which no adequate reasons are given, of seeking (at best) to persuade and not to reason.

Therefore, we have to look critically into Postel's argument by assessing and reviewing her description of reality, the normative and evaluative basis of her argument and the coherence of her moral judgement. How does she justify her claims? What follows from her extension of the moral community? Which other ethical theories are able to justify alternative moral practice? All questions assume that it is possible to argue rationally about ethics; yet this is contested, as well as the possibility to come to a universal ethics that Postel seems to favour. Finally, one can ask if ethics is at all able to arrive at alternative moral practices – or if this is beyond the legitimate scope of what ethics can achieve. These are only some of many challenges water ethics has to deal with in order to argue for a different water practice. Water ethics as an application-oriented ethics seeks to give reasons for its claims and thus needs to address its meta-ethical assumptions (O'Neill 2009).

3.3 Ethical cognitivism and non-cognitivism

Is it possible to *argue* about morals at all? This is a basic meta-ethical question. Yet, if we want to do water ethics, we should at least have an idea how we want to do it. (Ethical) cognitivism and non-cognitivism give different answers respectively. Cognitivist approaches assume that moral judgements are truth-apt, that is, that it is possible to evaluate moral sentences or judgements as correct or incorrect according to epistemological criteria. Cognitivists defend the perspective that moral statements are valid independently of the person making them.[5] Non-cognitivists deny the possibility that we can rationally argue about morals. For them, what is morally right cannot be a question of cognition, but must be one of feelings, personal judgements of taste or societal conventions. In this reading, it is impossible to *argue* about ethics. As non-cognitivism runs into self-contradictions, it cannot convince (Quante 2013). By giving reasons, it denies the possibility of justifying norms. Yet in doing so, it acknowledges reason as a final authority to judge actions and thereby chooses a cognitivist position. Besides, non-cognitivists have to face the challenge of finding (non-argumentative) ways to convince other people without manipulating them (Ricken 2013, pp. 56–9).

So, with that in mind, what can we make of Postel's idea of water ethics? As she argues for a different practice, we can assume that she favours a cognitivist approach. The decision for either approach has consequences for 'leaders, movements, advocates, and activists' when 'pulling society onto higher moral ground' (Postel 2010, p. 224): do they *argue* for another practice *by giving reasons*? Or do they appeal, request or plea for another practice by more or less manipulative means? In the understanding of this chapter, water ethics is cognitivist. It believes that we can argue about normative claims with regard to water. That means we can give reasons for values and norms.[6]

3.4 Universalism

Postel rejects present normative paradigms with regard to water because they are not 'universal' (Postel 2010, pp. 222, 224). In the context of her essay, we can read her in two ways. First, Postel believes that it is possible to argue for norms with universal validity. In this reading, she criticizes present water approaches that rest on maxims that cannot be universalized. Second, by criticizing the status quo for only recognizing human water needs but not those of other living beings and ecosystems, Postel argues for an extension of the moral community. In this reading, 'universal' refers to the scope of the moral community, that is, to the questions who deserves moral consideration and why, what follows from the attribution of a moral status and who is included (Düwell 2002, p. 418). The issue of moral universalism (in both readings) features prominently within water ethical discourses (see Schmidt & Peppard 2014).

3.4.1 Universalism and relativism

The meta-ethical debate between universalism and relativism concerns primarily the validity of moral principles. Universal approaches claim that it is possible to argue for criteria of what is morally right that are valid independently of time and space. In contrast, relativistic approaches deny this and insist that ethical judgements are inevitably connected to cultural and historical contexts (Quante 2013). Relativistic approaches hold the meta-ethical view that moral judgements are only valid relative to the standards of an individual or a group holding these standards. Normatively, they claim that others should not interfere and be tolerant. Obviously, the meta-ethical and normative assumptions contradict each other. Relativists need to argue for a universal understanding of tolerance that would conflict with their own meta-ethical claim. At the same time, relativists have no argumentative means against groups that do not want to be tolerant. There are also empirical arguments saying that the variety of different moral practices shows that universal ethics cannot do justice to this diversity. While it is important for ethics to consider diverse values and norms within a field of action, the diversity of values and norms does not prove that universal ethics are impossible.

Ethical universalism is highly contested because it is believed to be ignorant of different social and cultural water contexts (see e.g. Schmidt & Peppard 2014). In this critique, the same principle would be forced onto reality everywhere. Yet, one does not 'apply' ethics in a deductive sequence that will always come to the same moral judgements regardless of contextual conditions. We need to distinguish between moral principles and their argumentative elaboration in specific social circumstances. Moral judgements depend on prescriptive *and* descriptive premises. Therefore, different moral practices with regard to water can still be grounded in a universal norm, for example, that everybody ought to have enough clean and accessible water. Due to different social circumstances, different moral judgements are possibly based on the same universal moral principle (Quante 2013, pp. 151–5; Ricken 2013, pp. 37–8; Mieth 2002, pp. 65–71). For instance, enacting the right of water in a region with a developed infrastructure or plenty of water obviously poses other challenges than in a region without infrastructure or low water resources. Even in areas with hardly any water, the right is of great value if it can tell us how to organize access to and allocation of water. Lack of water does not prove the right to water wrong but serves as an extreme example of the difficult challenges we face when enacting moral principles. Ethical universalism is not blind to different social contexts.

3.4.2 Moral community

We will now deal with the second interpretation of 'universal' in Postel's paper. She calls for an extension of the moral community 'to rivers, plants, fish, birds, and the ecosystems of which they are a part' (Postel 2010, p. 224) because present anthropocentric approaches have led to detrimental consequences of global environmental change such as water scarcity. Just as in other environmental ethical discourses, in water ethics too the question is relevant who is part of the moral community (Krebs 1999), that is, to whom do moral agents have direct moral obligations. In ethical debates, some argue that the moral community cannot be smaller than humanity (Tugendhat 1997). Beyond that, physiocentric (i.e. non-anthropocentric) approaches argue that being member of humanity cannot be a sufficient criterion to determine who belongs to the moral community. Among these approaches, sentientists see the ability to experience positive and negative feelings as the crucial feature. Therefore, all creatures that can experience these feelings (usually by having a central nervous system) are members of the moral community. Biocentrists take a wider perspective and believe that life is the determining criterion. Finally, ethical holists hold that all nature (living and not-living) deserves consideration and should be included into the moral community (see Krebs 1999).[7] All approaches aim to give reasons for their respective boundary of the moral community. It is obvious that the decision for either one has great implications for water ethics: is it about human water needs? Also about animals and plants? Do ecosystems such as rivers also have a right to their water? All approaches are contested for different reasons that cannot be discussed here comprehensively (see Krebs 1999). Krebs rightly states that bio- and ecocentric approaches gain plausibility because environmental discourses take place between two stereotypical extremes: these are 'instrumentally truncated anthropocentrism' equated with the instrumental value of nature on the one hand, and 'absolute physiocentrism' equated with the absolute value of nature. If we reduce (water) ethics to these two oppositions, many differentiations get lost. According to Krebs, a different discursive field seems more productive:

> *Enlightened anthropocentrism* does not reduce nature to an instrument for human pleasures but accords nature various kinds of eudaemonic intrinsic value: aesthetic intrinsic value, *Heimat* value, sacredness. *Extensionalist physiocentrism* does not accept the absurdity of absolute values which are of value to nobody but extends elements of human moral culture, notably respect for the well-being of others, to nature. Since virtually the entire appeal of absolute physiocentrism rests on the failure of instrumentally-truncated anthropocentrism to account for the richness of our attitudes and feeling toward nature, absolute physiocentrism loses its appeal once we explore the intermediate terrain of enlightened anthropocentrism and extensionalist physiocentrism.
>
> (Krebs 1999, p. 137)

Postel calls for ethical holism because the status quo, by using nature (and water) instrumentally, causes water scarcity. With this, she precisely seems to reproduce the dichotomy that Krebs criticizes. The point I want to make here is that water ethics does not need to be physiocentric if it wants to argue for a different human attitude to and handling of water.

3.5 Is-ought-fallacy

Postel argues that all living beings, ecosystems and rivers need water; therefore, they should have it. With this, she gets close to deducing a normative conclusion from empirical premises. Since David Hume, this issue has been discussed as the is-ought-fallacy, that is, the idea that it is logically not possible to come to moral judgements from descriptive premises alone (Potthast 2015, pp. 131–3).[8] From the very fact that all living beings, rivers and ecosystems need water, one cannot deduce that they should have it. One needs prescriptive premises giving reasons for the claim that all animate and inanimate beings have a right to water. In our example, Postel justifies her claim with a form of ethical holism, yet without further discussing this approach.

For water ethics, this means that we have to become conscious of hidden normativities and make them accessible for ethical reasoning. Otherwise, there is always the danger of the is-ought-fallacy. The concept of moral judgements can help us to identify hidden assumptions but also to constitute argumentatively statements on what we ought to do.

3.6 Moral judgements

A moral judgement is the conclusion of a descriptive *and* a normative premise. That is why they are called mixed judgements – and why application-oriented ethics is always inter- and transdisciplinary. The underlying argumentative figure is the practical syllogism.

Constituting moral judgements needs *descriptive* premises about what is the case in a specific field of action such as water. Reviewing these premises require different forms of knowledge, that is, more than scientific knowledge alone. For reconstructing and reviewing *normative* premises, some ethical expertise is imperative (Düwell 2008)[9] including the ability to philosophically evaluate and give reasons for this evaluation and to recognize moral problems at all that we want to solve argumentatively (Dietrich 2007, pp. 40–4).

There are meta-ethical considerations such as the is-ought-fallacy calling for the separation of descriptive and normative premises (Potthast 2015). However, from this, one cannot conclude that there are descriptively examinable, value-free facts that in subsequent argumentative steps need to be evaluated normatively. Especially when statements describe social circumstances, they are linked to normative contexts and can therefore never be value-free. This model of moral judgements allows for analysing terms and conceptions with regard to their descriptive and normative components and examining their validity. Meanwhile, it becomes apparent that, against the backdrop of new or differently assessed normative references, different facts become relevant for the evaluation of a situation. Likewise, new facts can reveal further relevant values and norms (Dietrich 2007). When discussing descriptive and normative premises, the model considers the facts and values of all affected people and tries to systematize them in order to find the best argumentative status and a conclusion (Mieth 1993; Funtowicz & Ravetz 2015).

We can also find moral judgements within concepts or terms such as biodiversity that do not only describe a certain empirical form of biological diversity but also entails an evaluation that this diversity is good (Potthast 2015). We describe these terms as 'thick moral concepts' or 'epistemic-moral hybrids'. Thick moral concepts comprise descriptive as well as normative or evaluative elements, that is, facts and value judgements, which are closely intertwined in common language usage ('thick') but which nevertheless can be separated analytically (Dietrich 2004, p. 21; Ricken 2013, pp. 62–5). This hybrid might prompt meta-ethical as well as epistemological objections. Meta-ethical objections refer to the is-ought-fallacy. The epistemological objection is related to the meta-ethical one. In conventional understanding, science is about objective facts. As the biodiversity concept is intertwined with normative statements, its scientific soundness might be questioned. The argument I want to make made here is that there are scientific concepts that evolve and can be understood against the background of a certain normative theory. Yet, they are neither naturalistic fallacy nor political ideology. In epistemological terms, they can be described as 'epistemic-moral hybrids' (Potthast 2015). Regarding scientific concepts in such a way initiates an inter- and transdisciplinary discourse on concepts themselves, their meanings and implications. More precisely, it enables reflection on underlying norms and values.

The model of moral judgements is a valuable tool for water ethics. On the one hand, it helps to analyse terms and concepts of water governance and science with regard to hidden normative assumptions (e.g. with regard to Integrated Water Resources Management). On the other hand, it highlights the interrelation of facts and values. Finally, it raises awareness of inter- and transdisciplinarity.

3.7 Interim conclusion

Above in section 3, I introduced different meta-ethical considerations (morals-ethics, cognitivism-non-cognitivism, universalism-relativism, moral community, is-ought-divide, moral judgements) that water ethics inevitably face and to which it needs to find an answer (O'Neill 2009, pp. 221–3). Water ethics is more than arguing for another water practice but also – and more importantly – giving reasons for one's argument and reflecting on its argumentative status. We need some ethical expertise to argue for a different moral water practice, but also knowledge on practical circumstances in order to build moral judgements. Therefore, water ethics as an application-oriented ethics is always inter- and transdisciplinary. Against this background, water ethics faces two tasks. On the one hand, water ethics critically assesses arguments in order to make implicit normativities transparent and thus open to rational debate. This allows improving a normative argument and identifying evaluative and normative perspectives that have so far been excluded. It is in this respect that water ethics is an emancipatory exercise. On the other hand, it aims to argue what we ought to do by building moral judgements based on descriptive and prescriptive premises. While so far we have dealt with the first task, we will now look into the second.

4. Global water ethics

Together with an emerging water ethical discourse, the insight is growing that we need a water ethics that addresses global challenges. Water relates to and is affected by a global circulation of goods, global capitalism or global environmental changes posing particular moral questions. Meanwhile, particularly with the human right to water or the UN Sustainable Development Goals, there is a global discourse on norms and values concerning water.

Arguing for a global water ethics is challenging in many ways. We want to argue for (different) values and norms guiding water practices worldwide and give reasons for it. At the same time, we want to argue for universal principles that need to, and can be, enacted for different social and cultural as well as physical circumstances. In the following, I will approach the topic of global water ethics from two sides: by sketching a normative frame for a global water ethics and by arguing for value-reflexive governance representing a low-threshold entry into normative debates on water.

4.1 Sustainable water governance: a normative approach

From the beginning, water ethical discourses were related to sustainable development (Grunwald 2016), which has served as a reference point for both criticism of, and alternatives to, the status quo (Meisch 2014, 2016a). The relationship between sustainable development and water ethics is not logically consistent; however, there is an 'elective affinity'. Both deal with ideas of social nature (e.g. 'hydro-social cycle', Swyngedouw 2009) and ask for more just relations in these produced hydraulic environments.[10] If we see sustainable development as a socio-ethical concept, it is more general and encompassing than water ethics, which deals only with some, though important, aspects of this development.[11]

Basically, sustainable development is an ethical concept claiming that all humans today and in the future have the right to live a good life and find the necessary natural and social preconditions to live such a life. Therefore, these natural and social preconditions need to be protected and improved.[12] This understanding might be challenged for being (too) anthropocentric. There are at least two answers to this challenge. First, the presented understanding of sustainable development is in principle open to conceptual extensions of moral community to all (sentient) animals, living beings or ecosystems. Respective arguments need to specify what a 'good life' means with regard to the respective extended moral community and how to solve resulting normative conflicts consistently. Second, even in an anthropocentric approach, the non-human world is morally relevant (as we have seen above in section 3.4.1). As argued earlier, there is a rich field of ethical deliberation on human–nature relations between enlightened anthropocentrism and extensionalist physiocentrism. Debates on sustainable development and what humans need and owe each other might even contribute to and enrich this deliberation. Thus, we have to ask what humans need in order to live a good life. This is a valuable task in itself and a necessary task to spell out what needs to be done in order to achieve sustainable development. Yet, it also reveals that for a good life many different relations to water become visible. So, when arguing for a water practice, we have to explicate and justify what we ought to do.

By referring to the political philosophies of Alan Gewirth and Martha Nussbaum respectively, I will sketch what this might mean.[13] Nussbaum and Gewirth give an answer to the question of what humans need and have legitimate claims to in order to live a good life (Nussbaum 2006; Gewirth 1978). Both regard human dignity as the normative foundation of human rights and aim to justify universal moral obligations in order to protect these rights. Resulting normative duties encompass the protection and support of necessary preconditions of human life. While Nussbaum seeks criteria of a truly human life, Gewirth refers to prospective agency and what humans need in order to be prospective agents (freedom, well-being). Both try to specify what humans need for a truly human life or prospective agency respectively and justify universal moral obligations protecting and supporting these preconditions. As these universal moral obligations neither refer to the realization of specific goals nor the maximization of goods or utilities (Claassen & Düwell 2013, p. 494), both approaches remain open to diverse ways of living. Finally, both highlight negative as well as positive rights. Humans can claim that others do not interfere with their life and that others support them. Thereby, Nussbaum and Gewirth indicate a scope of legitimate state action, that is, supporting and protecting the preconditions for a life of dignity as defined by the respective approaches. It follows from both approaches that a democratic political order is a precondition for a life of dignity. With regard to her list of capabilities, Nussbaum precludes a hierarchy of capabilities and therefore does not offer a principle to rank them. As Gewirth starts from a supreme principle, he can build a hierarchy of rights (relative to the needs of agency).

Both approaches are able to make strong claims for securing human water rights, for example, the right to safe drinking water and sanitation, a basic condition not enjoyed by more than one billion people worldwide (Leese & Meisch 2015). By means of this example, we can understand that sustainable development always has an individual *and* a social ethical perspective: it asks, on the one hand, what individuals can legitimately claim in order to live a good life and, on the other, which social organization is the best to ensure that they can live such a life.[14]

Thus conceptualized, sustainable development contributes to deliberations on an enlightened anthropocentrism. For instance, Nussbaum[15] argues that people need to be able 'to live with concern for and in relation to animals, plants, and the world of nature' or 'to laugh, to play, to enjoy recreational activities'. For the moment, we can leave aside what this means in specific situations. What becomes clear is that both capabilities in different ways also help to determine different waters humans need for a good life. While the first would refer to different landscapes and the water bodies in them, the second refers to activities such as swimming and respective (social and natural) preconditions. More could be said about religious feelings, etc. Gewirth (2001), for example, acknowledges the right of present and future generations to appreciate the natural world and the necessary protection of other living beings. From this, we can also elaborate water ethical claims that need to be enfolded in specific social circumstances.

All this is inevitably still very general – a feature water ethics shares with other fields of application-oriented ethics (O'Neill 2009, pp. 223–4). Water ethics wants to suggest to people involved in a water practice 'how certain sorts of activities might generally be well undertaken' (O'Neill 2009, p. 224). In this line of thought, I wanted to show a way of sketching an ethical frame for global water governance that argues for universal claims but remains open to different conceptions of a good life and the water needed for it and particular social and physical local circumstances.

4.2 Value-reflexive governance of water

Postel argues that we need a debate on human values in order to determine 'the haves and have-nots within a healthy ecosystem' and establish a different water use. Other authors are also critical of present governance approaches that miss acknowledging the manifold value bindings that humans have with regard to water. In the following, I want to argue that paying more attention to the value dimension of water can make water governance more legitimate and democratic (Ingram 2011; Meisch 2014). Moreover, as we will see, it provides an opportunity for engaging people in normative debates on water in a low-threshold way. Yet, using a value approach in order to determine what humans need and can claim within the carrying limits of ecosystems rests on many challenging assumptions.

Against this background, the idea of value-reflexive governance was introduced (Meisch 2014; Meisch et al. 2012). It starts by suggesting a concept of value that allowed for *discussing* different values in the first place. There is an extensive debate on values that cannot be repeated here; yet it shows that the concept is notoriously contested (see Schnädelbach 1983; Joas 2001). Values originate in contexts of action and are connected to specific experiences in which people evaluate something (e.g. actions, persons, institutions, etc.) with regard to its (moral) desirability. Later, these evaluations become detached from their original contexts and more abstract (nouns). Then, they can travel to different social situations. As values originate in experiences, their meaning and name vary. The same value might mean something else in different cultures, and different cultures might have different names for the same valuation. Therefore, values always have a contingent side to them. Values serve as reference points for evaluation. People refer to them in order to determine what is worth desiring. Values have an active and passive, rational and emotional side to them. As they originate in experiences, they have something to do with the identity of people making them. They feel bound to their values and have strong feelings, for example, when the values are violated. At the same time, people can actively refer to values in order to legitimize actions and rationally discuss their value-based decisions. In contrast to norms that have a clear structure and tell people what to do, values are rather ambiguous and serve as motivators or orientation for action. The gap between values and specific actions needs to be bridged by norms.

Value-reflexive governance of water does not pretend to produce *value-based* governance. This would either be too indeterminate, as all governance is necessarily based on values, or it would be inappropriately moralizing when it does not give ethical justification for its own value claims. Value-reflexivity assesses which values are relevant in the debate on water, which or whose values are excluded and what is actually meant by a given value. It also inquiries into the relation of different values to each other. Obviously, there is the cognitivist assumption that it is possible to argue about values in spite of their historical and cultural contingency (Joas 2008). Value-reflexive governance would then have to constitute moral judgements on what to do on the basis of a systematically and hierarchically ordered set of values. On the one hand, this is a matter of ethical competence and, on the other, of political necessity. Value-reflexive governance of water serves as a low-threshold entry into ethical debates. As such, the concept is emancipatory in that stakeholders should participate in political processes affecting them. Starting with their values means introducing what is important to them and what is related to their identities and experiences into an ethical discourse on what to do with regard to water. This cannot stop with mere value talk. All stakeholders' views have to be taken seriously in deliberations on what to do, but they also have to be assessed with regard to ethical justification. After all, one cannot conclude from the empirical existence of a value (set) that it is ethically justified. The resulting challenge consists in finding an ethical frame that acknowledges different perspectives on values and a good life, formulates universal ethical principles to structure value debates and requires stakeholder participation. I regard the concept of sustainable development as such a frame.

5. Lessons for the Water Ethics Charter

After addressing what water ethics is and what it can do, I will ask of what use that might be for a Water Ethics Charter. Thus, this concluding section raises some questions that are relevant for the Charter:

- What is the Water Ethics Charter about? Educating morals or inviting ethical discourse? In the first case, the charter knows what is better and wants to help others to live accordingly (see section 3.2). In the second, it might provide a frame with 'red lines' and opens up a process in the course of which people discuss their values and are helped to arrive at moral judgements (see sections 3.6. and 4.2).

- Does the charter rest on the assumption that discussing normative and evaluative issues carries with it a gain in rationality? Or is it about appealing to emotions and feelings (see section 3.3)?
- How many thresholds are there for stakeholders to enter ethical debates? How much ethical expertise does the charter presuppose (see sections 3.6 and 4.2)?
- What are underlying or implicit normative and evaluative judgements? Does the charter make these transparent (see sections 3.5 and 3.6)?
- Does the charter make explicit statements about the moral community? What are the oppositions that set the discursive scene (see section 3.4.2)?
- What is the relation of the charter to the political systems? A charter is a self-commitment tool for a more ethically reflected governance of water bodies. Why can the task of bringing about this governance not be achieved by normal politics? There might be good reasons for a charter. Does the charter reflect its social ethical foundations (see sections 3.1 and 4.1)?
- How does the charter solve value and norm conflicts respectively? Is there a supreme normative idea that helps to set priorities? Postel, as well as the charter, name different normative principles such as the precautionary principle, the subsidiarity principle or public trust. How can they be brought into a coherent set (see sections 3.4. and 4.1)?

These are some questions that might be helpful in drafting a charter. This chapter aims to give advice and probably a hint of answers to them. A Water Ethics Charter can indeed be a tool for a deliberation on water use and governance.

6. What is and to what end do we study water ethics?

The chapter started with the puzzling observation that there is a young water ethics while human water dealings are old. One would rather expect an old ethical discourse. There might be three reasons for this. First, talk about a new water ethics might result from a terminological confusion of ethics and morals. In this line of thought, the water ethical discourse would be about establishing different water practices based on more or other values and norms and less about new and different ways and methods of ethically reflecting these practices. Second, timing might be an issue, that is, why this water ethical discourse is emerging now. For sure, it was triggered and is driven by the rejection of political developments (e.g. neoliberalization, commodification of water) (Meisch 2016b) and scientific paradigms (e.g. scientific naturalism, positivism, techno-science) (Meisch 2016c), whose underlying values and norms are regarded as harmful and inappropriate for addressing water scarcity. Third, the novelty of the water ethical discourse might be the result of a lacking interdisciplinary exchange between those scientific disciplines that dominate water now and others that had a say on water before and that are today aligned to the arts and humanities. Insights appear to be new and in need of reflection for some that are not at all new and without thought to others. This chapter addresses the first of these three issues and briefly touches on the second.

So, to what end do we study water ethics? It is evident that in the social field of water former certainties have been lost. Academic water ethics can address general ethics issues as I have done in this chapter. However, there are also many socio-ethical and epistemic issues that need to be dealt with. Water ethics is just as much an environmental ethics as it is a political ethics or an ethics of science.

Acknowledgements

I want to thank Uta Müller, Lieske Voget-Kleschin, Rafael Ziegler, Johannes Lundershausen, Leonie Bossert and Gero Bauer for their comments and feedback.

Notes

1 The academic literature also speaks of 'applied ethics'. I use the term 'application-oriented ethics' to avoid a false analogy. We do not apply ethical principles in the same way the natural sciences apply empirical or theoretical principles (see sections 3.4.1 and 3.6) (see O'Neill 2009, pp. 223–5).

2 Water-Culture Institute: Water Ethics Charter 2016, accessed 30 December 2015, www.waterculture.org/Water_Ethics_Charter.html.

3 This chapter does not aim for a review of the existing water ethical literature (see e.g. Schmidt & Peppard 2014; Kowarsch 2011; Armstrong 2006, 2009).

4 I do not discuss here whether and to what extent the understanding of utilitarianism in Postel's text and the wider water ethical discourse is appropriate and pays tribute to the theoretical richness of utilitarianism.

5 However, if one agrees that ethics can be the subject of cognition, different questions ensue. One of these is the problem of moral realism. If one believes that moral statements can be subject to cognition and exist independently of the person making them, then one has to cope with the question of the nature of these objects of cognition. Do they exist out there in the real world and only have to be recognized? This results in many difficulties that cannot be discussed here (see Düwell 2008; Krebs 1999).

6 I do not consider here if there is such thing as an eternal moral value of water or any moral significance of water that exists independently of humans. I do not believe that this is the case. I do not believe either that moral claims with regard to water are purely subjective.

7 See Chapter 7 of this volume for a biocentric approach to water ethics.

8 A meta-ethically related, but not identical problem, is that of the naturalistic fallacy formulated by George Edward Moore: he does not refer to the level of logical argumentation, but to the question of how we can determine what is really good (Potthast 2015, pp. 133–4). Moore claims that one cannot determine what is morally good from non-moral properties. In this line of thought, it is not possible to conclude that because it is natural that all animate and inanimate beings need water, it is desirable that they should have it.

9 Generally speaking, we can distinguish two ways of finding normative premises (Ricken 2013). First, foundationalist approaches of justification apply a supreme moral criterion to a context and come to a moral judgement in a deductive sequence. We have seen before that this approach is problematic. Second, more coherentist approaches determine the manifold normative references in a field of action, argumentatively examine their contents and bring them into a coherent set of statements. This approach acknowledges that in different social contexts there are different relevant normative references that have to be considered. It also allows for formulating hierarchical relations between different normative statements (Mieth 1993, 2002).

10 What has not been discussed in this chapter but needs to be considered by water ethics is the debate between (ethical as well as epistemic) constructivism and realism and the consequences for water ethics and governance.
11 Meanwhile, there are relevant water ethical questions that can be part of other fields of application-oriented ethics such as technology ethics (infrastructure projects) or cultural ethics (water in the arts).
12 This understanding of sustainable development is wider than the one of the Brundtland report that only speaks of 'basic needs'. While the Brundtland definition is the most famous one, we might ask why we should accept it as the authoritative one and whether there are good reasons to reformulate it.
13 Elsewhere, I have elaborated this thought in more detail (Meisch 2016a). There, I particularly refer to Claassen and Düwell (2013).
14 Due to the limited space of this chapter, I do not discuss marketization and commodification of water (Meisch 2016b). The issue of social organization of water deserves and needs more space. Just as ethics in general, the conceptualization of sustainable development developed here can serve as a critic of the status quo as well as a basis for arguing for an alternative. Deliberation on global water governance would start with the legitimate claims of people and then ask for the most appropriate institution and social mechanism to protect and support these claims. As humans need water regardless of their ability to pay for it, markets and their underlying rationale of making profits might not be the best solution to ensure that everybody has the water he or she needs to live a good life. This might rather be a call for democratic states capable of acting, that is, arguing for limits of markets (Meisch 2015).
15 I do not want to claim here that Nussbaum herself represents an enlightened anthropocentric approach, which at least in *Frontiers of Justice* (Nussbaum 2006) she does not do. However, we might build on her approach when elaborating enlightened anthropocentrism.

References

Armstrong, A 2006, 'Ethical issues in water use and sustainability', *Area*, vol. 38, no. 1, pp. 9–15.

Armstrong, A 2009, 'Viewpoint – further ideas towards a water ethic', *Water Alternatives*, vol. 2, no. 1, pp. 138–47.

Böhme, G 1997, *Ethik im Kontext – Über den Umgang mit ernsten Fragen*, Suhrkamp, Frankfurt am Main.

Claassen, R & Düwell, M 2013, 'The foundations of capability theory: comparing Nussbaum and Gewirth', *Ethical Theory and Moral Practice*, vol. 16, no. 3, pp. 493–510.

Dietrich, J 2004, 'Ethisch-Philosophische Grundlagenkompetenzen: ein Modell für Studierende und Lehrende', in M Maring (ed.), *Ethisch-Philosophisches Grundlagenstudium. Ein Studienbuch*, Lit Verlag, Münster, pp. 15–32.

Dietrich, J 2007, 'Was ist ethische Kompetenz? Ein Philosophischer Versuch einer Systematisierung und Konkretion', in R Ammicht Quinn, G Badura-Lotter, M Knödler-Pasch, G Mildenberger & B Rampp (eds), *Wertloses Wissen? Fachunterricht als Ort ethischer Reflexion*, Klinkhardt, Bad Heilbrunn, pp. 31–51.

Düwell, M 2002, 'Moralischer Status', in M Düwell, C Hübenthal & M Werner (eds), *Handbuch Ethik*, Metzler, Stuttgart, pp. 417–23.

Düwell, M 2008, *Bioethik. Methoden, Theorien und Bereiche*, Metzler, Stuttgart.

Düwell, M, Hübenthal, C & Werner, M 2002, 'Ethik: Begriff – Geschichte – Theorie – Applikation', in M Düwell, C Hübenthal & M Werner (eds), *Handbuch Ethik*, Metzler, Stuttgart, pp. 1–23.

Düwell, M, Hübenthal, C & Werner, M 2011, 'Einleitung. Ethik: Begriff – Geschichte – Theorie – Applikation', in M Düwell, C Hübenthal & M Werner (eds), *Handbuch Ethik*, Metzler, Stuttgart, pp. 1–24.

Funtowicz, S & Ravetz, J 2015, 'Post-normal science', in S Meisch, J Lundershausen, L Bossert & M Rockoff (eds), *Ethics of Science in the Research for Sustainable Development*, Nomos, Baden-Baden, pp. 101–12.

Gewirth, A 1978, *Reason and Morality*, University of Chicago Press, Chicago, IL.

Gewirth, A 2001, 'Human rights and future generations', in M Boylan (ed.), *Environmental ethics*, Prentice Hall, Upper Saddle River, NJ, pp. 207–11.

Groenfeldt, D 2011, 'Book review of Brown and Schmidt (Eds), 2010. Water ethics. Foundational readings for students and professionals', *Water Alternatives*, vol. 4, no. 1, pp. 106–9.

Groenfeldt, D 2013, *Water Ethics. A Values Approach to Solving the Water Crisis*, Routledge, London.

Grunwald, A 2016, 'Water ethics – orientation for water conflicts as part of inter- and transdisciplinary deliberation', in R Hüttl, O Bens, C Bismuth & S Hoechstetter (eds), *Society – Water – Technology. A Critical Appraisal of Major Water Engineering Projects*, Springer, Cham, pp. 11–29.

Ingram, H 2011, 'Beyond universal remedies for good water governance: a political and contextual approach', in A Garrido & H Ingram (eds), *Water for Food in a Changing World*, Routledge, London, pp. 241–61.

Joas, H 2001, *The Genesis of Values*, University of Chicago Press, Chicago, IL.

Joas, H 2008, 'Value generalization. Limitations and possibilities of a communication about values', *Journal of Business Ethics*, vol. 9, pp. 88–96.

Kowarsch, M 2011, *Diversity of Water Ethics – A Literature Review*, accessed 15 January 2016, www.hfph.de/forschung/institute/gesellschaftspolitik/forschung/wirtschaft-und-ressourcen/wassermanagement/2nd-working-paper-sept-2011.pdf.

Krebs, A 1999, *Ethics of Nature. A Map*, De Gruyter, Berlin and New York.

Leese, M & Meisch, S 2015, 'Securitising sustainability? Questioning the "water, energy and food – security nexus"', *Water Alternatives*, vol. 8, no. 1, pp. 695–709.

Meisch, S 2014, 'The need for a value-reflexive governance of water in the Anthropocene', in A Bhaduri, J Bogardi, J Leentvaar & S Marx (eds), *The Global Water System in the Anthropocene: Challenges for Science and Governance*, Springer, Cham, pp. 427–37.

Meisch, S 2015, 'Knowing one's food – making food a public issue', in D Dumitras, I Jitea, & S Aerts (eds), *Know Your food. Food Ethics and Innovation*, Wageningen Academic Publishers, Wageningen, pp. 306–11.

Meisch, S 2016a, 'A fair distribution within the Anthropocene: a normative conception of sustainable development', in P Pattberg & F Zelli (eds), *Environmental Politics and Governance in the Anthropocene. Institutions and Legitimacy in a Complex World*, Routledge, London, pp. 62–78.

Meisch, S 2016b, 'Wasserethik – im schmerz geboren', in M Maring (ed.), *Zur Zukunft der Bereichsethiken. Herausforderungen durch die Ökonomisierung der Welt*, KIT Scientific Publishing, Karlsruhe, pp. 381–399.

Meisch, S 2016c, 'Resilience and water governance in the Anthropocene'. Paper presented at the 3rd European Workshops in International Studies, Tuebingen, 6–8 April.

Meisch, S, Beck, R & Potthast, T 2012, 'Towards a value-reflexive governance of water', in T Potthast & S Meisch (eds), *Climate Change and Sustainable Development. Ethical Perspectives on Land Use and Food Production*, Wageningen Academic Publishers, Wageningen, pp. 413–18.

Mieth, D 1993, 'Norm und Erfahrung. Die Relevanz der Erfahrung für die ethische Theorie und die sittliche Praxis', *Zeitschrift für Evangelische Ethik*, vol. 37, no. 1, pp. 33–45.
Mieth, D 2002, *Was wollen wir können? Ethik im Zeitalter der Biotechnik*, Herder, Freiburg im Breisgau.
Mieth, D 2011, 'Sozialethik', in M Düwell, C Hübenthal & M Werner (eds), *Handbuch Ethik*, Metzler, Stuttgart, pp. 517–20.
Nussbaum, M 2006, *Frontiers of Justice*, The Belknap Press, Cambridge, MA.
O'Neill, O 2009, 'Applied ethics: naturalism, normativity and public policy', *Journal of Applied Ethics*, vol. 26, no. 3, pp. 219–30.
Pieper, A & Thurnherr, U 1998, *Angewandte Ethik: Eine Einführung*, C.H. Beck, München.
Postel, S 2010, 'The missing link: a water ethic', in P Brown & J Schmidt (eds), *Water Ethics. Foundational Readings for Students and Professionals*, Island Press, Washington, DC, pp. 221–6.
Potthast, T 2015, 'Ethics and sustainability science beyond Hume, Moore and Weber – taking epistemic-moral hybrids seriously', in S Meisch, J Lundershausen, L Bossert & M Rockoff (eds), *Ethics of Science in the Research for Sustainable Development*, Nomos, Baden-Baden, pp. 129–52.
Quante, M 2013, *Einführung in die Allgemeine Ethik*, Wissenschaftliche Buchgesellschaft, Darmstadt.
Ricken, F 2013, *Allgemeine Ethik*, Kohlhammer, Stuttgart.
Schmidt, J & Peppard, C 2014, 'Water ethics on a human-dominated planet: rationality, context and values in global governance', *WIREs Water*, vol. 1, no. 6, pp. 533–47.
Schnädelbach, H 1983, *Philosophie in Deutschland 1831–1933*, Suhrkamp, Frankfurt am Main.
Stoecker, R, Neuhäuser, C, & Raters, M-L 2011, 'Einleitung', in R Stoecker, C Neuhäuser & M-L Raters (eds), *Handbuch Angewandte Ethik*, Metzler, Stuttgart, pp. 1–11.
Swyngedouw, E 2009, 'The political economy and political ecology of the hydro-social cycle', *Journal of Contemporary Water Research & Education*, vol. 142, pp. 56–60.
Tugendhat, E 1997, 'Wer sind alle?', in A Krebs (ed.), *Naturethik – Grundtexte der gegenwärtigen tier- und ökoethischen Diskussion*, Suhrkamp, Frankfurt am Main, pp. 100–10.
Ziegler, R & Kerschbaumer, L 2012. 'Wasserethik – eine Übersicht', *Greifswald Environmental Ethics Papers* No. 3, University of Greifswald, accessed 22 June 2016, www.botanik.uni-greifswald.de/130.html

4 Beyond general principles

Water ethics in a Deweyan perspective

Martin Kowarsch

> (1) By the law of nature these things are common to mankind – the air, running water, the sea, and consequently the shores of the sea. [. . .] (2) All rivers and ports are public; hence the right of fishing in a port, or in rivers, is common to all men.
>
> (*Codex Justinianus* (529), Liber II, 1(1)–(2))

1. More policy-relevant and reliable water ethics

Water management involves lots of difficult ethical questions (see e.g. Armstrong 2006; Brown & Schmidt 2010; Groenfeldt 2013; and the introductory Chapter 1 by Ziegler & Groenfeldt in this volume).[1] Affecting the quantitative availability and the quality of water resources as well as natural water cycles (which are already significantly influenced by humans) can involve diverse far-reaching trade-offs and synergies for human beings and other living things. While these ethical dimensions seem to be increasingly acknowledged by many experts, stakeholders and practitioners involved in water management debates, it is not so clear what moral philosophers – particularly in light of their often controversial and sometimes overly generic ethical guidelines – can actually contribute to water management processes. Related to that, it is also unclear on which trustworthy methodology for water ethics this could be based.

Water ethics address a considerable diversity of aspects of water management, ranging, for example, from purely procedural to particular material issues, and from rights to utility or well-being; they can focus on the concerns of only a few current stakeholders or rather include a broad range of living beings now and in the foreseeable future. The formulations of water-related ethical issues are already frequently highly disputed and controversial, let alone the appropriate ethical responses to these issues. Some even assume that ethical claims are always purely 'subjective' in epistemological terms and merely a matter of individual preferences (Douglas 2009).

Moreover, although a human right to water, for example, seems relatively widely recognized in political debates as a normative guideline for water management, it is disputed what such a right actually means for various more specific ethical questions of water management (see also Chapter 10 in this volume).

This includes, for instance, how to allocate water equitably between, say, the current generation and remote future people, or between upstream and downstream riparians. Does nature have a right to water, too, and what would this mean more precisely in particular contexts of water management? Which use of water should have priority, for example, between particular communal, spiritual or industrial uses of water? Should water be priced and water supply be privatized? More importantly, it is sometimes unclear how these human rights or other ethical claims translate into (or challenge) particular economic, technological, institutional, legal and other water management options. Thus far, moral philosophers often do not specify the practical means required for realizing the suggested ethical claims. These means are, however, precisely what decision-makers are most interested in – in terms of concrete options for action in water management processes. Furthermore, identifying these means is also necessary for critically evaluating the ethical claims themselves, as will be argued below.

The mere existence of disagreement about water ethics claims and unclear means for achieving them is no reason per se to question their validity. However, such disagreement and unclear means make it hard for decision-makers to determine an appropriate role for ethics in water management, that is, to see how water ethics could inform water management processes in a highly policy-relevant, credible and legitimate manner. An approach to water ethics is needed that renders water ethics claims more justifiable and reliable while also acknowledging, and adequately discussing, divergent viewpoints among academics or other stakeholders in water management processes. This is to ensure legitimacy of the ethical advice for water management. Additionally, to increase policy relevance, the aspired approach to water ethics should significantly contribute to the identification and evaluation of specific and relevant water management solutions. This is a highly complex inter- and transdisciplinary challenge.

In the past few years, I have often been asked (in my capacity as moral philosopher), by colleagues from the natural sciences, geography and economics, to provide a general and clear-cut normative-ethical guideline for all cases of water management. I will argue here that this is *not* the most useful approach to water ethics, since different sorts of more specific, conflicting interests and values usually need to be balanced in each case, while often facing much uncertainty and complexity. Rather, this chapter proposes a particular pragmatist approach to ethically reflecting on water-related policy-making and management decisions. This approach puts emphasis on specific contexts and the various practical implications of particular water management options, without, however, joining the camps of moral relativism or radical contextualism. Developing such an approach first requires a discussion of fundamental philosophical issues of ethical epistemology, including how to determine ethical problems, and how to come to reliable hypotheses to convincingly address these ethical problems (section 2). Section 3 explains this approach in more detail, including how to appropriately identify and frame key ethical issues of water management; based on that, it develops a model for how to provide ethical advice to water management processes. Section 4 compares the proposed pragmatist approach, mainly with the popular ambition to develop general ethical principles or values for water management. Section 5 briefly evaluates economic approaches predominating in water management in light of the proposed approach to water ethics. Section 6 concludes.[2]

2. Deweyan pragmatism as ethical epistemology

The core philosophical idea underlying the approach to water ethics developed in this chapter is that (ethical and other) goals and means are interdependent via their practical implications, when being evaluated. This ends–means interdependency is a central assumption particularly for philosophical pragmatism in the tradition of John Dewey (1986, 1988; Putnam 1999), as a philosophy of (social) science and a theory of (e)valuation. The essence of this kind of pragmatism:

> is to trace and evaluate the practical consequences of hypotheses, be they scientific, ethical or just verbalized gut feelings in ordinary life. Hypotheses are possible means to solve any kind of practical problems; the 'rightness' of hypotheses (of any kind) depends on their potential to solve problematic situations, as experienced by us. Though this may sound disturbing for some natural scientists, the pragmatist method builds on the successful principle of experimentation that is fundamental to natural science.
>
> (Edenhofer & Kowarsch 2015)

Based on this, Dewey proposes a pattern of inquiry that should underlie all kinds of more specific research methodologies and that can be applied to a broad range of issues, including philosophical-ethical problems. This general pattern consists of five major steps (Dewey 1986, pp. 105–22; Brown 2012; Kowarsch 2016b):

- noticing a challenging or worrisome situation, e.g. the feeling that a situation is 'unjust' (Sen 2009);
- thoroughly analyzing the problematic situation as a whole, and its causes, constituents, contexts, etc. Moreover, identifying objectives (or broader goals) as potentially problem-solving conditions in this situation;
- developing tentative hypotheses for potential means of any kind (including ethical hypotheses for specific water management cases) for achieving these objectives;
- critically evaluating these hypotheses and possibly re-evaluating the objectives in light of assumed practical implications of the means (in terms of direct effects, unwanted side effects or co-benefits);
- further evaluating the hypotheses for means and the objectives alike in light of the actual practical consequences that these means have once they are employed in reality.

In order to ethically judge about (and to weigh) practical means-implications, ethical principles, values, etc. (e.g. human rights) are helpful as evaluation criteria (which are not necessarily different from predominant ethical approaches, given that Dewey only provides a fundamental ethical methodology). However, in contrast to many other ethical approaches, Dewey emphasizes that these generic evaluation criteria themselves need to be derived from earlier inquiries along this pattern. Moreover, they are incomplete and insufficient to guide the evaluation of means-implications, because also experiences – for example, in terms of what A. Schweitzer called 'Lebensanschauung' (1923) – and the related 'transaction' are required, as Dewey calls the need for deep personal involvement particularly in ethical inquiries (see Kuruvilla & Dorstewitz 2010 for an explanation). This can lead to additional or refined criteria for ethically judging practical means-implications. Consequently, ethical principles, values, etc. are always fallible from a Deweyan perspective and new contexts may require their revision. Section 4 will explain more detail in this regard.

Furthermore, a successful Deweyan inquiry into complex social-ethical issues presupposes a dialogue and collaboration between experts (including moral philosophers) and diverse stakeholders; experts alone cannot have a full overview of all relevant possible means and means-implications, and cannot easily judge after the fifth step whether or not a problematic situation is actually 'resolved' from the perspective of (some of) the stakeholders involved.

To illustrate how Dewey's idea can guide ethical inquiries, imagine a hypothetical case somewhere, say, in Australia, where freshwater particularly for agricultural irrigation becomes increasingly scarce. Conflicts between different groups of water users arise and many jobs in local agricultural companies are at stake. Noticing such an ethically worrisome situation can be regarded as the first step of Dewey's pattern of inquiry. The second step may, for instance, result in the identification of recently intensified agricultural activities, increasing seasonal variability in precipitation patterns, global climate change, etc. as key factors for the worsening of local water scarcity. It also results in determining the basic ethical goal for this case of water management which is to ensure sufficient and affordable freshwater supply for all people and business companies in this particular region. The third step may lead to the hypothesis that technology to increase blue water availability across seasons in this region – for example, building a huge dam nearby to create a new reservoir – may substantially mitigate the trade-offs regarding water use for a while.

However, after the fourth step, with the analysis of potential practical implications of the proposed dam, this hypothesis has to be slightly revised perhaps, for example due to rather high financial costs of the originally envisaged kind of dam which leads to the additional claim of more governmental subsidies for the dam, or because the originally envisaged kind of reservoir could have endangered a particular ecosystem nearby, requiring an enlargement of the basic goal to include also environmental protection. The fifth step – done after actually building the dam – may reveal that additional societally relevant, adverse side-effects have been neglected in the preceding *ex-ante* analyses. For instance, there might be negative consequences for native inhabitants who have used a lake nearby for some traditional religious practices in past centuries, but now this lake has almost disappeared due to underestimated changes in the regional water cycle resulting from the dam project. This involves the need to go back to the second step to revise the problem framing and goal determination (i.e. the second step of the inquiry), as well as the need to find better means to resolve the ethically problematic situation.

How can an inquiry along this pattern, including the experience and transaction involved, generate reliable, convincing and perhaps objective results? Many philosophers have criticized (different strands of) pragmatism for being 'too pragmatic' and giving up the idea of infallible, everlasting knowledge or absolutely true ethical claims (Ryan 2004; Cochran 2010). They assume that pragmatism leads to moral relativism or radical contextualism. While it is indeed true that all hypotheses remain fallible according to Dewey's pragmatism, and while there is certainly no direct access to the world out there that would come without any particular value-laden perspective, Dewey's pragmatism clearly does not lead to moral relativism and radical contextualism. The following brief explanation of how a Deweyan approach can result in objective hypotheses focuses on hypotheses that are repeatedly useful for different (similar) problematic situations (see more detail in Kowarsch 2016b):

> The point is that the problematic situations and their resolutions through scientific inquiry sometimes resemble other situations regarding key characteristics. In this case, the successful results of a pragmatist inquiry could perhaps be applied to these other, similar situations, and perhaps be acceptable for many people. In this sense, [ethical] hypotheses that have repeatedly turned out to be sound and reliable in terms of their practical consequences for the solution of the problematic situation at stake can serve as the premises for further inquiries.
>
> (Edenhofer & Kowarsch 2015)

To summarize, employing the Deweyan pattern of inquiry in order to address philosophical-ethical issues has two major, highly interlinked implications. *First*, the appropriate determination of the ethical problem at stake is crucial for the success of an inquiry; a problem well put is half solved. This requires continuous revision of the problem framing in light of newly identified practical consequences of the means that are employed to resolve the situation. An ethical problem is only 'solved' according to Dewey when the broader – often highly complex – undesirable situation largely disappears for those involved.

Second, ethical goals (or principles, values, etc.) for water management alone cannot justify the means – and the goals cannot be evaluated independently from the available means. Rather, possible means and the goals alike must be critically assessed in light of the diverse actual or potential practical implications of the means. This can change previous evaluations of both the means and the goals dramatically, while 'changing one's values is [. . .] frequently the only way of solving a problem' (Putnam 2004, p. 98). Ethical claims are thus never infallible or represent everlasting truth. Rather, in the Australian example above, the original approach to water ethics (in the second step) which had solely focused on economic well-being had to be substantially revised and amended during the course of this specific ethics inquiry by including also cultural and religious factors as well as ecological, procedural and intergenerational aspects. One could also imagine cases where the initial fundamental ethical claims have to be completely abandoned because of the extremely negative side-effects of the means required to realize these ethical claims.

Having explained in a nutshell some key aspects of a Deweyan-pragmatist approach to ethical issues, the question remains how Dewey's approach relates to other pragmatist approaches. While no meaningful literature review can be provided here due to space constraints, it should at least be clarified that 'pragmatism' includes a broad range of diverse philosophical approaches. Many currently think of Richard Rorty in particular when it comes to pragmatism. However, Rorty's and Dewey's philosophies differ significantly: Rorty chose a radical constructivist approach, in contrast to Dewey (Pihlström 2004). Furthermore, even within the literature explicitly referring to the Deweyan approach, many different things are highlighted (see e.g. Kuruvilla & Dorstewitz 2010). Moreover, in contrast to existing works on 'environmental pragmatism' (e.g. Parker 1996) as well as on 'pragmatist' water ethics (e.g. Norton 2010), this chapter particularly emphasizes, and elaborates on, the ends–means interdependency towards a new model for the role of ethics in water management processes. This will become clearer in the following section.

3. Applying Dewey: cartography of alternative water management pathways

Let's translate the Deweyan theory of (ethical) inquiry into a methodology for water ethics that is relevant to decision-making processes, focusing on problem framing and the exploration of solutions.

First, who defines, and how, the problem at the beginning of an inquiry into particular issues of water ethics? Major issues of water management include that, in some regions, water availability (or quality) decreases while demand for freshwater is often increasing (WWAP 2009). In addition, ecosystems lose their resilience due to direct or indirect human intervention (Falkenmark & Folke 2002). Floods and droughts pose further water-related threats. Major *drivers* include, for instance, climate change and variability; land use changes and hydraulic engineering; demographic factors; technology innovation and implementation; economic development and lifestyle changes; income distribution and allocation; and the quality of governance and management on different governance levels.

While these basic problems and key drivers are well-understood and ethical concepts such as 'realizing the human right to water' or 'ensuring ecosystem resilience' may serve as general ethical goals, water ethics must not neglect the fact that water management usually affects diverse stakes and stakeholders in different ways, and affects several (partly interdependent) policy fields simultaneously – as implied by the long list of drivers. Many studies (e.g. UNEP 2012) point out the complex interdependencies between water management and other policy fields or ethical priorities (climate change, agriculture and food, forestry, land ownership, health, biodiversity and environmental protection, economic well-being, distributional issues, culture, etc.). These studies therefore suggest that water management be embedded in the broader context of sustainable development with its *multiple* objectives entailing lots of trade-offs and synergies. The recent debate on the global Sustainable Development Goals (SDGs) perfectly confirms this viewpoint, highlighting the water–land–food–energy–biodiversity nexus and thus implying multiple ethical 'constraints' for water management.

Thus, given the policy complexity and interdependencies, adequate problem framing for water ethics is very challenging, while there are considerable differences between different cases of water management. As the Australian example above illustrated, (revised) problem framing for water ethics must take into account *all* of the ethically highly relevant aspects that are interlinked with the specific water management case at stake. These aspects are identified based on experience, transaction and ethical claims (principles, values, etc.) – as tentative evaluation criteria and 'heuristic tools' – resulting from previous inquires (see section 4). Then, these ethically relevant aspects have to be translated into objectives or broader goals for the water management case (such as in the Australian example).

In order not to miss important problematic aspects of a particular case, water ethics studies would ideally also involve the different stakeholders affected to discuss the appropriate problem framing with them. Such a collaborative problem framing should be done in an iterative manner, that is, if necessary revising the problem framing in light of diverse means-implications identified in the fourth or fifth step of a Deweyan inquiry. To conclude, an appropriate identification of *ethical* problems of water management in specific cases presupposes a dialogue with different affected stakeholders as well as a thorough understanding of relevant insights from the natural and social sciences and technology on the respective water management case and its complexity.

Second, closely related to what was already said, how can appropriate responses be identified more precisely? How can justifiable ethical hypotheses be generated – in terms of hypotheses resolving a problematic situation? There are a number of possible water management options that are related to the drivers and affected policy fields mentioned above. These include, for example, climate change mitigation and adaptation, laws to avoid particular land use changes, increased participation of affected people, governmental investment in technology (R&D), large-scale geoengineering, genetically modified seeds, constrained consumption, forced migration, reducing population growth, building higher dams or denying access to water for some people or corporations. It is obvious that a few of these possible management strategies are ethically questionable given their adverse side-effects. There are trade-offs, for example, between climate change mitigation (e.g. more hydro power and biomass needed) and protecting water cycles (for which extensive hydro power and biomass production are highly problematic). It is therefore again crucial for moral philosophers to collaborate with both stakeholders and other scientific disciplines to learn about possible options (particularly including policies) as well as especially their diverse practical effects including side-effects and co-benefits. For example, new studies show that the potential for irrigation efficiency improvements in agriculture has been underestimated so far. This may slightly reduce the pressure regarding other, perhaps ethically more problematic measures (Jaegermeyr et al. 2016). Also the Australian example above showed that some ethical (e.g. distributional) issues may disappear by implementing certain technologies or policies, while new ethical problems may be created when employing a particular technology or policy.

Given this, and based on an adequate problem framing, moral philosophers working on water ethics would ideally contribute to an inter- and transdisciplinary exploration of alternative policy pathways (i.e. future scenarios implying different sets of more specific policy objectives realized over time) for specific water management cases, again jointly with stakeholders and decision-makers. This exploration should include the identification and evaluation of policies or other options for realizing these pathways, and in particular the analysis of their positive and negative practical implications, including interdependencies across policy fields. According to Dewey's theory of inquiry, the methodological reason for exploring different *alternative* objectives and options is that this is the only reliable way to determine the best ones, because one first has to understand their different practical implications. Moreover, also for democratic reasons, and because typically so much uncertainty remains, *alternative* policy pathways should be explored rather than presenting only one option (Edenhofer & Kowarsch 2015).

There are three kinds of particular potential contributions from moral philosophers to such a pathway exploration.

- ***Develop ethically acceptable means for water management***: many questions of, for instance, distributional justice, arise in water ethics, including conflicts between upstream and downstream riparians, between different lifestyles and water uses (communal versus industrial), or between the current generation and remote future people (sustainability). These issues are related to many procedural questions (e.g. questions of participation, public debate, or the question of duty-bearers: the state or the individual, etc.). Moral philosophers should – and actually already do – provide preliminary ideas for how these issues can be addressed, that is, hypotheses for promising distribution schemes; proposals for procedural design of policy processes, etc., as means for achieving the goal of overcoming the problematic situation of distributional or procedural conflicts. While such work is a traditional focus and strength of large parts of moral philosophy, it should better focus on more specific cases and their contexts.
- ***Select scenarios to identify means-implications***: moreover, moral philosophers should make recommendations for which policy objectives or means to select to achieve a deeper analysis in the pathway exploration; not all of the countless alternative pathways that are possible in the future can be

assessed interdisciplinarily and in detail. For this purpose, moral philosophers, together with colleagues from the social and natural sciences, may translate major ethical disagreements into proposals for alternative scenarios. It would for instance be enlightening to learn about the different practical implications of alternative water allocation schemes (or other distributional ideas) for a particular case of water management. Biewald et al. (2015), for example, analyze ethical disagreement about future global food distribution and its implications for regional water availability:

> What and how much people eat influences their environments and societies because the extent of agricultural production is closely related to the usage of land, water and food prices. This is why exogenous model assumptions about the amount and composition of human food intake matters.
>
> (Biewald et al. 2015, p. 86)

The authors develop a:

> scenario where political actions regarding food waste management and health education in developed countries, as well as political decisions taken to provide a sufficient amount of food in developing countries, will lead to a globally equal diet of 2600 kcal per day and person in 2045 [. . .] The model results show that in the equal diet scenario in 2045, global crop production is about 40% lower than in the reference scenario, and global crop land decreases by approximately 10%. The release of pressure on agricultural systems through a global common diet also reduces global water scarcity (the average water shadow price decreases by almost 30%) and food prices (food price index is reduced by 15%) [. . .]. This effect can be derived from the reduction in the direct demand for crops, as well as from the lower demand for crops as livestock feed.
>
> (Biewald et al. 2015, p. 86)

- ***Evaluate the means-implications***: finally, moral philosophers should ethically evaluate the means-implications. More fundamentally, they should develop and provide evaluation criteria based on which of the side-effects or co-benefits of response options can be identified at all, as well as evaluated. This presupposes adequate problem framing as explained above.

These proposals for the ideal role of water ethics in terms of problem framing and pathway exploration are certainly highly ambitious and impossible to achieve for individual studies provided by small author teams. Rather, these ideas clearly call for larger, integrated scientific assessment processes on water management in which moral philosophers are seriously involved. Ideally, such assessments are sophisticated processes at the science–policy interface where a large number of diverse experts, in collaboration with other stakeholders, assemble and synthesize available knowledge from different fields and scientific disciplines to inform decision-making in a policy-relevant, credible and legitimate manner (Kowarsch 2016a). In such assessments, alternative pathways should particularly represent more fundamental, divergent policy narratives and viewpoints, including alternative sets of policy priorities. While individual ethical studies can only contribute jigsaw pieces, these jigsaw pieces are the necessary elements on which the broader inter- and transdisciplinary mapping of policy alternatives for water management is based. Moral philosophers interested in water management issues should thus not only focus on particular, isolated arguments, but also strive for being involved in – or even initiating – such indispensable assessment processes to more comprehensively and reliably inform decision-making.

4. Be careful with generic water ethics

Having explained the Deweyan approach to water ethics, what's wrong with the predominant approaches to water ethics? I argued in section 2 that despite existing disagreements, inquiries into ethical issues may – under good enough epistemic conditions – indeed come to reliable and trustworthy ethical claims, going beyond procedural issues. There is thus no need to generally restrict ethical argumentation to procedural aspects only, or to even adhere to moral relativism, including radical contextualism (for an overview of and examples of these different approaches to water ethics, see Brown & Schmidt 2010).

Another predominant approach to water ethics is to focus on isolated, abstract and generic principles (or values) for water ethics. Amartya Sen (2006, 2009) describes these as ideal ('transcendental') approaches to water ethics. 'In Sen's view, ideal approaches argue for particular ethical principles, more or less irrespectively of how they can be realized in practice. They also ignore the more specific implications of the realization of these principles' (Kowarsch & Edenhofer 2016).

An example would be the ethical claim that every person should have equal rights to use freshwater in a particular watershed, for example, in terms of the human right to water already mentioned at the beginning. While this generic claim may be compelling per se, it is hard to apply to specific situations of water management, not only because it does not specify indicators and means for realizing this claim, but also because it neglects all the trade-offs and other complexity of many water management cases as described in the previous section. In this sense, this claim is not only generic, but also abstract and isolated from other, highly interlinked important ethical aspects of the problem at stake. While being a promising concept, also the idea of regarding water as a common good (see the *Codex Justinianus* quoted at the beginning of this chapter) faces the challenge that it is unclear how this claim can be realized given the complexity of current water management issues.

Let me explain why generic, isolated principles or values have only limited value for decision-making processes and why they can even be misleading, although the proposed pragmatist approach to water ethics also, to some extent, relies on such ethical principles or values.

Section 1 already mentioned the fact that the more precise means for realizing these ethical principles or values, as well as particularly their implications, are most interesting for decision-makers; their collaborative exploration in integrated assessments (see previous section) would help enable social learning about the available policy alternatives and their pitfalls. This may help better address policy-relevant questions (rather than ethical problems that are primarily of interest within given academic discourses) and may provide solutions that are more practically based and better contextualized, inter alia through also exploring their practical implications. In contrast, if water ethics only focus on generic principles or values, this may entail limited policy relevance. At best, this can provide some general constraints for water management together with many other factors that would have to be taken into account. In many cases, however, it would not inform us about trade-offs or synergies with these other factors. These trade-offs and synergies can only be revealed if practical means-implications are explored.

More important, however, is a methodological argument against at least some kinds of generic, isolated ethical claims; this argument draws on the ethical epistemology introduced in Section 2. Given the assumption that ethical claims can only reasonably be substantiated and justified in light of their practical consequences, that is, their potential to resolve ethically problematic situations, all *a priori* approaches to water ethics as well as other approaches neglecting the ends–means interdependency via the means-implications are seriously flawed from a philosophical perspective. The same is true for all approaches to water ethics that claim to generate infallible ethical claims, neglecting the fact that contexts and conditions can change (not least language and meaning) which entails the need for revising such 'eternal' claims. The laborious pattern of inquiry proposed above might result in more reliable and justifiable ethical hypotheses than alternative approaches because of its iterative nature and inclusion of all kinds of relevant aspects of a particular ethically problematic situation. From a pragmatist point of view, general normative guidelines that usually result from previous ethics inquiries are always highly tentative as long as they have not (*a posteriori*) proven to be highly reliable in multiple situations and (Deweyan) inquiries.

Even if general ethical hypotheses are well justified, their appropriate and valuable role in water ethics is not to fully and sufficiently guide water management, but rather to serve as starting hypotheses, that is, as an assistance, in the analysis of new and specific, ethically problematic situations and the development of prima facie ethical hypotheses. For example, moral philosophers sometimes rightly point out issues of intergenerational justice in specific cases of water management, because future people cannot make themselves heard. Thus drawing on well justified general ethical concerns is useful here. If used as starting hypotheses and 'cumulated experience', general principles or values for water ethics can be very useful, but are neither sufficient to guide water decision-making, nor are they infallible. At best, general claims provide orientation rather than rules (Grimm 2010). This is not only due to the need for operationalization and interpretation of general ethical claims in specific situations, but also due to the pragmatist conviction that new and ethically problematic situations can require new, or revised, ethical guidelines. Otherwise, there is a danger that one simply attempts to apply fixed and general normative principles to specific situations without noticing the actual ethical problem in a given situation. This can be called 'the sticky finger problem' (Grimm 2010).

To illustrate how different ethical principles and values highlighted by different ethical approaches could – as fallible starting hypotheses ideally stemming from previous Deweyan inquiries – provide valuable orientation for ethical inquiries into water management cases, some prevalent concepts of justice are summarized in Table 4.1. These concepts help identify different ethically relevant aspects of water management. Although not being a comprehensive overview, it may inspire the development of a more comprehensive tentative list of multiple, societally relevant evaluation criteria for water management.

Another, more general concept of justice called the 'triangle of justice' (Kowarsch & Gösele 2012) presents a consistent set of tentative principles that aim to integrate many of the aspects listed in Table 4.1. The triangle of justice claims to be based on the widely accepted list of different human rights. Given the attempt to cover a broad range of ethically relevant aspects highlighted by different predominant ethical approaches (although it neglects biocentric concerns such as those raised, e.g. by Schweitzer 1923), this concept of justice may be suited to serve as the basis for developing starting hypothesis for water ethics inquiries. Employing the triangle of justice would mean focusing on how water management (globally and intergenerationally) affects basic needs fulfilment,

Table 4.1 Different conceptions of justice in water ethics

Water ethic standpoint	*Core ethical aspects of water issue*
Utilitarianism	How is water management affecting aggregated utility in terms of economic welfare (consumption, now and in the future)?
Libertarian freedom	How does water management affect liberal rights of people, and their freedoms (now and in the future)?
Liberal equality	How does water management affect equality (now and in the future) and human rights?
Social contract and discourse	Under which procedural conditions are decisions made?
Priority, threshold, feminist	How does water management affect the poor and oppressed? How is power used in water management, particularly by men?
Socialist, Marxist	How does water management affect the working class, distributional and property rights issues (now and in future)?
Physiocentric, biocentric	How does human use of water influence non-human living beings and ecosystems (now and in future)?
Communitarian, religious	How are local traditions and cultural-religious values affected?

Source: own.

opportunities and procedures, as three major dimensions of justice. The core claim is that everyone should equally have the inviolable freedom to live a flourishing life (that is, to choose from a sufficient range of life concepts), connected with – but not dependent upon – the duty for everyone to grant the same for other people and to actively support them concerning the realization of their rights, as far as possible without violating one's own rights. More specifically, the 'freedom to live a flourishing life' is interpreted along three (*tri*-angle) dimensions of being human (leading to the three dimensions of justice). First, physical-psychological; second, talents, interests, capabilities and actions; third, being involved in processes and decisions. The numbers of the dimensions have nothing to do with priority of any dimension. All of these three dimensions of justice are irreducible to each other in terms of their meaning and justification, but they are also highly interrelated. Economic opportunities, for example, can help to fulfil one's own basic needs. Basic needs fulfilment in general is necessary for most of the opportunities, and having effective access to basic goods highly depends on fair procedural conditions. Priority should always be given to those who are far from fulfilment of their rights.

To conclude, many decision-makers appreciate guidelines that are both general (i.e. valid for a broad range of cases) and easily applicable to specific cases. That may partly explain the huge success and power of neoclassical economics, but also the appeal of classical utilitarianism: despite their differences, both are general theories that are relatively easily applicable to all specific cases where values can somehow be quantified. While some philosophers decided to focus on general principles and values first and then see whether or not they are applicable and useful in practice, the above Deweyan model suggests doing it the other way round: first start with ethical solutions specially tailored towards specific cases of water management and then see to what extent they can be generalized. Then, such more generic (fallible) principles and values can be highly valuable as starting hypotheses for ethical inquiries into water management cases.

5. Water ethics and economics: a strained partnership

Motivated, inter alia, by the critique sometimes raised on pragmatism that it allegedly adheres to economic cost–benefit rationality (Ryan 2004), let us finally discuss the interesting relationship between Deweyan water ethics in the above sense and standard economic approaches to water management.

In fact, ethical argumentation does not simply introduce an additional and competitive aspect beside economic reasons, etc. when assessing a situation; ideally, ethical argumentation weighs up all these economic, socio-cultural, environmental and other aspects, as argued above (sections 2 and 3). A crucial implication of the Deweyan approach to water ethics, with its emphasis on the ends–means interdependency, is that there are no facts without values; factual and value judgments are highly entangled (Putnam 2004; Edenhofer & Kowarsch 2015). This is also valid for the economic studies on water management, as shown by Biewald et al. (2015). It is thus impossible to avoid all of the complex ethical issues mentioned thus far by focusing on natural or social science studies on water management alone. Deliberately or unintentionally, such normative assumptions can make economic studies policy-prescriptive and can reduce their legitimacy at the science–policy interface if not made transparent. Economists should not only make the normative assumptions in their studies more transparent, but moral philosophers could also encourage economic modellers to improve their coverage of the many ethical aspects of water policy, including social equity, intergenerational justice and ecological sustainability (Biewald et al. 2015).

These basic assumptions about the relationship between water ethics and economics, and the explanation of the Deweyan approach proposed above, should clarify that there are considerable differences between, on the one hand, a standard cost–benefit analysis – or more sophisticated variants such as the popular 'integrated water resources management' approach (see also Chapter 8 by Aldaya et al. in this volume) – and on the other hand, the proposed Deweyan approach to water ethics. *First*, the range of relevant means-implications taken into account goes far beyond what economists are typically interested in. It includes both quantitative and qualitative (co-)effects of all conceivable kinds that are relevant to those affected by a particular problematic situation, including distributional, cultural, spiritual and other aspects. *Second*, in contrast to cost–benefit analysis and classical utilitarianism, the goals of an ethics inquiry in terms of the evaluation criteria (for economists this is mainly 'utility') are not fixed. They can change in light of severe side-effects of means, for example. *Third*, the approach proposed here emphasizes the need to explore and compare alternative scenarios representing different, societally highly relevant, disputed viewpoints. For this approach, economic efficiency, growth or utility are only a few among many, partly competing policy priorities that need to be explored in terms of their practical means-implications.

The proposed Deweyan approach allows a better appreciation of the ethical potential of economic tools in water management than other water ethics (see also Priscoli & Wolf 2009; Sarni 2011, ch. 3) – particularly given the fact that globally, the water problem is becoming primarily a quantitative problem of water for agricultural irrigation and food security. While some moral philosophers seem to resist economic approaches per se, because they fear that this would endanger particular ethical values or principles, economic tools are neither good nor bad per se, but have to be evaluated in terms of their practical implications. While I agree that many economic approaches have undesirable technocratic or utilitarian implications, market-based instruments, including water pricing, have proven to be extremely effective, efficient and fair tools *if designed properly*. For instance, taxing away parts of the rents made by private resource-owners, including those in the water sector – based on the normative idea that natural resources basically have to be regarded as common goods (which does not exclude private property per se) – can be a very interesting tool for realizing the commons idea without the need for communism. On the spending side, governments can use the tax income for investments in infrastructure, health programs, education, environmental protection, etc., which would mostly benefit the poor. This is but one example indicating how moral philosophers should perhaps better appreciate the potential of economic tools. The ethically decisive questions are not whether water pricing, etc. is ethically good or bad, or whether water is a private or a public good per se, but rather, under which conditions are water pricing, privatization, etc., desirable. Like a knife, you can use them for good and for bad purposes.

6. Enabling deliberative, democratic and sustainable management of the commons

To conclude, one of the most important tasks for water ethics is the assessment of practical implications of water management options jointly with other disciplines and different stakeholders. This needs to be done in an iterative manner that also involves the revision of the initial problem framing if necessary. There should be no sacred cows among the initial ethical claims and possible means of water management when they are evaluated in light of the practical implications. It is mistaken to believe that abstract ethical claims can be infallible and would suffice for ethical guidance of water management. However, generic ethical principles and values can be helpful as fallible starting hypotheses for water management. The Water Ethics Charter (see also Chapter 15 by Groenfeldt in this volume) is a good example. It provides a promising attempt to develop a set of well justified starting hypotheses needed to guide water ethics. But we should not stop there. Rather, to appropriately inform water management, water ethics have to embrace complexity in an inter- and transdisciplinary manner in assessments of particular water management pathways, even if this entail a laborious and sometimes painful iterative learning process. No one is better trained to do this than moral philosophers. The thoughts presented above suggest that many of the diverse aspects that different moral philosophers focus on in water ethics are somehow relevant; all of these aspects might be elements of a bigger picture that need to be mapped by integrating these partly conflicting aspects. This will not necessarily resolve all ethical disagreement – that is probably impossible – but will at least enable deliberative learning processes regarding the governance of freshwater as a common good of humankind and other living beings.

Notes

1 Philosophical water ethics, as I understand them, discuss moral guidelines (i.e. principles, values, etc.) for decision-making and public policy regarding the management of freshwater resources.

2 The present chapter elaborates and refines the arguments made in Kowarsch (2012).

References

Armstrong, A 2006, 'Ethical issues in water use and sustainability', *Area*, vol. 38, no. 1, pp. 9–15.

Biewald, A & Kowarsch, M (equal contributions), Lotze-Campen, H & Gerten, D 2015, 'Ethical aspects in the economic modelling of water policy options', *Global Environmental Change*, vol. 30, pp. 80–91.

Brown, MJ 2012, 'John Dewey's logic of science', *HOPOS*, vol. 2, no. 2, pp. 258–306.

Brown, PG & Schmidt, JJ (eds) 2010, *Water Ethics: Foundational Readings for Students and Professionals,* Island Press, Washington, DC.

Cochran, M (ed.) 2010, *The Cambridge Companion to Dewey*, Cambridge University Press, Cambridge.

Dewey, J 1986, 'Logic: the theory of inquiry', in *John Dewey. The Later Works, 1925–1953, Volume 12: 1938*, ed. Jo A. Boydston, Southern Illinois University Press, Carbondale and Edwardsville, pp. 1–527.

Dewey, J 1988, 'Theory of valuation', in *John Dewey. The Later Works, 1925–1953, Volume 13: 1938–1939*, ed. Jo A. Boydston, Southern Illinois University Press, Carbondale and Edwardsville, pp. 189–251.

Douglas, H 2009, *Science, Policy, and the Value-Free Ideal*, University of Pittsburgh Press, Pittsburgh, PA.

Edenhofer, O & Kowarsch, M (equal contributions) 2015, 'Cartography of pathways: a new model for environmental policy assessments', *Environmental Science and Policy*, vol. 51, pp. 56–64.

Falkenmark, M & Folke, C 2002, 'The ethics of socio-ecohydrological catchment management: towards hydrosolidarity', *Hydrology and Earth System Sciences*, vol. 6, no. 1, pp. 1–9.

Grimm, H 2010, *Das moralphilosophische Experiment: John Deweys Methode empirischer Untersuchungen als Modell der problem- und anwendungsorientierten Tierethik*, Mohr, Tübingen.

Groenfeldt, D 2013, *Water Ethics. A Values Approach to Solving the Water Crisis*, Routledge, London.

Jaegermeyr, J, Gerten, D, Schaphoff, S, Heinke, J, Lucht, W & Rockström, J 2016, 'Integrated crop water management might sustainably halve the global food gap', *Environmental Research Letters*, vol. 11, p. 025002.

Kowarsch, M 2012, 'Wie sollte die Nutzung des knappen Gutes Süßwasser geregelt werden?', in Mathias Maring (ed.), *Globale öffentliche Güter in interdisziplinären Perspektiven*, ZTWE, series, vol. 5. Karlsruhe: KIT, pp. 161–78, accessed 18 June 2016, http://digbib.ubka.uni-karlsruhe.de/volltexte/1000027069.

Kowarsch, M 2016a, 'Policy assessments to enhance EU scientific advice', *Nature Climate Change*, vol. 6, no. 1, pp. 15–17.

Kowarsch, M 2016b, *A Pragmatist Orientation for the Social Sciences in Climate Policy. How to Make Integrated Economic Assessments Serve Society,* Boston Studies in the Philosophy and History of Science 323, Springer International Publishing, Switzerland.

Kowarsch, M & Edenhofer, O 2016, 'Principles or pathways? Improving the contribution of philosophical ethics to climate policy', in C Heyward & D Roser (eds), *Climate Justice in a Non-ideal World,* Oxford University Press, Oxford, pp. 296–318.

Kowarsch, M & Gösele, A 2012, 'Triangle of justice', in O Edenhofer, J Wallacher, H, Lotze-Campen, M Reder, B Knopf & J Müller (eds), *Climate Change, Justice and Sustainability: Linking Climate and Development Policy*, Springer, Dordrecht, pp. 73–90.

Kuruvilla, S & Dorstewitz, P 2010, 'Climate transactive rationality for public policy and administration', *Policy Sciences*, vol. 43, no. 3, pp. 263–87.

Norton, BG 2010, 'The rebirth of environmentalism as pragmatic, adaptive management', in PG Brown & JJ Schmidt (eds), *Water Ethics: Foundational Readings for Students and Professionals*, Island Press, Washington, DC, pp. 179–96.

Parker, KA 1996, 'Pragmatism and environmental thought', in A Light & E Katzin (eds), *Environmental Pragmatism*, Routledge, London, pp. 21–37.

Pihlström, S 2004, 'Putnam and Rorty on their pragmatist heritage: re-reading James and Dewey', in EL Khalil (ed.), *Dewey, Pragmatism, and Economic Methodology*, Routledge, London, pp. 39–61.

Priscoli, JD & Wolf, AT 2009, *Managing and Transforming Water Conflicts*, Cambridge University Press, Cambridge.

Putnam, H 1999, *The Threefold Cord: Mind, Body and World. John Dewey Essays in Philosophy*, Columbia University Press, New York.

Putnam, H 2004, *The Collapse of the Fact/Value Dichotomy and Other Essays*, Harvard University Press, Cambridge, MA.

Ryan, FX 2004, 'Five milestones of pragmatism', in EL Khalil (ed.), *Dewey, Pragmatism, and Economic Methodology*, Routledge, London, pp. 15–26.

Sarni, W 2011, *Corporate Water Strategies*, Earthscan, London.

Schweitzer, A 1923, *Kultur und Ethik*, Beck, München.

Sen, A 2006, 'What do we want from a theory of justice?', *The Journal of Philosophy*, vol. 103, no. 5, pp. 215–38.

Sen, A 2009, *The Idea of Justice*, Harvard University Press, Cambridge, MA.

UNEP (United Nations Environment Programme) 2012, *Fifth Global Environment Outlook. Environment for the Future We Want*, United Nations Environment Programme, Nairobi.

WWAP (World Water Assessment Programme) 2009, *The United Nations World Water Development Report 3: Water in a Changing World*, United Nations, Paris.

5 Incorporating ethics into water decision-making

David Groenfeldt

Introduction

This chapter discusses the link between values and decision-making in water management. Many, and probably most, values in water management remain implicit, yet those tacit values can determine critical decision outcomes. This situation is certainly not optimal; indeed, it is dangerous. Management decisions about a resource that is vital to life are being made with only a cloudy notion of what our values, and hence our management objectives, truly are (Keeney 1992). This type of 'blind' management has long been the norm in the world of water resources development. When water supplies are abundant, and water demand is relatively low, management can be sloppy and inefficient, without dire consequences. But those days are long gone. Responsible water management in our overcrowded and increasingly fragile planet needs higher standards. We need to know what our value priorities are, and then develop responsible strategies for meeting those priorities. That, in my view, would constitute 'ethical' water management: bringing our water behaviour in line with our values.

What is ethical decision-making?

From the Ten Commandments and Buddhism's Four Noble Truths to the UN Declaration of Human Rights, human societies have long recognized the need for identifying and documenting (whether on stone tablets or online reports) broadly shared value principles. Indeed, shared values are the glue that holds societies together. Rendering those values and ethics explicit is a way of keeping that social glue strong.

In modern complex societies, the shared values of society are incorporated into such a plethora of laws and regulations that the underlying values are often obscured and can easily be overlooked. Open any newspaper and you can find stories about people, or corporations, using the legal framework to support greedy behaviours which their own society recognizes as unethical, but which are nonetheless legal. That is why such stories are published in the newspapers; they highlight the surprising inconsistencies between what the laws permit and what reasonable people would condone. In the words of German Chancellor Angela

Merkel, chastising the American intelligence services for tapping her private phone for many years, 'Just because you can doesn't mean that you should.'

The point here is that legal frameworks, policies, regulations and even detailed operating guidelines and procedures cannot substitute for human judgment and considered decision-making. This is the reason that business managers have embraced the concept of 'value-focused thinking' (Keeney 1992, 1996). Deciding on the best course of action depends on knowing what values you, or your community, or your company wish to advance. Are you interested only in maximizing economic returns? Or do you also place a value on social justice? If you are a coffee retailer concerned about social justice, you will want to ensure that the farmers who grow your coffee, and the labourers who process the coffee, are working under good conditions. You might willingly sacrifice some of your financial returns (economic value) for the benefit of social returns (social value).

In some situations, there is not even a trade-off between economic value and social value. The ice cream company, Ben & Jerry's demonstrated that social concerns (values) can also contribute to profitability, because their customers also placed a value on the social implications of their ice cream purchase (Vogel 2005). An additional reason for integrating social values into company decisions is that the employees become more motivated and enthusiastic in 'values-driven' companies (Barrett 2006).

A key feature of values-based decisions is the notion that the values we hold dear are not necessarily obvious even to ourselves. The first step in working with values is a process of self-discovery, identifying what our values are and making them explicit, and deciding which ones are core/fundamental values and which ones are less important. In the case of a company, or any collective organization (the analogy with water would be the water stakeholders), the group needs to arrive at a workable consensus about the values the whole group can endorse, and also enforce.

In the business literature, the emphasis is on harmonizing everyone's values into a common 'corporate culture' that everyone understands. The values are typically documented in the form of a code of ethics, which reminds the employees about their responsibility to act according to the corporate values and with personal integrity. Ethical behaviour can then be defined as adherence to the company's code of ethics (Croft 2013).

Within the water sector, codes of ethics have become an important tool for promoting 'integrity', a term defined by the Water Integrity Network[1] as having four dimensions: transparency, accountability, participation and 'integrity' (i.e. non-corruption). While the concept of integrity emphasizes core processes, its meaning can also include water management outcomes such as social justice and sustainable water ecosystems.[2]

Extending the concept of integrity to include water management outcomes, and not only the internal processes of water management, is important. Values are not necessarily ethical, so integrity in adhering to the organization's values is no guarantee of ethical outcomes. For example, a business that seeks to maximize profits regardless of the social or environmental impacts, might inculcate an organizational culture based on purely financial values. Indeed, this is precisely how Wall Street banks precipitated the 2008 economic collapse in the United States. The bank managers were acting in accordance with corporate values that emphasized loyalty to the financial interests of their investors. These widely shared values of placing investor profits ahead of any sense of responsibility to the larger society is now recognized (in hindsight) as unethical.

To transform abstract values into proper ethics requires a process of 'ethical reflection' (Harris 2008). More specific than 'thinking' and more holistic than 'analysis', ethical reflection is the process by which 'the ground or basis for a belief is deliberately sought and its adequacy to support the belief examined' (John Dewey, cited in Harris 2008, p. 385). Ethical reflection can be framed as a form of strategic thinking oriented around outcomes. '[Both] ethical reflection and the strategy process focus people's attention on the preparation and justification of future actions by raising the question: What do we want to achieve?' (Behnam & Rasche 2008, p. 80). Closely related to the notion of ethical reflection is 'moral imagination' (Werhane & Moriarty 2009, pp. 2–3) which introduces an active, creative dimension to the reflection process:

> Moral imagination includes an awareness of the various dimensions embedded in a particular situation – in particular, the moral and ethical ones. It entails the ability to understand one's situation from a number of perspectives. Moral imagination enables managers to recognize a set of options that may not be obvious from within the overarching organizational framework; evaluate these options from a moral point of view; and actualize them. . . . Moral imagination is the ability to discover and evaluate possibilities within a particular set of circumstances by questioning and expanding one's operative mental framework.

What sorts of ethical reflection, or moral imagination, can we discern within the water sector? Practical water decisions, for example about whether to build a dam, or what rate-structure to impose on urban water users, are invariably based on certain values, particularly economic and political ones, but usually including social and environmental considerations as well, at least to some extent. An analysis of the values motivating water decisions is rarely made explicit, but can be inferred from the benefit and cost categories used in project feasibility studies and economic cost–benefit analyses.

Thus, the topic of values is not ignored in conventional water governance and management, but I am suggesting two common shortcomings: (1) values are insufficiently considered, that is, some values are left out of consideration, while other values are given too much or too little consideration; and (2) the values which are considered are not subjected to enough critical reflection, much less moral imagination, as to whether realizing those values will lead to ethically desirable outcomes. Perhaps this sounds like nit-picking. Isn't it enough to consider the values even in a modest way? How much reflection is actually needed? If the proposed dam will generate electricity, provide irrigation water, and help control floods, do we really need to ask still more questions?

The answer that has emerged ever since the environmentalism of the 1970s is that social and environmental impacts need to be assessed and compared against the economic costs and benefits of the proposed intervention (Lawrence 2013). Social and environmental impact assessment has become an integral part of the feasibility and planning stages for major water infrastructure projects and policy initiatives. Values of various sorts are also addressed in impact analysis, but most commonly through an economic lens, for example, what is the monetary value of retaining access to a community forest (Chaudhury 2009)? The purpose of impact assessments, and the analysis of economic as well as social and cultural values that are factored in, is to avoid harmful collateral impacts.

Minimizing negative impacts on cultural values about sacred forests, or the environmental impacts of an ecosystem-changing high dam, is intuitively preferable to ignoring such impacts. But this type of analysis can be provided through standard economic tools. What further guidance can ethics analysis provide, whether through ethical reflection, moral imagination or some other conceptual tool? Identifying the full range of values operating in a particular context – not limiting our analysis to the readily monetized economic values, but including the fuzzier environmental, social and cultural values – provides an inventory of what not to do. These are the values that we want to protect so far as possible, yet at the same time, we have the promise of major economic benefits from a proposed project (e.g. a dam) that will certainly impact negatively on this inventory of values. How should we proceed? What's the ethical path?

Avoiding harm is clearly an important principle to keep in mind. Former US President Obama referred to this principle (speaking about his foreign policy strategy) with the maxim: 'Don't do stupid stuff.' Both economists and ethicists could readily agree. But in addition to the precaution motivated by fear of doing the wrong thing, there is another reason for inventorying the values ahead of project decisions, and that reason is 'creativity'. Once all the desirable values are identified, rendered explicit, what should we do with them? It is not practical to accept every value as sacrosanct; that would halt every development plan. The values need to be explored, subjected to open discussion and debate, about which values are core principles and which values are sub-topics, or corollaries. Only then can values get translated into behavioural implications, or 'ethics'.

For example, as we will discuss in more detail below, the Rio Grande river in New Mexico is valued as a living, sacred being by the native Pueblo Indians. The sacredness of the river is incontrovertible; the river, as a sacred being, must be respected. The operative value is not that the river is sacred. That is not a value, but rather a fact. The value is that because it is sacred, it must be respected. Does that mean 'no dams'? Perhaps, but perhaps not. There is room for creativity and moral imagination. Respect for the sacred river becomes a design parameter. It can be viewed as a design constraint, or it can also be viewed as design guidance for solution-making.

Research on decision-making processes has shown that approaching decisions through systematic consideration of values (including distinguishing between primary, root values and secondary, dependent values) yields better and more innovative outcomes than does a decision-making process based on considering alternatives (Selart & Tvedt Johansen 2011). When conflicting stakeholders use the language of values, they will not necessarily agree on the specifics of what water policies to adopt, but they can often agree on the design parameters that need to be considered in any ultimate design solution.

An ethics framework for water decisions

Our starting point is the premise that values and ethics are bound up in water decisions whether consciously or unconsciously. The initial challenge is to identify the operative values, which might be explicit or, more commonly, implicit. Standard methods for deducing the implicit ethics include interviews, questionnaires and focus groups. Once the values are identified, those values can be subjected to reflection and assessment as to whether these are the desirable values to advance, or whether a different set of values would be preferable. This is an important step, because values are not fixed, but can and do change, and the revised values are expressed through revised ethics rules. How values change, and under what conditions, is a complex topic, but we know for certain that values do not stay the same forever. Guiding the process of value change is critical to the success of climate resilience strategies (Bauer 2015).

To harness the power of values, and begin to apply those values in service of sustainable water management, it helps to know our starting point (what the values are) and our management aims. The framework presented in this section does not reveal the current or desired values; that is a topic for the water stakeholders to decide. But the framework can help the water stakeholders, whether government agencies, civil society groups, farmers, business owners, etc. to identify what their values and ethics are. The framework can help with the process of 'values-focused thinking' and ethical reflection about those values and their implications.

The water ethics framework we will consider comprises five main categories: (1) environmental, (2) economic, (3) social, (4) cultural and (5) governance. These categories refer simultaneously to values, in the sense of priorities about what's important, and to ethics, in the sense of normative principles for behaviour: given these values, how should we act? What should we do and not do? This five-part framework has been adapted from the literature on multifunctional agriculture (Wilson 2007) and specifically the 'house of functions' developed by Fleskens et al. (2009), which distinguishes five slightly different dimensions: ecological, productive, economic, social and cultural. In adapting these dimensions for the water sector, I have combined productive and economic into one (since this distinction makes more sense for agriculture than for water), and have added the dimension of 'governance' which is critical to water sustainability. The water framework presented here also shares the meta-frame of multifunctionality, that is, the notion that multiple objectives can, and indeed should, be addressed simultaneously (see discussion below). This ethics framework can be applied at the level of whole water ecosystems (a river, lake, spring, aquifer or estuary), or to particular resources or categories of resources: the water itself, fish, biodiversity, etc. (Figure 5.1).

Crosscutting these categories are some additional value principles, including the following:

- *Water as a commons*. All people, including future generations, have a vested interest in the water, and from a moral perspective, we can consider everyone as joint owners of water resources. What this really means is that we share a mutual responsibility in caring for the water commons. This principle can also be extended to non-human life forms (e.g. fish and other riparian life) and to the physical health of water ecosystems (e.g. river channel morphology) and the ecology of the overall water basin.[3]
- *Precautionary principle*. Given the high stakes of tampering with water ecosystems, directly or indirectly, we have an ethical responsibility to take precautions.
- *Resilience*. The capacity to rebound after a disturbance is increasingly important for both rivers as well as human systems (e.g. cities along the rivers) as storms and droughts become more powerful due to climate change. Favouring approaches that enhance resilience becomes an important ethical consideration.
- *Multifunctionality*. Synergies among different values and functions can give more total aggregate value of all types (Netherlands Enterprise Agency 2016). The concept is essentially that 'the whole can be greater than the sum of the parts'. Our aim should be, for example, to enhance ecological resilience while also serving important social and cultural functions, contributing to local economies, and strengthening political engagement (governance value).

Environmental values

A central principle of environmental ethics is 'respect for nature' (Taylor 1986), and within the water sector, a 'reverence for rivers' (Leopold 1977). More specifically, the principle of maintaining robust ecological function implies the intrinsic

	Environ. values/ethics	Economic values/ethics	Social values/ethics	Cultural values/ethics	Governance values/ethics
Governance of Freshwater Eco-systems					
Governance of Freshwater					

Figure 5.1 Domains of freshwater governance (left) and domains of values/ethics (top).

Source: own.

value of natural functions, beyond any utilitarian value to people. The value of biodiversity faces a similar tension. While nearly everyone would agree that some amount of nature's abundant biodiversity should be saved, even if it has no clear utility to humans, how much biodiversity is justified? For a deep ecologist, all biodiversity might be considered sacrosanct.

Pollution policies bring out the distinguishing features of environmental values particularly starkly. How much ocean pollution, of what type, is ethical or not? Should pollution be considered a civil (pay a fine) or a criminal (go to prison) offense? When a coal-powered ferry discharges coal ash into the North American Great Lakes, rather than paying the cost for disposal on land, is this an ethical crime against nature, or does it depend whether the local aquatic ecosystem is obviously damaged? It becomes easier to define good ethical behaviour than to define the boundaries of bad, unethical behaviour. Aldo Leopold (1949, p. 2) summarized the ethics we should aspire to in this famous quote, 'That land is a community is the basic concept of ecology, but that land is to be loved and respected is an extension of ethics.'

Economic values

The twin faces of economic ethics are efficiency and frugality, ensuring that water resources, whether water, fish, hydropower or recreation, are utilized with minimum wastage and for maximum utility. The concept of 'environmental flow' is important for protecting river ecosystem stocks, but the concept of a 'minimal environmental flow' (and similarly the concept of 'maximum sustainable yield') is dangerous because it invites operational targets too close to the threshold of unsustainability (Overton et al. 2014; Pang et al. 2014). Targets need to be viewed through the lens of the precautionary principle.

Economically useful ecosystem services of water ecosystems include water, of course, but also biodiversity, riparian habitat for both aquatic and terrestrial species, and mitigation of floods and droughts through 'nature's infrastructure' (CBD 2013). There is a moral imperative to govern rivers effectively to protect these services, and manage these resources. There is usually a logic for doing so that extends beyond monetary rationalism. For example, protecting biodiversity and wildlife habitat can simultaneously contribute to economic, environmental, social and cultural aims, while also advancing governance best-practice. Aggregate utility, and in this sense, economic efficiency, can often by enhanced through strategies of multifunctionality (Netherlands Enterprise Agency 2016). Indeed, economic ethics can serve to 'complete the circle' in designing new initiatives that educate and inspire people to become engaged in water ecosystem governance and find creative synergies with Nature. Examples range from ecotourism to aquaculture to recreation.

Social values

The most famous social value operating in the water sector is the idea of access to safe water and sanitation as constituting a human right, which was reaffirmed by the UN General Assembly in 2010. This ethical principle motivates programs to help underserved communities to realize their right. Access to water is fundamental because we all need water to live. But access to *safe* water brings in an additional principle of health. Water should allow not only for survival, but health.

A similar logic applies to water and employment. Water provides jobs through agriculture, industry and indirect value chains (WWAP 2016). Agriculture, which uses most of the world's water, also provides the most jobs. This is a good thing for both economic values, and social ones. More jobs are better than fewer jobs, *ceteris paribus*, but what agricultural policies will provide the best mix of jobs, economic security and well-being, and for which segments of the population? Women? Migrants? Minorities? Youth?

Agriculture is inherently multifunctional, providing social and environmental services along with food (Mölders 2014), and certain forms of agriculture can provide many more of these non-production services than others (Marsden & Morley 2014). The social benefits from water are complex and interlinked with economic, cultural and other variables. Ethics offers a way of navigating through the range of potential values that agriculture (and agricultural water) could provide, to identify healthy options.

Cultural values

Culture assigns meaning to rivers, and rivers shape cultures by offering a palette of subsistence and semiotic resources. Indigenous communities that have evolved in symbiosis with a river consider themselves to be inextricably linked to that ecosystem, whether it is the Mekong, the Amazon or the Rio Grande. Rivers provide not only food and water, but cultural identity. This is most starkly evident in traditional and Indigenous communities, but also applies to contemporary city life. Imagine Paris without the Seine, or Rome without its fountains. And though the River Tiber plays a very weak role in Roman culture today, that might soon change, as Rome[4] and many other cities rediscover the cultural meanings of local rivers.

Indigenous peoples' right to their culture was formally recognized in the 2007 UN Declaration on the Rights of Indigenous Peoples. The broader principle is that culturally distinct communities have a right to pursue their own cultural happiness, and that this right is accorded to the group (the cultural community) as well as to the individual. On a global level, traditional societies constitute a form of cultural heritage in which all people, including future generations, have a stake. Modern societies also derive important and necessary meaning through their cultural heritage, and contemporary cultural expressions. Cultural and psychological identity are fundamental to human well-being, and depend upon healthy relationships with both people and nature (Kellert 2012).

Governance values

The arrangements by which governments or other local authorities (e.g. landowning corporations or individuals) make decisions about water resources inevitably favour certain governance values and disfavour others. What stakeholders are recognized and given voice? How are Indigenous stakeholders and civil society organizations engaged in decisions about water resources? How transparent and accountable are the governance bodies? What are the geographic boundaries and scope of the intended governance jurisdiction, and how are the boundaries established? In the case of urban water governance, for example, is the upper watershed also part of the governance responsibility?

Commonly accepted values of water governance include the principle of subsidiarity (management at the lowest practical level), favouring local management of local resources, and democratic representation of all stakeholder interests in water decisions. The scope of governance should include not only resource governance but also 'knowledge governance' and decisions about what research topics are pursued and how the data are used. Commercial corporations utilizing or impacting water resources have social and environmental responsibilities which the governance system should address.

The principles of 'integrity' can be applied as standards for water governance actors, whether public or private, namely (a) transparency, accountability, and representation of all stakeholder interests including Nature (environmental values), diverse cultures and people, including future generations. Governance should also include effective and fair enforcement of rules and regulations. Finally, water conflict mechanisms need to be designed into the governance arrangements.

The case of Cochiti Dam

In the Southwestern US state of New Mexico, Cochiti Dam, on the main stem of the Rio Grande, takes its name from the Pueblo of Cochiti, founded roughly 1000 years before the dam was built. The US Army Corps of Engineers decided in the 1950s that a dam on the Rio Grande at this particular location, within Pueblo territory, was needed to protect the city of Albuquerque, about 60 km downstream. Constructing what at the time was the largest earth-filled dam in the world, on lands belonging to an Indigenous community, for the purposes of reducing the threat of floods downstream which (we now know) were actually necessary for the ecological health of the river, met with little controversy in the 1950s and 1960s when the planning was taking place.

Cochiti tribal members were divided over the prospect of the dam, with recent World War II veterans convincing the more conservative tribal members that the development which would surely be sparked by the dam would bring modern conveniences and opportunities. Construction of the dam, just 2 km from their village, proved traumatic. In the words of one resident:

> One of the most emotional periods in our history was watching our ancestors torn from their resting places, removed during excavation. The places of worship were dynamited, destroyed, and desecrated by the construction. The traditional homelands were destroyed. When the flood gates closed and waters filled Cochiti Lake, to see the devastation to all of the agricultural

> land upon which we had walked and had learned the lessons of life from our grandfathers destroyed before our eyes was like the world was coming to an end. And all we could do was watch.
>
> (Pecos 2007, p. 645)

Today, more than a half century later, tribal residents of Cochiti Pueblo still have tears in their eyes when they speak of the dam's construction and the wrenching decision that allowed it to go forward. Why did the tribe agree to a plan that was so deeply inconsistent with their traditional values? In another essay (Groenfeldt 2016), I address this question using the concept of 'semiotic hegemony'. Basically, the tribal members' cultural logic was upset by the dominant cultural discourse of American society, and specifically the modernization message of the Army Corps. Cochiti tribal members were literally unable to 'think straight' because their cultural references were so conflicted.

Could 'value-focused thinking' (Keeney 1996) have resulted in a different outcome? My contention is that clarity of values, coupled with ethical reflection about the relative priority of the various values, might well have led to a rejection of the dam proposal. We can't know, of course, and the cultural currents of the 1950s and early 1960s were powerfully optimistic about a materialist modernity. There was little support in the non-Indigenous world for spiritual beliefs about the natural world. It is precisely this type of smugness, or what Daniel McCool (2012) calls 'hubris' (referring to the attitudes underlying the big dam projects of the American West), which systematic ethical analysis and reflection might be able to counter.

How would an ethics-based approach be applied to the decision of whether or not to allow the Cochiti Dam? The first step would be to identify core values related to the Rio Grande, and then to reflect on the behavioural implications of those values (see Table 5.1). The five categories outlined above (environmental, economic, social, cultural and governance) provide a framework to ensure that multiple kinds of values are considered in the analysis. These categories could be expanded or revised; the important feature is that a broad range of values be represented.

Cochiti water values

What do Cochiti tribal members 'value' about their local environmental context within the Rio Grande river basin? This might seem an odd question to the people concerned. The river's healthy condition was always a fundamental assumption of the community, and to be a healthy river, it needs to flow. The ethical implication is that the river should be left alone; it should be allowed to flow.

The river has, for many centuries, provided the economic base for the community through floodwater farming and fishing (Lange 1959) and irrigation. An irrigation diversion dam constructed in the 1930s as part of a larger government-sponsored irrigation district, provided reliable irrigation for Cochiti farmers. There is no indication that Cochiti leaders objected to the irrigation infrastructure at the time.

Social values about water emphasized sharing under the political authority of the tribal leaders, who allocated irrigation water and maintained drinking water sources. Culturally, the river was viewed as a spiritual being and an object of veneration (Swentzell 2012). In terms of governance values, the river governs itself, or if humans are to be involved, Cochiti tribal sovereignty should be respected (Pecos 2007).

Army Corps water values

The Army Corps entered the scene in the 1950s with a proposal for a major dam, which promised to improve the reliability of irrigation water, but its main benefit would be to protect the downstream city of Albuquerque from flooding, thus not directly benefitting Cochiti. The cultural values and implied ethics underlying this proposal are outlined in Table 5.2.

The environmental value of the river is its flow of water, the aquatic habitat it provides and the water resilience, which the natural river supports through feeding the local groundwater aquifers; however, these natural functions also entail threats from flooding. The ethical implication is that the river should be controlled, to take advantage of its benefits (the water) without the costs (the floods). Economically, the dam would result in reduced flood costs, and income from recreational fees (from camping, swimming and boating in the reservoir) while the huge capital costs of the dam were provided by the federal government and not included in the economic analysis.[5] The ethical implication was that these new services would render the river more useful than it would be in its natural state.

The social values underlying the dam proposal was that Cochiti Pueblo should be a good neighbour to downstream communities by allowing the dam to be built. And there is a deeper cultural value as well. Rivers should be controlled; they should not be allowed to remain wild. Both before and after construction of the dam, the Army Corps engaged in a massive program of 'rectifying' (straightening) the river channel to remove the meanders (Phillips et al. 2011). Finally, the governance values of the Army Corps feature centralized authority (i.e. the Army) based on rational planning.

Lessons from Cochiti

Clarifying the values and ethics about water ecosystems – the Rio Grande in this case – can be helpful for any set of stakeholders, whether a government agency

Table 5.1 Cochiti Pueblo: Values → Ethics

	Cochiti values	*Cochiti ethics*
Environmental	River should flow	Leave the river alone
Economic	River already useful (fishing, irrigation)	Use the river as it is
Social	Share water/common good	Share the benefits
Cultural	River as friend	Pray to the river
Governance	River governs itself	River should govern itself

Source: own.

Table 5.2 US Army Corps: Values → Ethics

	US Army Corps values	*US Army Corps ethics*
Environmental	Water is a resource; river is a 'floodway'	River needs to be controlled
Economic	River is a means to another end (reservoir, recreational facility)	Make the river more useful through engineering
Social	River has potential for social benefit through flood control and creating reservoir	Share the benefits
Cultural	River as threat and potential resource	Modify the river (neutralize the threat)
Governance	River needs to be tamed	People, and USACE specifically, should exert control over river

Source: own.

like the Army Corps, or an indigenous community. Whether such a process would have made a difference some 50 years ago is not known; clearly it was a different time with a different gestalt taking shape in post-World War II American society. The social context today is certainly more favourable to reflection and respect for cultural diversity. I think it is safe to presume that were the decision to be made today about building Cochiti Dam, it would not be built, though the reasons might be based more on economics than culture or environment.

Indeed, it is the singular reliance on economic analysis that ethics can help to overcome. Certainly the economics are important, and in the case of Cochiti Dam, it is likely that a purely economic-oriented cost–benefit analysis would have stopped the dam at the early planning stages. Economics is an invaluable tool, but it is not everything that needs to go into water decisions. What I am suggesting is that multiple value categories need to be taken into account in water decisions, because water is connected to so many facets of life, over and above the economic, that decisions need to consider as many of those facets as possible.

An analysis of core values in Cochiti Pueblo would have revealed very strong values and ethics on issues of little concern to the Army Corps. However, it would not have been necessary to convince the Army Corps of the value of religious shrines or other sacred sites that would be (and were) destroyed by the dam's construction. The Cochiti community did have the legal authority to block construction if they had chosen to do so. This raises the central lesson (in hindsight, of course) about the Cochiti experience. Tribal leaders were themselves split on the desirable course of action. Reflecting on traditional values and ethics could not yield a clear answer because those values were in flux at the time.

While the Cochiti case is extreme in the wrenching choices confronted by the community, the feature that is common to any analysis of values is that identifying and articulating the values is a complicated task. The entire community needs to be involved, directly or indirectly, in acknowledging their shared values. There needs to be 'buy-in' from the whole community, and a common understanding.

Furthermore, the values of the community need to be communicated to the outside world both to inform the world about the community's stance, and in so doing, to reinforce that stance. In the Cochiti case, the primary outside audience would have been the Army Corps of Engineers, as well as local government agencies, businesses and private developers. It is not a matter of trying to convince the outside that the values of the community are the right ones, that they should accept the sanctity of the river, but rather to alert the outside society that this particular community regards the river as sacred. It is a way of pre-empting an influx of culturally 'hegemonic' influences seeking to convert this Indigenous community to neoliberal economic thinking (Groenfeldt 2016).

The potential value of ethics-based decision-making processes lies in careful analysis and reflection about the values which the community desires to live by, and undertaking this evaluation in advance of specific proposals to do this or that. It is very likely that through a process of debating the values/ethics, there will be some surprises. Just as psychological therapy can bring out unrealized emotions and feelings, a reflective process about values and ethics can also yield new understandings.

Yet even clear values and carefully considered ethical positions is only the first stage of what might need to be a three-step process. The second step would be for the community to document the consensus water values and ethics in the form of a 'water ethics charter' or some kind of written statement of key principles in terms that can be readily understood both inside the community as well as outside. The purpose would be to remind the community what they have already decided, while at the same time educating the outside world about these distinct principles.

The third step in this process of reclaiming their traditional water ethics would be to publicize the water ethics document (the charter) both within and beyond the community's boundaries. The best defence against unwanted proposals from the outside is proactive communication about the ethical principles for water which the community has adopted. By creating a document with the values and ethics systematically recorded, those values attain a legitimacy as well as a transparency, which only written words can provide. The list of values and ethics is transformed into a charter that can serve as a guide for future decisions about policies, projects and programs.

Conclusions: from local to global, and back

The capacity to make ethically informed decisions about water development depends on knowing what your values and ethics are (or what you would like them to become), and I am suggesting that it also requires a process of documenting and publicizing those values and ethics, to create an operating space for applying the ethics. But how sustainable is that ethical operating space? Would the outside world respect the ethical principles of the community, or would neoliberal forces send economic missionaries to convert the Natives back to conventional water orthodoxy? Certainly it would be supportive for a community engaged in documenting its water ethics to be part of a larger network of other communities engaged in the same process. And it might also be helpful to have a common frame of reference about water ethics principles and categories.

These two support functions – an information network and a common frame of reference – are the ideas behind the Water Ethics Network, and the initiative to develop a global Water Ethics Charter. Together these efforts can contribute to an enabling environment for water ethics:

- *Water Ethics Network.* Established by Water-Culture Institute in 2011, the Water Ethics Network (http://waterethics.org) serves as an information-sharing platform on water ethics. With a website, newsletter and social media sites, the aim of the Network is not only to provide information, but at the same time to legitimize water ethics as both an important topic for research and analysis, as well as an operational approach to water governance and decision-making.
- *Global Water Ethics Charter.* A recommendation of the 6th World Water Forum held in Marseille, France in 2012, a coalition of organizations, initially led by the French Water Academy, Water-Culture Institute and UNESCO, launched an initiative to formalize a set of universal principles about water ethics. This task proved to be more complex than anticipated, and the initiative has since been folded into the Water Ethics Network. The current draft of the Water Ethics Charter is posted on the Network website, and will continue to undergo revisions under the direction of an international Steering Committee. Whether a 'final' version of the global charter is ultimately produced or not, even the draft versions can serve as a common reference that can guide local water ethics initiatives. For details about the global charter, see Chapter 15 in this volume.

The growing community of practice on water ethics and ethics-based water decision-making holds a powerful potential for creating an enabling environment for water-related ethical reflection. Realizing this potential, and establishing water ethics as a legitimized, mainstream set of theory and practice within the water profession, hinges on the interaction of the global and local levels of water ethics. A draft of a global water ethics charter may sound rather weak, but it can nonetheless serve the important function of a point of reference for stimulating local water ethics charters. A local community water ethics charter might seem disconnected from the technically sophisticated realm of national and global water policies, but it can draw legitimacy from the global water ethics charter. And both the global and local initiatives can build collective strength through networking with communities, cities and even corporations engaged in applying ethical principles to water policies (see the chapters in this volume by Smith, about the Water Justice Framework; by Härlin, about Berlin's Water Charter; and by Sym, about the Water Stewardship Standard). Local decisions about water are not made in an ethical or political vacuum. Making tacit values explicit, reflecting on the ethical implications, and referencing international standards and best practice can facilitate the process of ethics-based reflection and decision-making.

Notes

1 www.waterintegritynetwork.net, accessed 15 July 2016.
2 For a concise definition of 'water integrity', see the 2013 Delft Statement on Water Integrity', www.waterintegritynetwork.net/wp-content/uploads/2015/03/WI-DelftStatement_EN_2013.pdf, accessed 15 July 2016.
3 The concept of geoethics is being elaborated by the recently formed International Association for Promoting Geoethics, www.geoethics.org, accessed 15 July 2016.
4 See the website of Tevereterno, 'A multidisciplinary cultural project for the revival of Rome's Tiber River', www.tevereterno.it, accessed 15 July 2016.
5 It was standard practice for the Army Corps at the time to consider only the benefits, and not the costs. This was changed in the 1990s but the culture of minimizing economic considerations has continued.

References

Barrett, R 2006, *Building a Values-Driven Organization: A Whole System Approach to Cultural Transformation*, Elsevier, Boston, MA.

Bauer, J 2015, *Forging Environmentalism: Justice, Livelihood, and Contested Environments*, Routledge, London.

Behnam, M & Rasche, A 2009, 'Are strategists from Mars and ethicists from Venus? Strategizing as ethical reflection', *Journal of Business Ethics*, vol. 84, pp. 79–88.

CBD 2013, *Natural Solutions for Water Security*, Convention on Biological Diversity, Montreal, accessed 11 June 2016, www.cbd.int/idb/doc/2013/booklet/idb-2013-booklet-en.pdf.

Chaudhury, M 2009, 'Valuing services: an analysis of the ability and willingness to protect forest-based environmental services in the Asia-Pacific region', FAO Regional Office for Asia and the Pacific Working Paper No. APFSOS II/WP/2009/16, Bangkok.

Craft, J 2013, 'A review of the empirical ethical decision-making literature: 2004–2011', *Journal of Business Ethics*, vol. 117, pp. 221–59.

Fleskens, L, Filomena, D & Eicher, I 2009, 'A conceptual framework for the assessment of multiple functions of agro-ecosystems', *Journal of Rural Studies*, vol. 25, pp. 144–55.

Groenfeldt, D 2016, 'Cultural water wars: power and hegemony in the semiotics of water', in C Ashcraft & T Mayer (eds), *The Politics of Freshwater: Access, Conflict and Identity*, Routledge, London.

Harris, H 2008, 'Promoting ethical reflection in the teaching of business ethics', *Business Ethics: A European Review*, vol. 17, no. 4, pp. 379–90.

Keeney, RL 1992, *Value-Focused Thinking: A Path to Creative Decision-making*, Harvard University Press, Cambridge, MA.

Keeney, RL 1996, 'Value-focused thinking: identifying decision opportunities and creating alternatives', *European Journal of Operational Research*, vol. 92, pp. 537–49.

Kellert, S 2012, *Birthright: People and Nature in the Modern World*, Yale University Press, New Haven, CT.

Lange, CH 1959, *Cochiti: A New Mexico Pueblo, Past and Present*, University of Texas Press, Austin.

Lawrence, D 2013, *Impact Assessment: Practical Solutions to Recurrent Problems and Contemporary Challenges*, 2nd edn, Wiley, Hoboken, NJ.

Leopold, A 1949, *A Sand County Almanac*, Oxford University Press, London.

Leopold, L 1977, 'A reverence for rivers', *Geology*, vol. 5, pp. 429–30.

Marsden, T & Morley, A (eds) 2014, *Sustainable Food Systems: Building a New Paradigm*, Routledge, London.

McCool, D 2012, *River Republic: The Fall and Rise of America's Rivers*, Columbia University Press, New York.

Mölders, T 2014, 'Multifunctional agricultural policies: pathways towards sustainable rural development?, *International Journal of the Sociology of Agriculture and Food*, vol. 21, no. 1, pp. 97–114.

Netherlands Enterprise Agency 2016, *Reinventing Multifunctionality: Combining Goals, Sharing Means, Linking Interests*, Netherlands Enterprise Agency, The Hague, viewed 15 March 2016, accessed 18 July 2016, http://english.rvo.nl/sites/default/files/2016/03/Reinventing%20Multifunctionality.pdf.

Overton, IC, Smith, DM, Dalton, J, Barchiesi, S, Acreman, MC, Stromberg, JC & Kirby, JM 2014, 'Implementing environmental flows in integrated water resources management and the ecosystem approach', *Hydrological Sciences Journal*, vol. 59, nos 3–4, pp. 860–77.

Pang, A, Sun, T & Yang, Z 2014, 'A framework for determining recommended environmental flows for balancing agricultural and ecosystem water demands', *Hydrological Sciences Journal*, vol. 59, nos 3–4, pp. 890–903.

Pecos, R 2007, 'The history of Cochiti Lake from the Pueblo perspective', *Natural Resources Journal*, vol. 47, no. 3, pp. 639–52.

Phillips, FM, Hall, GE & Black, ME 2011, *Reining in the Rio Grande: People, Land, and Water*, University of New Mexico Press, Albuquerque.

Selart, M & Tvedt Johansen, S 2011, 'Understanding the role of value-focused thinking in idea management', *Creativity and Innovation Management*, vol. 20, no. 3, pp. 196–206.

Swentzell, R 2012, 'Pueblo watersheds: places, cycles and life', in J Loeffler & C Loeffler (eds), *Thinking Like a Watershed: Voices from the West*, pp. 28–44, University of New Mexico Press, Albuquerque.

Taylor, P 1986, *Respect for Nature: A Theory of Environmental Ethics*, Princeton University Press, Princeton, NJ.

Vogel, D 2005, 'Is there a market for virtue? The business case for corporate social responsibility', *California Management Review*, vol. 47, no. 4, pp. 19–45.

Werhane, P & Moriarty, B 2009, *Moral Imagination and Management Decision Making*, Business Roundtable Institute for Corporate Ethics, Charlottesville, VA, accessed 30 June 2016, www.corporate-ethics.org/pdf/moral_imagination.pdf.

Wilson, GA, 2007, *Multifunctional Agriculture: A Transition Theory Perspective*, CAB International, Wallingford.

WWAP (World Water Assessment Programme) 2016. *The United Nations World Water Development Report 2016: Water and Jobs*, UNESCO, Paris.

6 Transcending water conflicts

An ethics of water cooperation

Angela Kallhoff

Introduction

Philosophical approaches to water ethics have been developed against the backdrop of a broad range of problems. They include shortages in water supply and overall scarcity of water resources, the pollution of water reservoirs, the dangers that water provides by flooding and the need to protect societies from growing sea levels. In particular, ethics addresses the question of how persons should relate to water as a non-renewable and particularly valuable resource (Delli Priscoli et al. 2004; Brown & Schmidt 2010). Due to the broad variety of themes, it is obligatory to narrow down the scope of the investigation first. This contribution focuses on how conflicts that result from multiple use of a limited water resource can be resolved in a peaceful way. It explains ways to transcend resulting conflicts by means of an ethical framework. A common interpretation of how overuse of a scarce resource spurs conflict among various water users is the interpretation of the 'tragedy of the commons' (Hardin 1968) as a necessary outcome. This narrative states that over-exploitation and degradation are the destiny of natural resources, because they are scarce, but lack clear-cut access-barriers. In this chapter, I shall portray constructive alternatives to this interpretation of water conflict. I shall argue that some proposals from water ethics are particularly helpful in transcending water conflicts and in arguing water cooperation as a reliable alternative.

This article has five sections. The first section introduces a paradigm case of conflicts that result from multiple use of a limited water resource. Section two recalls the interpretation of severe water conflicts as the 'tragedy of the commons' – which is a familiar interpretation of water conflicts. In order to address this interpretation thoroughly, two research questions will be distinguished. Section three develops a normative framework for water cooperation. It profits from normative insights that have been reasoned in environmental ethics: the values of environmental conservation, environmental justice in terms of distributive fairness and guidelines of mutual and intercultural respect. Section four argues that an interpretation of cooperation as 'joint agency' is particularly helpful in introducing a concept of cooperation that can be applied to multiple use of a shared natural resource. Different from cooperation as a process of mutual exchange, this approach favours 'joint action' as a goal-driven activity that is sustained by various actors who are committed to realizing a joint goal. This approach also resonates with a concern of authors in the field of water management (Delli Priscoli & Wolf 2009). They say that an approach to water ethics needs to address joint goals in shaping water resources. In particular, special attention needs to be paid to shared values underlying the formation of goals. In section five, I shall draw a conclusion from this proposal.

1. A scenario of water conflicts

Imagine the following situation. A village is located at the borders of a lake. The lake serves many different interests, including services as a sink for waste. For years, the practice of utilizing the lake as a waste dump has not been questioned. The inputs have been marginal as compared to the vast amount of water in the lake. No one has really taken notice of the problems resulting from polluting the lake. But one day the situation begins to cause trouble. The lake starts to stink; it poisons a river. As a consequence, the practices of using the lake as a waste dump are called into question.[1] At that point, a basic problem in using the lake becomes apparent: even though the lake serves many different interests, it is necessary to coordinate the profit-seeking activities. Moreover, it is necessary to reduce exploitation and to prohibit harmful practices of water use. In order to prevent tipping points beyond which the water is not usable any longer, the villagers come together and discuss the problem and possible solutions.

A first step in developing a solution is an interpretation of the occurring problems. One might argue that the problems that the inhabitants of the village face result from over-exploitation of a shared resource. Yet, this interpretation is only part of a much more complex situation. A lake is a common pool resource. Common pool resources are specified by two characteristics: they do not have effective entrance-barriers; as a consequence, profiteers are free to profit from that resource. Simultaneously, many beneficiaries compete over that resource. In particular, a frequently occurring situation is the following: 'Some of the most serious environmental problems occur when the same resource is used both as a source and as a sink: for example, when the same stretch of river is used both as a water supply and as a sewer' (Jamieson 2008, p. 14). It is not only scarcity of a shared natural resource that spurs conflicts and that might lead to the degradation of that resource. Instead, uncoordinated use of a resource has the effect that some types of usage are made impossible by other types of usage.

In my view, the villagers are well-advised to realize a model of cooperation that forestalls degradation by uncoordinated use of the resource and that coheres with claims of fairness in profiting from the shared resource. This includes a focus on some core values and the attempt to translate them into some basic normative guidelines of water usage. In particular, the water users should act together in order to realize future developments as a practice of 'active codesigning with nature' (Delli Priscoli & Wolf 2009, p. 121). Instead of regarding effects on water resources and types of water usage as given, and instead of regarding effects on water as side-effects of their activities, they need to regard themselves as authors of rules that shape future practices of water usage. That means that they do not only interpret their shared activities as driven by a distinct vision of the future development, but also that it is based on values that help them to shape and to coordinate activities. Together, they wish to realize a good access-regime that coheres with their values. They also know that in order to succeed in achieving a good situation, they have to work together. They will rely on values of environmental conservation of that resource, on values of a fair distribution of burdens and shares in that resource, and on values of cultural identity that are attached to certain usages of water. In subscribing to an ethos that includes these values, water users form a group. They also develop a shared vision of scenarios that can accommodate the values accordingly.

This proposal for framing 'water cooperation' as a value-based type of joint action is based on two premises. First, it argues that cooperation in terms of 'joint action' is indeed the best model for resolving and transcending water conflicts. Second, it presupposes that there is some common ground regarding basic values that help to shape the decisions of the group members. In particular, this proposal amounts to the following strategy among the villagers:

> As a group, the villagers first discuss their shared values that help to shape their goals regarding the common pool resource. Basically, they pay tribute to the fact that they are all in some way dependent on the water resource. Moreover, they also agree that shared goals have priority over private goals. In addition, they also reach consensus about the priority of environmental conservation, about fairness in distributing profits and burdens from water usage, and about respect for various cultural values attached to the water resource. They also figure out which concrete values are best to shape their common action. They then agree about cooperative schemes of water usage that shall be realized.[2]

A minimal agreement on shared values includes three aspects. *First*, the conservation of the lake is prioritized. All water users agree that heavy pollution and heavy over-exploitation of the lake is a worst-case scenario. They agree that it is reasonable to keep basic eco-systemic functions of the lake intact. They also discuss how resilience of the lake can be supported and how eco-functions that already are under pressure can best be restored. *Second*, the villagers agree that the profit-seeking practices need to be framed and guided by long-term strategies that cohere with claims of fairness. This includes a range of principles of fairness. Water fairness includes at least three aspects: (a) priority to urgent needs; (b) distributive fairness in sharing burdens and profits from the lake; (c) respect for cultural and aesthetic values. *Third*, the villagers agree that they will realize the agenda and achieve the goals as joint actors. This means that each of them subscribes to fair practices of water usage.

Obviously, this proposal for resolving water conflicts is ambitious. It rests on the idea that potential beneficiaries from a natural resource also subscribe to a set of shared values. Water resources are interpreted as shared resources that deserve environmental protection. Fairness provides a range of restrictions in using water resources. Moreover, the proposed scenario abstracts from many additional conflicts and problems in water supply. In order to focus on one type of conflicts, it abstracts from non-anthropogenic pollution, from extreme scarcity due to climate change, and from water markets, among others. In this chapter, there is not the space to argue each claim in water ethics accordingly.[3] Instead, some core ideas regarding an approach to joint action and regarding basic values of a water ethics shall be outlined. Before doing that, it is helpful to contrast an approach to water cooperation with a familiar interpretation of the necessary degradation of natural common-pool resources. This is the 'tragedy of the commons'.

2. The tragedy of the commons recalled

The usage of a natural common pool resource by various water users has been interpreted as a paradigm case of the 'tragedy of the commons' (Hardin 1968). This narrative recalls the necessary degradation of a shared natural resource that is accessible by each beneficiary who wishes to use it. Moreover, the 'tragedy' means that a socially desirable goal is not achievable, even though this is not willed, but the outcome of a hazardous situation.

One interpretation of this narrative states that natural resources will necessarily be over-exploited by uncoordinated behaviour. As common pool resources, lakes, rivers and groundwater reservoirs do not possess natural entrance-barriers (Buck 1998; Kaul et al. 1999). Simultaneously, competition over water reservoirs is high. As life-sustaining and simultaneously multi-functional goods, many diverse users are interested in optimizing their profits from a common pool resource. One immediate consequence is that over-exploitation is likely to occur. As for water reservoirs, the overall situation is worsened by climate change that contributes to desertification and to growing scarcity of water supply, even in formerly water-rich regions of the world (Feldman 2007). The situation is also worsened by the additional problem of long-term effects of profit-seeking behaviour and by an exponential growth of need in some areas of the world that suffer from over-exploitation, in particular in megacities.[4] In addition to over-exploitation, in-pool pollution is another threat to common-pool resources. Due to the quality of water as a cohesive medium, one disastrous incident of pollution can set the whole natural resource at risk. Moreover, uncoordinated interests contribute to practices which are not mutually compatible: a natural resource cannot simultaneously be used as a sink and as a resource for life-supporting activities (Jamieson 2008, pp. 14–22). As long as access to natural common pool resources is not restricted, conflicting interests result in the destruction of that resource.

The proposal to frame multiple use of a water resource as water cooperation rejects this narrative of a 'tragedy'. The degradation of water resources is not regarded as a necessary and as a tragic event which results from competition and a lack of entrance-barriers. Even in circumstances in which institutions that regulate access have not been established yet, it is possible to develop regimes of fair access that cohere with conservation of that resource. As for the theoretical resources the biggest difference between water cooperation and the 'tragedy of the commons' is the following: whereas the background idea of the interpretation as a 'tragedy of the commons' is basically a concept of profit-seeking behaviour of self-interested individuals, the interpretation of water cooperation builds on capacities to negotiate and to realize joint goals. In particular, it regards it as possible for individuals and for groups of people to act together against the background of a set of shared values.

In the reminder of this chapter, I wish to argue in favour of water cooperation as a reliable alternative to the alleged tragedy. Therefore, I start with rendering explicit some justificatory burdens. The story of the 'tragedy of the commons' helps to highlight the burdens of proof of a theory of water cooperation. I shall formulate them as two critical questions that the theoretical approach to water cooperation has to answer:

1 Are there any shared values that persons will be willing to subscribe to in water cooperation?
2 Why will people act together in profiting from water resources?

The first question is debated in ethics. Normative claims need to be justified. In particular, in order to frame cooperation, it does not suffice to portray and to examine values from a philosopher's point of view. Instead, they need to be related to the 'real will' of persons. Following John Rawls (2005), philosophy serves at portraying and justifying basic principles of justice. Yet, in answering severe reservations of Jürgen Habermas regarding the connectedness of this philosophical enterprise with the real will of the people, democratic procedures are obligatory in order to get the necessary reassurance that the principles are really willed by the people (Habermas 1996; Rawls 1996). Philosophers cannot realize the step from theory to practice. But they can argue that the proposed values are good enough to pass a basic contractual test: basic normative principles need to be good enough to be accepted for reasons that each person can subscribe to. In my view, the debate on 'environmental justice' can be seen in this light. Justice is a very basic concept in ethical terms. It can be spelled out in a range of principles of fairness that can be accepted by each person.

The second question about the motivation to cooperation has another twist. It addresses the very basic issue of *homo economicus* on the one side and persons endowed with a will to cooperate on the other side. Moreover, it also calls into question whether or not cooperation is possible at all. Instead of getting into a debate on the foundations of ethics, recent approaches to 'joint action' in social philosophy demonstrate that persons are not only willing to cooperate in order to achieve a shared goal, but also capable of realizing shared goals by working together (Bratman 2014; Gilbert 2003; Tuomela 2010). Authors in the field of social philosophy argue that 'collective action' is not as problematic a notion as also recalled in the narrative of the 'tragedy of the commons'. Instead, it is a usual state of acting in societies. An explanation of these facts can be applied to an interpretation of environmental cooperation, too.

As for water resources, another theoretical line of thought can be added. If there is any truth in the narrative of the 'tragedy of the commons' it is the following. A loss of water resources will indeed be a tragic event. Living beings on planet Earth need water to survive. Since ancient times, civilizations have been dependent on water resources. The approach to water cooperation does not only highlight the very need to develop distributive schemes that contribute to protecting a valuable shared natural resource. It also argues that it is time to first render the effects of our current lifestyles on water resources visible; it is then necessary to get beyond schemes of distribution and types of usage that privilege some and that remain uncoordinated. Water resources are part of the common heritage of mankind; they are the most basic natural common pool resources we have. Therefore, it is obligatory to address these resources not only as jointly possessed goods, but also as goods that need to be protected from further damage.

3. A framework for water cooperation

Following theories of joint agency, cooperation does neither have to be paralleled with a voluntary exchange of assets, nor is it congruent with a social contract.[5] Instead, cooperation is a goal-directed activity of various persons who contribute to realizing a shared goal. The group of persons that acts together realizes a shared goal jointly. Before spelling out this model of joint action more thoroughly, I shall first discuss the space for introducing values in cooperation and some most basic values that underpin shared agency in terms of 'water cooperation'.

Following Tuomela, joint action rests on a scenario in which persons form a group and subscribe to a shared goal (Tuomela 2010). What holds the group of persons together, and what gives reason to realize steps towards the realization of that goal, is not only the attractiveness of the goal. Instead, various persons are related to that goal via a shared 'ethos'. A group is literally constituted by the commitment of all group members to a shared ethos, whereby the ethos contains 'constitutive goals, values, beliefs, norms and standards collectively accepted by the members acting' (Tuomela 2010, p. 44). This ethos represents a joint commitment. Moreover, it provides persons with reasons to act in a certain way, even when acting as individual persons.

Theories of joint action explain the importance of a shared set of values, an ethos, both in forming a group and in realizing a shared goal. They regard this as a fact. Now, ethics can supply ideas of which values in which scenarios matter. In order to justify a range of values that people can subscribe to, it is important not to choose particularly controversial values, but basic values such as the worth of a resource as deliverer of eco-services or the value of fairness. This approach to cooperation is much more modest than the claim that persons are endowed with a sense of justice and therefore tend to cooperate as fair partners (Rawls 1996, pp. 19, 74–75). Instead, it rests on an interpretation of cooperation that motivates the willingness to contribute to a cooperative scheme in a distinct setting. As for the content of the values in water cooperation, this includes the value of integrity of the natural resource, fairness in distributing profits and burdens among water users, and respect for various cultures in using water (Kallhoff 2014). Even though distributive schemes have been debated in terms of water markets, the distribution of access to water is also a matter of highly developed cultural practices. This also needs to be taken into account when shaping access-conditions to shared water resources.

In the reminder of this section, I shall highlight the content of the three groups of values that water cooperation includes: environmental conservation and the value of integrity of natural resources, distributive fairness and respect for culture.

Environmental conservation and integrity of natural resources

Conservation of the shared resource and restoration of basic eco-functions if already forestalled are basic issues in water cooperation. Obviously, the health and the resilience of natural goods resonate with conditions of good practices in using that resource. In particular, resilience of water systems is limited. As for distinguishing various good practices from bad practices of water usage, not only the quality of the emissions into a lake, but also the quantity and the types of usage can be distinguished as – taken together – either beneficial *or* neutral *or* detrimental to the basic eco-functions of the resource. As for long-term goals in cooperation, it is reasonable for each participant of cooperation to subscribe to environmental conservation as a prioritized goal.[6] The underlying rationale is that multiple use of water can only be guaranteed when basic eco-functions that enable this multiple use are all kept intact. A lake or river that in contrast is under severe stress or that suffers from incidents of pollution and scarcity delivers only a restricted range of eco-functions (Ripl 2003). Basically, it is the option to multiple use that profiteers should be interested in. The underlying rationale is comparable to what Rawls (1996) argues in terms of a 'veil of ignorance': since future beneficiaries will not now which eco-functions they will need or desire, the best possible future scenario is a broad range of eco-functions. In addition, it is reasonable to restore eco-functions that have already been lost. In subscribing to conservation as a core value, some of the mechanisms that lead to the tragedy of the commons can already be overcome. As supporters of conservation of the resource, water users do not act as self-interested profiteers any longer. As a minimum, they comply with conservationist goals.

Distributive fairness

As for fairness in water cooperation, two basic moral insights play an important role. First, water is a common and a critical good. It has life-sustaining qualities. As such, each person has a right to access-conditions that meet basic needs. Recently, this intuition has been enshrined in a right to water (Gleick 1998;

Scanlon et al. 2004). At least, the right of each person to receive sufficient supply of clean water within a reasonable reach – among other things – is fair to claim. This first claim has already severe implications, particularly so in situations of scarcity. As for the scenario of cooperation that focuses on a shared resource, it defines one important area of fairness. It defines a fair share that each person as endowed with human rights deserves. Yet, two more spheres of justice need to be addressed.

Second, fairness in access to a common-pool resource also includes a fair share of profits from that resource and of costs in protecting that resource for the future as also for remedies. The allocation of 'fair share' of burdens and of profits in environmental goods is a complex matter. Climate ethics has been particularly engaged in debating principles of fairness regarding a natural common pool resource (Gardiner 2004; Kallhoff 2015). As for water cooperation, three principles are of particular importance:

- The *polluter pays principle* says that actors who are responsible for environmental hazards are also reliable for remedying that damage. Regarding natural resources that cannot easily be repaired and regarding which responsibility is often unevenly distributed, it is sometimes difficult to qualify the polluters. Yet, as a principle that supports cautious practices in using water, the polluter pays principle is helpful.
- The *beneficiary pays principle* says that persons who profit from a resource should also pay for it. In cooperative schemes over natural resources, the beneficiary pays principle has been interpreted in two different ways: first, it says that persons who profit from practices that have contributed to damage should now also pay for that damage. Different from that concept of restitutive justice, the beneficiary pays principle can secondly also be framed more directly: persons who profit from a resource should pay for environmental protection, too. This second claim can be translated into a scheme that helps to integrate environmental costs into water usage. Whereas the former principle addresses moral fault in the past that now contributes to benefits of profiteers from a resource, the second is suited for spurring fair schemes of water cooperation.
- The *ability-to-pay principle* claims that burdens should be allocated according to the capacity to contribute to environmental protection. As for the natural commons, it is important to first acknowledge that each person needs water and air. As a consequence, it is also fair to claim that each person contributes something to support fair access-conditions and a fair share of the burdens resulting from environmental protection. Yet, in order to allocate burdens in accordance with capacities, the principle also says that the most rich and economically potent profiteers should contribute significantly more.

Overall, water fairness means two things: each person should receive a fair share in a natural resource that is life-supporting. Following theories of sustainability, this also includes a fair share for future generations (Feldman 2007). As a consequence, natural common goods such as water reservoirs need to be protected from destruction and instead be procured for future generations. And the burdens from environmental protection, from remedies for environmental hazard and the burdens from building new protective institutions, need to be shouldered according to principles of fairness.

Respect for culture

A third important pillar of basic values in water cooperation is respect for culture and various cultural needs. Water is also part of a web of cultural practices and religious rites (Gyawali 2009; Macauley 2010). In particular, some water reservoirs are also part of the web of goods that deserve the status of the common heritage of mankind. Respect for cultural identity does not mean that each type of water practice needs to be accepted. Yet, sometimes it is particularly difficult to integrate or even transform cultural values towards a culture of mutual respect. Examples are given in a study on water values in Nigeria: ritual procedures and religious values might support a view of water as a religious and even saintly good which claims that water reservoirs should not be transformed in any way (Akpabio 2011). Sometimes, conflicts arising around the categorization of water as saintly are deep, as examples from Hinduism demonstrate (Gyawali 2009, pp. 13–17).

Authors in the field of water ethics have claimed a culture of respect for water. In particular, an approach to water ethics does not have to accept these approaches straight away. Instead, a minimum requirement is that '. . . a new ethic, even in our advanced technological age, should be based on finding a new balance of the sacred and utilitarian in water' (Delli Priscoli & Wolf 2009, p. 121). An ethos that supports group action includes respect for values of cultural heritage and for aesthetic values in addressing water. In particular, the diversity of values needs to be respected. As a consequence, there might be reasons to spare a segment of a river or a lake from usage. Perhaps, it is also necessary to prevent certain types of action from happening. Overall, it is important to regard cultural heritage as an important aspect of water cooperation.

To sum up, I have argued so far that a range of values serve as a yardstick in water cooperation. They include priority to environmental protection, environmental fairness and respect for diversity in cultural values. These values are part of the ethos of members of a group in water cooperation. They also contribute to shape the goals of cooperation in developing schemes of cooperation in addressing water as a shared resource.[7] I shall now explain what I mean by 'cooperation' and for this draw on theories of joint action that have been debated in social philosophy.

4. Theories of joint action as a theoretical background

In philosophy, goal-directed action has primarily been explored as events that are caused by single actors. Recently, authors in social philosophy explain that actions by single actors are only one mode of acting. They highlight another mode of acting, which explains the very basics of living together in societies in another light. Authors in social philosophy argue that acting together is a rather usual thing to happen. Indeed, without acting together on a daily basis, neither single persons nor societies would survive. Approaches to what has been called 'shared agency' explain how individuals come to 'work together', how they act collectively and what the processes of group formation and of collective agency imply (Gilbert 1996; 2003; List & Pettit 2011; Tuomela 2010). Some of these theories subscribe to collective agency as exemplifying all critical characteristics of an action undertaken by a single person: a group is considered as having intentions and a single goal. It is able to realize that goal, and therefore it also exemplifies the motive and wish to realize that goal (Searle 1995). Yet, different from a metaphysically thick concept of collective agency, authors also discuss the phenomenon of acting in a group as different from acting as a single person (Tuomela 2010, 2012). These debates have also contributed to a new interpretation of 'cooperation'.

Whereas cooperation has often been reduced to a market-like type of exchange, theories of shared agency regard cooperation as a collective, goal-driven and mutual enterprise. In this section, I wish to draw some lessons from theories of joint action for water cooperation. In my view, a model of joint action is helpful in explaining necessary elements in an approach to water cooperation. I shall restrict the discussion to some basic insights that allows us to apply joint agency to an approach of water cooperation.

In profiting from a shared resource, persons are already in a situation in which they interact in one way or another. In particular, not only acts of pollution or of spoiling have effects on other participants in the same situation. Instead, intricate water supply systems and irrigation systems as well as complicated rules of allocating rights of access in profiting from waterways demonstrate that profiting from water necessarily has an impact on the usage by other persons. In addressing a shared natural resource, two options are available: either not developing explicit schemes of cooperation, but simply letting things take their own course, or developing and negotiating schemes of cooperation that serve the interests of all parties. In such a situation, it is better to negotiate distributive schemes instead of not doing so.[8] The first lesson that can be learned from theories of joint action can be termed 'explicitness'. In order to develop fair schemes of cooperation, it is necessary to name the conflicts and to explore fair schemes of distribution and of access to a shared resource.

Theories of joint action state that 'joint action' has a structure that differs both from the behaviour of swarms and from acting as one single actor. In acting together, each person is mutually responsive to what other persons do; in particular, persons contribute to realizing a shared goal by each doing his or her part. In modelling conditions for best practices of environmental cooperation, this is an important insight. Successful irrigation systems and water supply systems work like that: they have feedback mechanisms that restructure profit-seeking behaviour at various points (Ostrom & Gardner 1993). Yet, this practical dimension needs to be supplemented by a normative one. In addition to the approach of 'governance' favoured by Elinor Ostrom (2003), it is possible to transcend conflicts not only by rules that groups of profiteers from a shared resource develop from the scratch, but by a shared commitment to some basic values. The second lesson from approaches to joint agency says that group action is favoured by a shared 'ethos'.

Following Tuomela (2010), acting as a group member differs from other types of cooperation in terms of a shared 'ethos'. As members of a group – that is not necessarily a long-lasting or natural group, but might also be a temporary task group (Tuomela 2010, p. 164) – persons engage in we-attitudes. They start reasoning and intending as group members. Moreover, a shared goal resonating with we-intentions is also action-guiding for group members – at least in a situation in which the goal is tied to an ethos that participants of the group subscribe to. 'In this case of collectively intentional collective cooperative action, the group members, in effect, has a collective intention expressible by 'We will cooperate to achieve goal G. . . . The group will at least try to bring it about that G is satisfied or promoted in accordance with its ethos' (Tuomela 2010, p. 166). The ethos gives expression to a collective commitment. As for water cooperation, the ethos of conservation, distributive fairness and respect for culture serves as a proposal for a shared ethos of profiteers from a water resource.

In addition, joint action depends on a clear expression of joint goals. A theory of shared agency does not presuppose that each single person realizes the same goal. Following the proposal of Bratman of a 'planning theory of joint action' (2014), persons act together when they try to realize a goal that fulfils two presuppositions: the goal itself is a joint action; and the goal cannot be achieved without realizing it together. In order to constitute a collective of persons acting together, it suffices to identify a group of persons who are related to each other in terms of 'modest sociability' (Bratman 2014, pp. 92ff.). Yet, it is obligatory to interpret the single actions as parts of that endeavour to realize a shared goal. As for water cooperation, it is necessary to envision a situation that all possible actors can subscribe to. As for this goal, the ethos supports goals that serve the group as a whole, not single actors alone. In particular, it is helpful to envision concrete patterns of exchange that resonate with the group ethos.

A final important comment addresses the problem of 'defection'. People even if convinced of a shared ethos are not saints. Instead, they might still try to cheat on co-operators; they may also still prefer a behaviour of free-riding. And what if a 'big polluter' is not willing to join the group of co-operators altogether? In my view, these aspects do not speak against a model of cooperation in terms of joint action. Instead, it speaks in favour of institutional instruments to reduce defection and to provide incentives for cooperation. Different from theories that address cooperation in a rather loose sense, a concept of joint environmental action highlights that both the goals and the moral commitments which support cooperation can be spelled out within a theoretical and philosophical framework. Instead of profiting from a shared resource from a single-minded point of view and instead of giving water politics into the hands of power politics, for which examples are plenty (Radkau 2002), I have argued for goal-driven and value-framed cooperation. This is not to say that conflicts do no longer arise. But it is to say that the values which should guide solutions of those conflicts can be regarded as basically reasonable values in addressing a resource jointly.

Conclusion

This chapter started with the example of a village that is situated at the banks of a lake. It has explored an approach to water cooperation that helps the villagers to develop ideas for future-looking types of water usage. In particular, environmental conservation needs to be prioritized, distributive schemes need to resonate with principles of fairness, and practices of water usage should pay tribute to cultural values attached to water usage. If successful, villagers will form a joint agency; they will shape their practices of water usage according to this 'ethos'. At the end of this contribution, I wish to highlight two consequences of this approach and also address as a third point the issue of a 'global common good'.

First, it is important to understand that water cooperation will not succeed unless actors also subscribe to a range of basic values. Different from patterns of market-style exchange, joint action is goal-driven; it is about processes that have been termed 'active co-designing with nature'. Yet, shaping the goals presupposes commitment to some basic values, including environmental integrity of that resource, principles of distributive fairness and respect for various cultural types of usage. The current exploitation of water reservoirs is far from fair; in particular, as long as it rests on either uncoordinated or purely pragmatically justified practices, such exploitation will continue to spur the 'tragedy of the commons'. Without a shared ethos, an alternative to over-exploitation and destruction will not be developed.

Second, the model of joint action I have proposed in this chapter is based on some very basic insights into how people succeed in acting together. In order to achieve a shared goal, people need to be clear about their common goals. They need to understand in which ways the shared goal resonates with values they have or they are willing to invest in. Following proposals from Tuomela, groups do not have to be pre-existent, instead they can develop as answers to shared sets of values. In particular, persons and institutions that subscribe to shared sets of values can form collective actors that address joint goals. Yet, in order to succeed, it is necessary to understand the very nature of the collective enterprise. As a first step, the shared goals need to be made explicit. As a second step, it is necessary to design institutions that spur collective action and that contribute to preserving natural common pool resources from being exploited by single and self-interested actors. This is of course one of the most challenging aspects of this theoretical approach, which would need much more space to unfold in terms of institutional settings that resonate with the normative proposals. This chapter has had the much more modest goal of outlining an alternative to random exploitation.

Third, as for water as a global commons, two comments help to apply the proposals to global agendas. Even though water conflicts are always shaped by many local factors, and even though water conflicts do not result from uncoordinated patterns of exploitation and from pollution alone, but also from environmental problems that affect water systems, a worldwide commitment to a water ethics approach would be helpful in two respects. Charters are not so much about really designing and channelling institutional change, but they contribute to empowering actors in supporting commitment to shared values. It makes a big difference when persons who invest in cooperative and fair schemes of water supply can fall back on a shared commitment or they need to defend their endeavours from scratch. Moreover, raising the issues of fairness and of justice is particularly important in addressing natural commons right now. Yet, instead of using justice like a sword that divides people in groups that are worse off and in groups that need to make good on former harm, it is helpful to integrate justice into a framework on which people can act jointly.

Overall, theories of joint action demonstrate that there is an alternative to a competitive model that rests on a concept of the purely self-interested *homo economicus* and to random exploitative patterns. As compared to theories of individual action, this research shifts the focus of attention in several respects. In particular, it opens new fields of research that are of eminent importance for re-thinking environmental cooperation. In addition, and different from social choice theory, processes of team formation – even comprising 'team reasoning' (Sugden 2003; Tuomela 2012) – are addressed as critical elements in working out shared agendas of joint action. It also needs to be studied in which ways the processes of forming groups includes the transformation of individual preferences along the lines of collective preferences (Poteete & Janssen 2010; Seemann 2009). Another remarkable new focus is the debate on the responsibility of groups and normative constraints of group agency likewise (List & Pettit 2011, pp. 153–169; Miller 2010). Overall, the role of shared goals is strengthened. Basically, water reservoirs need to be procured and to be protected from degradation and over-exploitation. I have argued that water ethics contributes to this shift of attention in that it contributes an analysis of basic values underlying the formation of joint goals.

Notes

1 I have chosen an example that is very close to the example of Peter Singer (2002, p. 27). Whereas Singer speaks of a 'giant sink' and compares this to the atmosphere as a waste dump for greenhouse gases, I shall take the lake as a literal example of a joint natural resource.

2 This claim is, of course, a rather challenging one. Yet, regarding water resources, there has always been some backing both in the theory of water laws and in the practices of sharing water resources. For examples, see Ripl (2003), Rose (1990), Sadoff and Grey (2002), Naff and Dellapenna (2002). Falkenmark and Folke (2002) go so far as to claim that their investigation on water conflicts 'clearly showed that at the international level, water appears to pose a reason for transboundary co-operation rather than for war, often preventing escalation instead of causing it' (p. 140).

3 For a more comprehensive approach to water ethics, see Kallhoff (2012, 2014, 2015).

4 This interpretation is particularly important in the context of climate ethics. For this 'real tragedy of the commons', see Gardiner (2001, 2011); for an interpretation of climate change as resulting to a significant degree from a comprehensive collective-action problem, see Jamieson (2014, p. 64).

5 The debate on collective action and on joint agency in particular is a vast debate in social philosophy. For the goals of this contribution, it suffices to select basic concepts and to

go back to the most basic contributions in that field of research: see Bratman (1993) and Gilbert (1996).

6 An example of an approach to restrict over-exploitation of groundwater use is given by Llamas and Martinez-Cortina (2009, pp. 193f) regarding a Spain Water Law of 1985 that restricts groundwater use.

7 Overall, this framework stands in stark contrast to normative guidelines in terms of 'precaution' and 'risk-aversive behaviour'. Even though it is also necessary to evaluate future scenarios against the background of risks, precautionary principles are not particularly helpful in defining future goals of cooperation.

8 In water supply, the frequently favoured alternative to governance structures that allow for normative constraints is the 'water market'. Yet, examples demonstrate that even though markets serve the goal to set incentives for economizing water supply, they do not support, per se, environmental goals (Bernsen 2011). Nor do they have positive effects on communal values. For further discussion, see the conclusion of this chapter.

References

Akpabio, EM 2011, 'Cultural notions of water and the dilemma of modern management: evidence from Nigeria', in M Kowarsch (ed.), *Water Management Options in a Globalised World. Proceedings of an International Scientific Workshop (20–23 June 2011, Bad Schönbrunn)*, 2nd rev. edn, Institute for Social and Developmental Studies (IGP) at the Munich School of Philosophy, pp. 156–71, www.htph.de/igp/proceedings 2011.

Bernsen, K 2011, 'Water pricing strategies in an integrated water services management framework', in M Kowarsch (ed.), *Water Management Options in a Globalised World. Proceedings of an International Scientific Workshop (20–23 June 2011, Bad Schönbrunn)*, 2nd rev. edn, Institute for Social and Developmental Studies (IGP) at the Munich School of Philosophy, pp. 78–86, www.htph.de/igp/proceedings 2011.

Bratman, ME 1993, 'Shared intention', *Ethics*, vol. 104, no. 1, pp. 97–113.

Bratman, ME 2014, *Shared Agency. A Planning Theory of Acting Together*, Oxford University Press, Oxford and New York.

Brown, PG & Schmidt, JJ (eds) 2010, *Water Ethics. Foundational Readings for Students and Professionals*, Island Press, Washington, DC.

Buck, SJ 1998, *The Global Commons. An Introduction*, Island Press, Washington, DC.

Delli Priscoli, J & Wolf, AT 2009, *Managing and Transforming Water Conflicts*, International hydrology series, Cambridge University Press, Cambridge and New York.

Delli Priscoli, J, Dooge, J, Llamas, R & World Commission on the Ethics of Scientific Knowledge and Technology 2004, *Water and Ethics: Overview*, UNESCO, Paris.

Falkenmark, M & Folke, C 2002, 'The ethics of socio-ecohydrological catchment management: towards hydrosolidarity', *Hydrology and Earth System Sciences*, vol. 6, no. 1, pp. 1–9.

Feldman, DL 2007, *Water Policy for Sustainable Development*, The Johns Hopkins University Press, Baltimore, MD.

Gardiner, SM 2001, 'The real tragedy of the commons', *Philosophy and Public Affairs*, vol. 30, no. 4, pp. 387–416.

Gardiner, SM 2004, 'Ethics and global climate change', *Ethics*, vol. 114, pp. 555–600.

Gardiner, SM 2011, *A Perfect Moral Storm. The Ethical Tragedy of Climate Change*, Oxford University Press, Oxford and New York.

Gilbert, M 1996, *Living Together. Rationality, Sociality, and Obligation*, Rowman and Littlefield, New York.

Gilbert, M 2003, 'The structure of the social atom: joint commitment as the foundation of human social behaviour', in F Schmitt (ed.), *Socializing Metaphysics: The Nature of Social Reality*, Rowman and Littlefield, Lanham, MD, pp. 39–64.

Gleick, PH 1998, 'The human right to water', *Water Policy*, vol. 1, pp. 487–503.

Gyawali, D 2009, 'Water and conflict: whose ethics is to prevail?', in MR Llamas, L Martinez-Cortina & A Mukherji (eds), *Water Ethics. Marcelino Botin Water Forum 2007*, CRC Press, Boca Raton, FL, pp. 13–24.

Habermas, J 1996, *Die Einbeziehung des Anderen: Studien zur politischen Theorie*, Suhrkamp, Frankfurt am Main.

Hardin, G 1968, 'The tragedy of the commons', *Science*, vol. 162, no. 3859, pp. 1243–48.

Jamieson, D 2008, *Ethics and the Environment. An Introduction*, Cambridge University Press, Cambridge.

Jamieson, D 2014, *Reason in a Dark Time. Why the Struggle Against Climate Change Failed – And What It Means for Our Future*, Oxford University Press, Oxford and New York.

Kallhoff, A 2012, 'Addressing the commons: normative approaches to common pool resources', in T Potthast & S Meisch (eds), *Climate Change and Sustainable Development. Ethical Perspectives on Land Use and Food Production*, Wageningen Academic Press, Wageningen, The Netherlands, pp. 63–8.

Kallhoff, A 2014, 'Water justice: a multilayer term and its role in cooperation', *Analyse & Kritik*, vol. 36, no. 2, pp. 367–82.

Kallhoff, A 2015, 'Klimakooperation: Kollektives Handeln für ein öffentliches Gut', in A Kallhoff (ed.), *Klimagerechtigkeit und Klimaethik*, Wiener Reihe. Themen der Philosophie, De Gruyter, Berlin and Boston, MA, pp. 143–67.

Kaul, I, Grunberg, I & Stern, MA (eds) 1999, *Gobal Public Goods. International Cooperation in the 21st Century*, Oxford University Press, New York and Oxford.

List, C & Pettit, P 2011, *Group Agency: The Possibility, Design, and Status of Corporate Agents*, Oxford University Press, Oxford and New York.

Llamas, MR & Martinez-Cortina, L 2009, 'Specific aspects of groundwater use in water ethics', in MR Llamas, L Martinez-Cortina & A Mukherji (eds), *Water Ethics. Marcelino Botin Water Forum 2007*, CRC Press, Boca Raton, FL, pp. 187–204.

Macauley, D 2010, *Elemental Philosophy. Earth, Air, Fire, and Water as Environmental Ideas*, State University of New York Press, Albany.

Miller, S 2010, *The Moral Foundations of Social Institutions: A Philosophical Study*, Cambridge University Press, Cambridge.

Naff, T & Dellapenna, J 2002, 'Can there be confluence? A comparative consideration of Western and Islamic fresh water law', *Water Policy*, vol. 4, pp. 465–89.

Ostrom, E 2003, *Governing the Commons. The Evolution of Institutions for Collective Action*, Cambridge University Press, Cambridge and New York.

Ostrom, E & Gardner, R 1993, Coping with asymmetries in the commons: self-governing irrigation systems can work, *The Journal of Economic Perspectives*, vol. 7, no. 4, pp. 93–112.

Poteete, AR & Janssen, MA 2010, *Working Together: Collective Action, the Commons, and Multiple Methods in Practice*, Princeton University Press, Princeton, NJ.

Radkau, J 2002, *Natur und Macht. Eine Weltgeschichte der Umwelt*, Beck, München.

Rawls, J 1996, *Political Liberalism*. With a new Introduction and the "Reply to Habermas", Columbia University Press, New York.

Rawls, J 2005, *A Theory of Justice*, Belknap Press, Cambridge, MA.

Ripl, W 2003, 'Water: the bloodstream of the biosphere', *Philosophical Transactions of The Royal Society of London, Biological Sciences*, vol. 358, pp. 1921–34.

Rose, CM 1990, 'Energy and efficiency in the realignment of common-law water rights', *The Journal of Legal Studies*, vol. 19, no. 2, pp. 261–96.

Sadoff, CW & Grey, D 2002, 'Beyond the river: the benefits of cooperation on international rivers', *Water Policy*, vol. 4, pp. 389–403.

Scanlon, J, Cassar, A & Nemes, N 2004, 'Water as human right?', *IUCN Environmental Policy and Law Paper*, vol. 51, IUCN–The World Conservation Union.

Searle, J 1995, *The Construction of Social Reality*, The Free Press, New York.

Seemann, A 2009, 'Why we did it: an Anscombian account of collective action', *International Journal of Philosophical Studies*, vol. 17, no. 5, pp. 637–55.

Singer, P 2002, *One World. The Ethics of Globalization*, Yale University Press, New Haven, CT.

Sugden, R 2003, 'The logic of team reasoning', *Philosophical Explorations: An International Journal for the Philosophy of Mind and Action*, vol. 6, pp. 165–81.

Tuomela, R 2010, *The Philosophy of Sociality: The Shared Point of View*, Oxford University Press, Oxford.

Tuomela, R 2012, 'Group reasons', *Philosophical Issues*, vol. 22, no. 1, pp. 402–18.

Part II

Global water ethics, local cases and a diversity of perspectives

7 Safe, just and sufficient space

The planetary boundary for human water use in a more-than-human world

Rafael Ziegler, Dieter Gerten and Petra Döll

1. Introduction

The global sustainability debate has been inspired by the iconic image of the blue planet as an entity with finite environmental limits. Most recently, these limits have been acknowledged via the discussion of planetary boundaries (PB), which considers nine critical Earth system processes – all strongly influenced by human activities – in a single conceptual framework (Rockström et al. 2009; Steffen et al. 2015). One of these planetary boundaries is concerned with an upper global limit to human freshwater consumption (Gerten et al. 2013). This boundary implies scientific-ethical challenges worthy of serious and sustained attention, and as such will be elaborated on in this chapter.

According to the general idea animating the concept of planetary boundaries, human societies as we know them have developed within the conditions of the Holocene over roughly the last 11,700 years. With industrialization, human activities have come to affect the functioning of the Earth system to such an extent that its capacity to persist in a Holocene-like state is threatened (Steffen et al. 2015). As a result, there is the risk that human societies will exit the 'safe operating space' delineated by the nine boundaries (Rockström et al. 2009). This risk diagnosis is worked out with PBs for biosphere integrity (consisting of functional diversity and genetic diversity), climate change, introduction of novel entities, stratospheric ozone depletion, atmospheric aerosol loading, ocean acidification, biogeochemical flows (consisting of phosphorus, nitrogen), freshwater use, and land-system change (Steffen et al. 2015). Following the precautionary principle, the planetary boundaries are not specified as thresholds or tipping points, but rather are located upstream of such possible thresholds or gradual aggregate impacts (i.e. in a safe distance, at the lower tail of a scientific uncertainty zone). This is to account for uncertainty in the precise position of critical developments and to provide society with an early warning sign (Steffen et al. 2015). The planetary boundary for human freshwater use (PB-W), which we focus on here, is defined as the maximum tolerable amount of total global freshwater consumption by humans for irrigation, livestock raising, domestic purposes, manufacturing and cooling of thermal power plants. The term 'freshwater consumption' here refers to the fraction of the abstracted freshwater volume that evapotranspires during use and is no longer available in liquid form to other human or non-human water users; notably water quality is not considered. PB-W is 'the maximum amount of freshwater that can be appropriated by humans, beyond which there is a high probability of (possibly abrupt) water-induced changes with large detrimental impacts on human societies' (Gerten et al. 2013, p. 551). Human freshwater use does not have a singular threshold at the global or continental scales, but aggregate effects of excessive local freshwater use would imperil the Earth system.

What type of threats does excessive human freshwater use pose to the functioning of the Earth system? To begin with, freshwater biota (including riparian vegetation) as well as coastal biota suffer from alterations of ecologically relevant streamflow regimes. Alterations include decreases of mean annual streamflow, changes in the seasonality, decreases (in case of water abstractions) or increases of low flows and decreases of flooding (Döll et al. 2009). On more than one sixth of the global land surface (excluding Greenland and Antarctica), human water use has decreased mean annual streamflow by more than 10% as compared to the situation without human water use (Döll et al. 2009). These streamflow alterations have contributed to the observed strong population decline of freshwater species (MEA 2005; Döll et al. 2009). While knowledge about quantitative relations between streamflow alterations and changes of freshwater ecosystems is very poor, there is broad scientific agreement that even small alterations can change the complex ecosystems (Richter et al. 2006; Poff & Zimmermann 2010). Streamflow alterations lead to changes in sediment transport and may cause salinization and inland movement of saltwater in delta areas. Excessive water use upstream may lead to a lack of water or even the drying out of downstream river stretches, thereby endangering crop and livestock production. In the case of groundwater depletion, with steadily falling groundwater tables, water supply for irrigated crop production or other human uses may become unavailable. Groundwater depletion contributes to rising sea levels, with an estimated 0.3 mm/yr during 2000–2009 (Döll et al. 2014). In addition, irrigation causes changes of spatial patterns and seasonal timing of atmospheric moisture flows.

The main driver of these threats is irrigated agriculture, which accounts for roughly 70% of the global rate of water abstraction, followed by industry (20%) and municipal water use (10%) for drinking water, hygiene, etc. (FAO 2004). With a high amount of evapotranspiration, irrigation accounts for almost 90% of global water consumption. Water consumption in irrigated agriculture amounts to an estimated 1100 km^3/yr, while water consumption in the other water use sectors is estimated to be 300 km^3/yr (Döll et al. 2014). In addition, more than 200 km^3/yr of evaporation loss is due to man-made reservoirs (Shiklomanov & Rodda 2003), such that global human water consumption now amounts to ~1600 km^3/yr. The currently proposed PB-W is 4000 km^3/yr (Steffen et al. 2015; Gerten et al. 2015). Therefore, the PB-W seems to suggest that we are currently well below the planetary boundary and therefore in the 'safe space'. What is the implication of such a finding for environmental philosophy? And what are the ethical presuppositions for arriving at this boundary estimate?

The planetary boundary proposal invites a reconsideration of Paul Taylor's argument, made in his seminal contribution to environmental philosophy (Taylor 1986). Arguments in environmental philosophy rely on the interrelation of fact and evaluation. On the one hand, we draw on what we know about the world – the knowledge made available by hydrology, ecology, evolutionary theory, etc. – while on the other hand, we draw on our ethical principles and intuitions. A philosophical worldview[1] is the result of an attempt to bring together the descriptive and the evaluative in a way that is comprehensive and complete, systematic and coherent (Taylor 1986, p.158f). Such a worldview informs and supports our attitude of living with others, and by implication the search for principles and standards that practically express the worldview (Taylor 1986, p. 41ff). Exploration of the planetary boundary framework should make the evaluative aspect explicit and explore its implications for operationalization and ultimately for water governance.

At first glance, the concern with a 'safe operating space' for humanity seems to express an anthropocentric perspective.[2] This chapter argues otherwise and proposes that the planetary boundaries framework can show respect for a more-than-human world. It explores this claim via a discussion of PB-W and biocentrism, a worldview that recognizes the intrinsic value of all living beings. Our discussion focuses on an ideal-theory exploration: what does a global perspective on freshwater use look like if humanity were to take a biocentric perspective? As we will see, this question is a demanding one, pushing the boundaries of a chapter. The conclusion provides some suggestions for further development in the non-ideal world.

For our exploration, we first note that the planetary boundaries framework includes ethics. The central place of the precautionary principle in the framework focuses the quantitative analysis of water use on unacceptable moral harm (section 2). But what is unacceptable moral harm? And whose harm is to be considered? The focus on a collectively safe operating space requires a qualification: not only safe for the whole – humanity according to an anthropocentric reading, the biocentric community according to the reading explored here – but also for the individuals that the whole consists of. This introduces distributive justice and the idea of a safe *and just* operating space (section 3); '. . . wellbeing depends on keeping total resource use below critical natural thresholds, but it equally depends upon every person having a claim on the resource they need to lead a life of dignity and opportunity' (Raworth 2013, p. 30). Evidently, considering the 'social foundation' (Raworth 2013) is very important to explore the link between the framework and global water governance discussions such as the human right to water and the water-related Sustainable Development Goals.[3] In a further step, section 4 critically revisits this social foundation and its relation to planetary boundaries. While broadly speaking the biocentric approach supports the call for a justice foundation of the framework, it challenges the idea that beyond a justice threshold of freshwater for humans and their needs, there is an additional empty 'just and safe space'. This space is already used by non-human biota. We highlight this additional consideration with the qualifier just, safe and *sufficient* space as it introduces a further challenge of *sufficient* human water use that is largely absent from the anthropocentric discussion of the framework. It highlights an alternative use of the PB-framework with arguably radical consequences for production and consumption. We discuss biocentric priority rules for thinking about water use conflicts once this more-than-human space is taken into account. Thus, this exploration of the freshwater boundary shows a biocentric way of developing the framework. Its focus on sufficiency contributes to a reflexive use of the planetary boundaries concept and of global water ethics.

2. Safe operating space, the precautionary principle and environmental philosophy

A central bridge linking empirical boundary specification and ethical questions in the planetary boundaries framework is the precautionary principle. 'The PB [planetary boundaries] framework is based on critical processes that regulate ES [Earth system] functioning. By combining improved scientific understanding of ES functioning with the precautionary principle, the PB framework identifies levels of anthropogenic perturbations below which the risk of destabilization of the ES is likely to remain low – a "safe operating space" for global societal development' (Steffen et al. 2015, p. 736[4]).

The precautionary principle is a key normative principle in global environmental politics, as testified by Article 3 of the United Nations Framework Convention on Climate.[5] A general definition of the principle is offered by UNESCO-COMEST: when human activities may lead to morally unacceptable harm that is scientifically plausible but uncertain, actions shall be taken to avoid or diminish that harm (UNESCO-COMEST 2005). The definition clarifies central aspects of the precautionary principle: a) morally unacceptable harm,[6] b) accompanying scientific research[7] and c) a call to action.[8] None of these are simple challenges for such a global issue!

Notably, climate ethics provides a lively debate around the question of limits and boundaries in relation to morally unacceptable harm, with contributions from the social sciences (for example, Stern's green new deal proposal: Stern 2010, p. 44) as well as from political philosophy (Gardiner et al. 2010). According to a human rights perspective, *morally unacceptable harm* refers to human rights violations caused by or plausibly attributed to climate change. The human rights perspective articulates a threshold that no one should fall under. For example, the human rights to water specifies a minimum of safe, affordable and acceptable water for drinking and hygiene that every human being is entitled to *qua* human being (independent of specific contracts, etc.). Entitlements generate a universal obligation to protect and respect this basic minimum standard (Caney 2010, p. 165[9]). In addition, the moral threshold enjoys lexical priority. 'Human rights generally take priority over moral values such as increasing efficiency or promoting happiness' (Caney 2010, p. 165).

The human rights perspective suggests thinking about unacceptable harm in terms of a moral threshold, and we follow this idea here. But while the human rights approach helpfully clarifies various aspects of a moral threshold, it is only concerned with harms to humans. Yet, in environmental philosophy, an open question is whose harm is to be considered? Harm only to human beings, or also harm to sentient beings (sentientism), or harm to all living beings (biocentrism), or damage (in a wide sense) to all entities (holism)?[10] The concept of dignity, which at least in the Western tradition plays an organizing role for the justification of human rights, can also be explored for sentient beings and living beings. In fact, the Swiss constitution recognizes the dignity of creatures (*Würde der Kreatur*), and explicitly includes in this animals, plants and other organisms. In a well-known justification of human rights, rights are understood as instruments to protect and promote central capabilities for leading a life in dignity (Nussbaum 2006). 'Dignity' here serves as an organizing term for various goods ('doings' and 'beings') that are claimed to be of constitutive importance for leading a dignified life. For moral agents, those capable of deliberation and choice, the freedom aspect is important; treating them as ends requires a recognition of their capacity to choose ends for themselves. Yet already within the species of humans, there are moral patients (synonym: moral subjects), that is, human beings (e.g. young children) who are not capable of deliberating and choosing but who still have a dignity and various goods that moral agents recognize. This observation suggests an extension of the moral community (Nussbaum 2006). The first ethical question at this point is: can we speak of 'good for *x*' where this good does not depend on the interest of a third party (instrumental value)?

> Perhaps the most ethically significant fact about moral subjects is that it is always possible for a moral agent to take a moral subject's standpoint and make judgments from its standpoint about how it ought to be treated. The standard implicit in such judgments is the furtherance or preservation of well-being of the subject, not of the one who does the judging.
>
> (Taylor 1986, p. 17)

By contrast, if we cannot take the standpoint of the respective being, the discussion of 'good for *x*' necessarily depends on an external standard we impose, and thus becomes an instrumental use. According to the biocentric position, moral subjects 'include all living organisms . . . inanimate objects (stones, grains of sands, puddles of water, snow, fire, ice, air) are not themselves moral subjects' (Taylor 1986, p. 18[11]).

If the scope of having a good of one's own includes all living organisms, does this also mean that we have to respect them as ends in themselves on an equal level with us? Are human beings not superior? At this step, Taylor proposes a worldview argument. Taking into account an evolutionary and ecological perspective, we can observe the joint, evolutionary membership of the human species in the Earth's community of life, an interdependent community that consists of diverse members each with a good of their own (Taylor 1986, Chapter 3). From this outlook, the idea of human superiority can be understood in a historical perspective, that is, as a result of earlier ontology that posits human beings on top of the chain of living beings ('*scale naturae*' in Greek philosophy and influential strands of early Western Christianity; White 1967) or that posits a mechanical world of *res extensa* and an independent world of *res cogitans* (Cartesianism). However, in Taylor's worldview argument, these earlier worldviews are no longer consistent with what we know about the world. The idea of human superiority cannot be simply posited; rather, the new default position is equal moral consideration of all beings that have a good of their own (even if these goods are (partly) species-specific, and thus also the respective notion of dignity and what it means to harm the respective being; Nussbaum 2006). To be sure, this biocentric argument is more complex than the brief sketch that we can offer here. It is also controversial (Nolt 2014, pp. 169ff). But a controversial status is a general feature of basic positions in environmental ethics. More important for our purposes, biocentrism offers one possible and coherent outlook for thinking about planetary boundaries in contrast to a purely anthropocentric reading.

In conclusion, a closer look at the precautionary principle leads to an ethical question in terms of environmental ethics: the question of moral inclusion, that is, who should be considered morally? This consideration pushes in the direction of further reflection for a transparent and reflexive discussion of the planetary boundaries framework.[12] In contrast to the conventional anthropocentric reading, we explore the framework from a biocentric perspective in section 4 below. But first we need to attend to a further qualification.

3. Just operating space

If we care about unacceptable damage to the Earth system that renders it less hospitable for the development of human societies, we also ought to care about the individual members of these societies, if they are unacceptably damaged as individuals, as groups, etc. Thus, the concern with a safe operating space implies a concern with justice, including the distributive consequences of industrialization in the Holocene. It is unsurprising then that the discussion of the planetary boundaries framework quickly yields the proposal of a safe *and* just space (Raworth 2013).

In the eco-space literature, of which the planetary boundaries framework is a more recent development at the global level, this point has been recognized in terms of securing a social minimum (Spangenberg 2003, p. 28). Likewise, it has been pointed out in the planetary boundaries discussion that in addition to an environmental ceiling (synonymous here with planetary boundary), we also need to discuss the social foundation in terms of the resources needed for leading a life in dignity (Raworth 2013; with respect to the Sustainable Development Goals discussion, see Griggs et al. 2014). The importance of this point is hard to overestimate: the warning of a possible exit from a hospitable space remains strangely biased if it does not take into account that the conditions in this space are already not hospitable for billions of people lacking adequate access to drinking water and sanitation, food or energy.

With respect to our focus on freshwater, this calls for a specification of unacceptable harm occurring if basic human needs for freshwater[13] are not met so as to arrive at estimates of the social foundation. Here we can only sketch what such an estimate might look like. A fully worked out specification is far beyond the possibilities of a chapter; moreover, a social process is required to determine the needs in each specific context.[14]

For drinking water and domestic use, Chenoweth (2008, quoted in Feitelson 2012) notes two approaches:

- Investigate the freshwater requirements for basic domestic functions (drinking, sanitation, washing, cooking). Gleick (1996) argues that 50 litres per capita per day, equalling 18.25 m^3 per capita per year, are needed for this.
- Investigate the amount of water abstractions by prudent societies that enjoy a high level of technological development. For this, Falkenmark (1986) specifies 274 litres per capita per day, about 100 m^3 per capita per year, as the required amount.

Norman Feitelson (2012), combining both approaches, arrives at an estimate of 60 m^3 per capita per year. This value is approximately equal to the per capita abstractions for public water supply in Germany (Statistisches Bundesamt 2013). With a global population of currently 7.3 billion people, this amounts to a freshwater requirement of roughly 440 km^3/yr, and of 134 km^3/yr according to the more modest proposal by Gleick (1996). Global freshwater abstractions for household use were estimated at 350 km^3/yr during the period 2003–2009, of which 60 km^3/yr was water consumption (Döll et al. 2014). Thus, considering both the water needs and water abstractions of households in comparison to the PB-W value of 4000 km^3/yr as specified in Steffen et al. (2015), there is no global freshwater problem as far as the physical availability for meeting this need is concerned. PB-W would not be transgressed if the basic human water needs of domestic water use were met for all human beings.

However, there are many places around the globe where this need is not met. The adjusted (just and safe) planetary boundary framework highlights that the challenge with meeting this need is not due physical freshwater limitation globally. In this sense, it opens space for further discussions and practical possibilities: as freshwater is not equally distributed worldwide, do water-rich societies have a responsibility to help water-poor societies meet their needs? Do they only have this responsibility when their way of living negatively affects, or puts pressure on meeting, water needs in water-poor societies? And linked to this last question: what about the responsibility of each society to develop ways of meeting the need adapted to their geographical-cultural context and within the possibilities of their watershed? The adjusted planetary boundary framework does not provide 'the answer' to these questions, but it helps to put them in a global context and thus to move beyond the frequently very place-based discussions in water ethics and water governance.

What about water for food production under irrigation, by far the largest water use at the global level? According to FAO estimates, about 3000 kcal per person per year need to be produced (e.g. consisting of 80% vegetal and 20% animal-based products) for healthy nutrition (see Mauser 2007; Gerten et al. 2011[15]). At current levels of productivity, this requires on average 1075 m^3 per person per year of green water (precipitation evapotranspired in both rain-fed and irrigated agriculture) and blue water (irrigation water evapotranspired in irrigated agriculture) (Gerten et al. 2011). With a population of 7.3 billion people, this amounts to roughly 8000 km^3 of green and blue water per year. This estimate consists of the green and blue water consumption on both cropland (including crops grown for feeding livestock) and grazing land (from which only 30% of evapotranspiration is assumed to contribute to food production). In the year 2000 about 13% of total water consumption of cropland and grazing land was from irrigation water (estimate derived by combining data of Siebert & Döll 2010 and Gerten et al. 2011). Thus, at current crop productivity levels, roughly 1000 km^3/yr of consumptive blue water use (which is what PB-W refers to) is required to allow healthy nutrition for 7.3 billion people. Note that from a biocentric perspective this is a conservative over-estimate of human freshwater needs as the FAO estimate includes 20% animal-based products. The actual green and blue water use for crop production and grazing is estimated to be rather similar to the estimated required value of 1075 m^3 per person per year (given the large uncertainties), with on average 1400 m^3 per person per year (estimate derived by combining data of Siebert & Döll 2010 and Gerten et al. 2011). Assuming, however, a fraction of only 10% instead of 20% animal-based food, the food-related basic water need would drop to only 768 m^3 per person per year.

Assuming that a PB-W of 4000 km^3/yr is not an underestimation, there seems to be no global freshwater problem as far as physical availability for meeting this need is concerned. However, with further growth in world population as well as changes in de facto food consumption habits, crossing PB-W (*sensu* Gerten et al. 2013, who calculated variants of the boundary value with an average estimate of 2800 km^3 yr^{-1}) is a possible scenario at current levels of technology and water consumption per food unit produced. As water use for food in the global North is much higher than in the global South (Mauser 2007, p. 180), the following questions arise again: do water-rich societies have a responsibility to help meet food-related water needs of water-poor societies? Do they have this responsibility only when their food habits negatively affect the possibility of meeting the needs in other societies (for example, large-scale land acquisition for food exports and water grabbing, virtual trade in water that does not enhance water efficiency but yields net exports of virtual water from water-poor countries; Mehta et al. 2012)? What is the responsibility of each society within its watersheds and physical water availability? Again, the adjusted PB-framework does not provide 'the answer' to these questions, but it helps to put them in a global context, which these global dynamics of trade and political economy require.

4. Planetary boundaries and sufficiency: the biocentric addition

If all living beings are to be considered as users of water in their own right and not just because freshwater biodiversity is good for humans, then how are we to think of the safe space and the numerous moral conflicts between different water users? Taylor proposes four main biocentric rules for moral agents (Taylor 1986, pp. 172–86):

1 Rule of non-maleficence: the duty not to harm any entity in the natural environment that has a good of its own.
2 Rule of non-interference: the duty to refrain from placing restrictions on the freedom of individual organisms, hands-off policy with regard to whole ecosystems and biotic communities, that is, providing 'safe space' for ecosystems due to minimal interference by humans.
3 Rule of fidelity: the duty not to deceive or mislead any animal capable of being deceived or misled, to uphold an animal's expectations, which it has formed on the basis of one's past actions with it, and to be true to one's intentions as made known to an animal when it has come to rely on one.
4 Rule of restitutive justice: the duty to restore the balance of justice between a moral agent and a moral subject when the subject has been wronged by the agent.

As it is difficult for humans *not* to violate these rules, Taylor also proposes priority rules so as to deal with conflicts between users. These contribute a genuine addition to the discussion of distributive justice within a planetary boundaries framework. The priority rules can be stated as (Taylor 1986, pp. 265ff[16]):

- a principle of self-preservation for meeting and securing one's basic interests,
- a principle of proportionality that gives priority to basic interests over non-basic interests in case of conflicts,
- a principle of minimum wrong for the pursuit of non-basic human interests as long as they concern culturally important human interests (Taylor 1986, p. 281) and as long as they are performed in a way that minimizes harm to non-humans – essentially this is the contested space where human water use can in principle exceed the sufficiency threshold,
- a principle of restitutive justice to make up for harm done under the prior principles – for example, the preservation or restoration of rivers as a compensation for the modification of rivers elsewhere for meeting basic human interests (Taylor 1986, p. 305).

The principle of self-preservation has a special justification for humans as moral agents: meeting basic needs is a way to ensure that the conditions of moral agency are met, and thus of ensuring a precondition of the 'respect for nature' demanded by environmental philosophy. If you starve or die from thirst, there is little space for moral action. As Taylor does not say much about the further moral and political implications of moral agency, it is noteworthy that his discussion of basic interests is conceptually very close to Nussbaum's discussion of central capabilities. This link suggests that for the political implications of moral agency, a biocentric approach to planetary boundaries could learn from and build on her work on justice (Nussbaum 2006).

In Taylor's view, interests are objects or events that serve or protect, to some degree or other, the good of a living thing (1986, pp. 270f). For human beings,

> basic interests are what rational and factually enlightened people would value as an essential part of their existence as persons. They are what people need if they are going to be able to pursue those goals and purposes that make life meaningful and worthwhile. Thus for human beings their basic interests are those interests, which when morally legitimate, they have a right to have fulfilled.
>
> (Taylor 1986, p. 272)

Thus, basic interests refer very much to those interests that need to be secured with a view to the moral threshold discussed in the last section. With the biocentric view, not only humans have such interests but also other living beings, each in their species-specific way. The distinction between basic and non-basic interests is significant for humans due to the unlimited way of shaping needs and wants via culture and in capitalism especially via markets. The significance of the distinction is much less clear for other living beings. We therefore here generically refer below to 'water needs of living beings', keeping in mind that for human beings this refers to their basic needs.

In sum, there is a way of dealing with the interests of different water users.[17] This is no doubt demanding, but not logically incoherent or in principle impossible; and with it comes a fresh way for thinking about the planetary freshwater boundary. We turn to this now.

Human appropriation of freshwater revisited

At first sight, the biocentric perspective suggests to move from a social foundation (Figure 7.1 A, modified from Raworth 2013) to a 'full' biocentric foundation that includes the freshwater needs of all living beings (Figure 7.1 B).[18] A second possibility to harmonize the planetary boundaries framework with the biocentric perspective is to re-interpret the PB-W. In the anthropocentric specification (Figure 7.1 A), all freshwater is a possible object of human appropriation, and PB-W is defined as the maximum amount of human water consumption that avoids detrimental impacts on human societies. In the biocentric specification, both human and non-human water users have the right to use water, and priority rules need to be applied to determine the environmental ceiling PB-W (Figure 7.1 C). Note that the environmental ceiling in Figure 7.1 C is lower than in Figure 7.1 A, as under the biocentric worldview more water is allocated to non-human water users, who have to be equally considered.

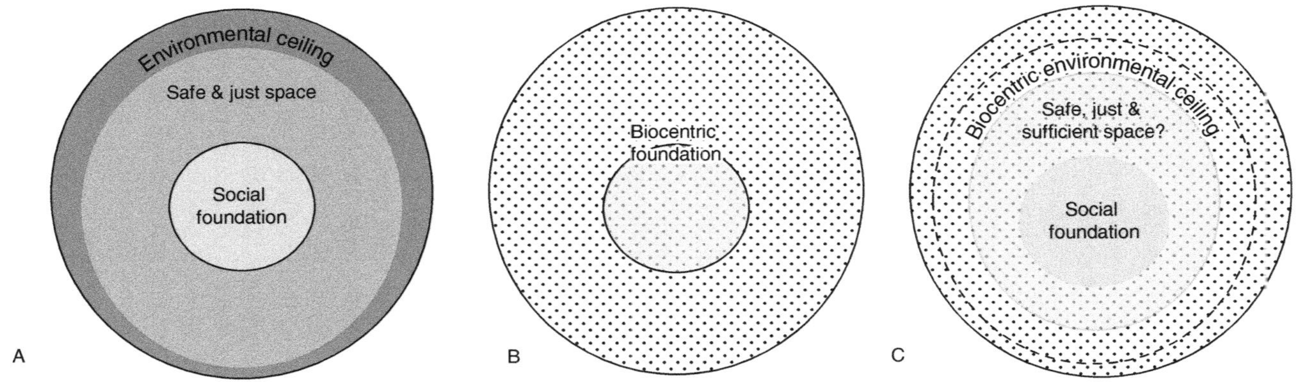

Figure 7.1 Varieties of perspectives on PB-Freshwater: A represents an anthropocentric perspective, B and C a biocentric perspective. All figures: the outer line of the circle describes the total amount of freshwater potentially useable by humans (for an estimate, see Gerten et al. 2013). The figures should not be read as representations of factual water use proportions; they seek to clarify the complex relation between a biocentric perspective and PB-W.

Source: own.

Figure 7.1 A shows a safe and just space for human water use between the social foundation (basic human water needs) and the environmental ceiling (PB-W) from an anthropocentric perspective (modified from Raworth 2013). The social foundation space refers to needs in relation to consumptive water use (as estimated in section 3 in relation to domestic water needs and food). Figure 7.1 B shows what happens if we extend the idea of the moral community: as all freshwater potentially useable by humans is used by some living being, the new biocentric foundation covers the entire space, and includes as a subpart the human social foundation (now with a dotted background to underline the community continuity). Strictly speaking, this is a partial representation of the biocentric foundation as there is freshwater use by living beings beyond the amount potentially useable by humans (for example, river run-off in remote areas). An illuminating tension results from the observation that the freshwater use of living beings is not necessarily consumptive. While water use of some plants causes evapotranspiration and it thus consumptive, the freshwater needs of fish refer to water as a habitat, with seasonally appropriate water availability and flow velocities and without barriers to migration. The total amount of potentially useable freshwater beyond the social foundation should therefore not be identified with water that is consumed by non-human living beings, but rather with water that is blocked due to a variety of non-consumptive water uses by non-human living beings. Figure 7.1 C considers the human basic water needs (social foundation) vis-à-vis the water needs of other living beings in the surrounding 'dotted' zone. From the biocentric perspective, excessive water consumption is still potentially dangerous for human beings, but in addition, the water is valuable for other species. This shift in normative significance implies that in the 'dotted' zone there is a blurring of 'danger for human beings' with 'water needs of other living beings'. The dotted area refers to a difficult to define 'basic water need of non-human living beings'. The new biocentric environmental ceiling is lower because it includes, further to anthropocentric risks, a respect for the water needs of other living beings (as a reference point, we keep the purely anthropocentric environmental ceiling as a dashed line). The respect for water needs of others refers to the extended sense of water use: humans can respect the water needs of a migrating fish like the huchen by not damming and abstracting water from the river. The sufficient space between the outer 'dotted' zone and the inner zone of the 'social foundation' refers to the possibility of human water uses beyond basic needs that do not negatively interfere with the water needs of other living beings. In the limit, the boundary of the social foundation is a sufficiency ceiling; there is no consumptive human water use that does not negatively interfere with the water needs of other living beings, and the social foundation is equivalent with a just, safe and sufficient space for humans. If there are synergetic uses or use potentials, then there is a (potential) *safe, just and sufficient space* beyond the social foundation.

There is some evidence that biocentric aspects are present in the conceptual and operational development of the planetary boundary framework so far. 'It has been recognized that terrestrial systems and aquatic ecosystems are legitimate water users for the sake of maintaining the ecosystems themselves *and* their contribution to services that support human well-being' (Gerten et al. 2013, p. 552, drawing from Falkenmark & Rockström 2004). In short, non-human water users are framed instrumentally (contributing services that support human well-being) and at least indirectly as potential ends in themselves ('for the sake of maintaining the ecosystem themselves'). In terms of operationalization, the maximum amount of freshwater is defined as accessible blue water, but 30% of accessible blue water is subtracted to avoid 'water stress' (to ensure maintenance of moisture feedback, wetlands and estuaries; Gerten et al. 2013, p. 554) and in addition environmental flow requirements are subtracted. Environmental flow requirements can be a way to recognize the water needs of other species in and around aquatic ecosystems. As Figure 7.1 C illustrates, a biocentric application of environmental flow requirements implies a shift in the normative significance of the framework: water consumption beyond the environmental ceiling is not just a likely danger for humans, it also most likely curtails the water needs of other beings. We can thus think of environmental ceilings not only in terms of risks to human beings but also in terms of a respect for other beings.

There are at least two ways in which a biocentric conception of the planetary boundaries framework for freshwater can be considered: the needs of moral agents in relation to the needs of other living beings (Figure 7.1 C); the shared needs of living beings in the bounded 'operating space' of a full planet (Figure 7.1 B). They are not exclusive but rather highlight different aspects of what such a conception has to focus on. To be sure, this only accounts for one aspect of these needs (i.e. in relation to consumptive water use), but this is a limitation of the planetary framework in general that is not specific to the anthropocentric–biocentric discussion. The social foundation, which Figure 7.1 C highlights, comprises the water needs of human beings, and thus of the moral agents in the Earth system.[19] It provides a very relevant focus from a biocentric perspective: the specification of basic needs (not wants) of moral agents in relation to freshwater. This way of exploring the planetary boundary framework from a biocentric perspective has the pragmatic advantage of fitting with the general planetary boundaries framework in a straightforward way.

Sufficient space: a biocentric addition to the debate

From the biocentric perspective, the basic interests and by implication the basic freshwater needs of all living beings deserve the respect of moral agents. For moral agents, the distinction of basic and non-basic interests and subsequent water needs is especially important. Moral agents, based on cultural and individual preferences as well as their knowledge (technological knowhow, etc.) are capable of consuming considerably more freshwater than would be required for meeting their basic needs. It is an open eudemonistic question whether consuming more than basic needs is enjoyable or really a nuisance (Paech 2012). Independently of this question, the biocentric perspective suggests a moral distinction:

- Claiming and meeting basic water needs is legitimate (principle of self-preservation), but requires estimates of freshwater needs for drinking, food, etc., that is, of the water-related social foundation of moral agents, and to this end ultimately a social process to determine this foundation.
- Human water use beyond this basic level is subject to the principle of proportionality, which prioritizes the basic interests of plants and animals over non-basic interests of humans.[20]

Thus, with the biocentric perspective the way of thinking about the safe operating space changes. Rather than focusing on the possible human freshwater appropriation that does not lead to negative impacts for human (the purely anthropocentric case), the focus is on the *sufficient human use* that leaves a maximum of water to non-human water users. Here there is an overlap with the discussion of the social foundation in the prior section – with the difference that beyond the social foundation there is no 'empty' space for humans to use, but freshwater that is used by other species. Thus, the operating space for human water use needs to not only be safe and just, but safe, just and sufficient (Figure 7.1). To be sure, human appropriation does not necessarily mean that water has to be maximally used in this space, even on the anthropocentric reading – but with the powerful narrative of 'growth within limits' (Crépin & Folke 2014, p. 58), this maximization reading is certainly there. The biocentric reading provides *an* alternative reading.

Environmental flow requirements reconsidered

PB-W is computed by subtracting environmental flow requirements, that is, the water needs of non-human water users, from total renewable water resources, both in an anthropocentric and in a biocentric perspective. In an anthropocentric perspective, humans need to leave some water for freshwater ecosystems as otherwise they would degrade, which would have some negative impacts on humans. But as 1) knowledge on sufficient water for ecosystems to avoid negative impacts for humans is to a large degree lacking and 2) more water for ecosystems restricts human water use and thus well-being, environmental flow calculations are highly value-laden. The PB-W value of 4000 km^3/yr was derived by assuming a 'fair-to-good' status of the freshwater ecosystem (Steffen et al. 2015; Pastor et al. 2014).[21] Following a biocentric perspective, a 'good-to-very good' status appears to be more appropriate. This would lower PB-W and require human water use to become more sufficient.

Conclusion

The precautionary principle provided an entry for our discussion of the planetary boundary framework and environmental philosophy. The principle calls for a clarification of morally unacceptable harm. So what harm does this refer to? And whose harm is to be considered?

With a view to moral agents, we first turned to a threshold notion of moral harm, the concept of dignity and of human rights as a political framework to further specify the idea of morally unacceptable harm. The first lesson of this discussion was that the planetary boundary framework, with its focus on 'a *safe operating space* for humanity', needs to be extended to also consider the members of which this whole consists. It is at this level that harm occurs and where many people are already harmed. Morally unacceptable harm occurs not only if the PB-W (or environmental ceiling) is exceeded but also if basic human water needs are not fulfilled (if the social foundation is absent). Such a *safe and just space* (Raworth 2013) in turn provides a conceptual link to other planetary boundaries where similar questions emerge (see, for example, Caney 2010 on harm in relation to dangerous climate change).

In a second step, we explored a biocentric perspective that considered all living beings, and not just humans, as subjects of potentially unacceptable harm. We concluded that the planetary boundary framework is not necessarily anthropocentric. According to a biocentric conception of the framework, we could move from a social foundation to a biocentric foundation to highlight the full-scope of the moral community. Alternatively, we can redefine the planetary boundary such that the water needs of non-human living beings are considered, not only with the aim to avoid harm to humans by degradation of the freshwater ecosystems but also taking into account the water needs of non-human living beings as intrinsically valuable beings. This shift from an anthropocentric to a biocentric interpretation of the PB-W could be achieved by aiming for a better ecological status of the freshwater ecosystem of the PB-W. For this, it is important to distinguish the water needs of moral agents (i.e. humans) from their water wants. According to the biocentric priority principles of self-preservation and proportionality, basic interests should be taken into account when thinking about water use conflicts. In the biocentric perspective, human appropriation of freshwater beyond basic needs is in principle subject to scrutiny: does it conflict with the basic interests of other members of the moral community? A sufficient human water use is asked for, and from a biocentric perspective, the operating space for moral agents should be *safe, just and sufficient*.

Inversely, the biocentric perspective (as well as water ethics) can benefit from an adjusted planetary boundaries framework. A central contribution is no doubt the effort to move beyond the (important but not exhaustive) local or regional level to also consider global dynamics. Economic globalization, population growth and climate change imply that water use cannot be considered as a merely regional, catchment-based question. The freshwater boundary provides one way to identify threats to the global community of living beings, which in turn can be linked back to local issues and responsibilities.

This chapter has delineated the conceptual space for a biocentric conception of planetary boundaries. Operationalizing such a conception with a view to a biocentric freshwater ceiling as well as estimating further freshwater basic needs are logical next steps. In addition, the link between the conception and water governance needs to be considered.[22] A further conceptual question is to move from the exploration of the ideal environmental ethics perspective as explored here (if humankind, especially its water managers, acted as if everyone took a biocentric perspective) to the non-ideal world where this is not the case.[23] We conclude with some considerations of why these steps should be taken and how they are likely to enrich the discussion of planetary boundaries, global water ethics, and ultimately the actions that the precautionary principle requires (see note 5).

Sufficiency and development. There is a tendency to interpret the planetary boundaries framework as 'growth within limits' especially in relation to the biophysical expansion of the human dimension, whereas others call for a more substantial shift to a focus on qualitative development in terms of well-being and fair distribution (Crépin & Folke 2014, p. 58). The biocentric perspective supports this second reading, and helps to identify priority areas. Current water use in agriculture together with an expected growth in population suggests as key priorities of biocentric politics and economics of planetary boundaries: a search for improved water productivity (including a discussion of improved spatial distribution) as well as on the demand side a reconsideration of water-intensive 'lifestyle choices', in particular diets that include a large amount of animal products. In the field of energy production, it suggests a need to reconsider hydropower not just as a potential renewable energy source for climate protection but also as a technology that strongly interferes with aquatic ecosystems (and in the case of large dams, also human settlement and cultural heritage). Accordingly, energy sufficiency as well as restitutive justice – for example, where dams are built or have been built – become important topics.

Realism. While growth within limits is no doubt one current worldview, it is by no means the only one. The biocentric worldview is a philosophical articulation of what has been identified as an integrative worldview (in contrast to traditional, modern, and postmodern worldviews; see de Hedlund-Witt 2014). It is not traditional as science plays a central role; it is not modern as science as such is not identified with progress; it is also not postmodern as the aspiration is ultimately global and affirmative of an ideal of universal values. It is held by members of Western societies but also resonates with views elsewhere. A prominent example is the new Mother Earth movement that reflects Latin American indigenous perspectives on the human–environmental system and recognizes intrinsic rights of nature.[24] The rights of nature have already been included in the constitution of some Latin American countries, also affecting water governance (compare the water use specifications in the constitution of Ecuador[25]).

The recognition of values and worldview differences is a first step toward a realist use of the framework: there are different value perspectives that should not simply be ignored in democratic practice. In addition, it has been argued that making values explicit is a 'low-threshold entry' to debates, and thus may improve the possibility to include citizens and other stakeholders in discussion (Meisch 2014, p. 433). The recognition of different worldviews also encourages self-reflexivity (de-Hedlund Witt 2014, p. 8322ff): if we come to understand how our way of thinking and acting is embedded in a worldview, we can also better appreciate other worldviews and even learn from them, the overlaps between these as well as the areas where discussion is called for.

Pluralism and overlaps. However, there is no need to focus on difference exclusively. Recalling environmental flow requirements, it is evident that there are biocentric and anthropocentric reasons for considerations of a planetary boundary for human freshwater use.[26] This is of practical importance for the application of PB-W in water governance in a world with different worldviews and should not be overlooked due to a misconceived understanding of various centrism (i.e. as if there is no place for a social foundation in a biocentric approach, and no role for freshwater beyond economic exploitation in an anthropocentric approach). Crossing planetary boundaries means detrimental impacts on the Earth system and its human and non-human living beings.

Notes

1 Our focus is on the conceptual question. In addition, and in many ways complementary, worldviews can be empirically studied. For example, Hedlund-de Witt distinguishes four worldviews in Western culture, based on empirical research (Hedlund-de Witt 2014, p. 8313ff): traditional, modern, postmodern and integrative. We return to this link between environmental philosophy and worldviews in the conclusion.

2 Historically, this evaluative distinction rests on a number of further ideas. Anthropocentrism in the Western tradition pulls from Christian cosmology, which puts human beings at the top of the natural entities in the 'Great Chain of Being', from a dualism of mind and nature (with humans attributed a non-material soul) and from a related further dualism of the mechanical material world subject to laws of nature versus an independent world of the mind. According to a widely and controversially discussed thesis in environmental philosophy, it is the combination of this worldview with technical invention and the geographic specifics of Western Europe that enabled the 'ecological crisis' (White 1967) – to which the planetary boundary framework seeks to respond at the global level.

3 SDG 6 seeks to ensure availability and sustainable management of water and sanitation *for all* (www.unwater.org/sdgs/en/, accessed 18 December 2015).

4 Steffen et al. further specify the precautionary principle: 'human societies would be unwise to drive the Earth system substantially away from a Holocene-like condition. A continuing trajectory away from the Holocene could lead, with an uncomfortably high probability, to a very different state of the Earth system, one that is likely to be much less *hospitable* to the development of human societies' (Steffen et al. 2015, pp. 737f, italics added). No further explicit definition of the principle can be found in Steffen et al., nor in Rockström et al. (2009).

5 According to Article 3 of the United Nations Framework Convention on Climate Change: 'the Parties should take precautionary measures to anticipate, prevent or minimize the causes of climate change and mitigate its adverse effects. Where there are threats of serious or irreversible damage, lack of full scientific certainty should not be used as a reason for postponing such measures.' The UNFCC refers to a principle adopted by the 1992 Rio Summit on Environment and Development: 'in order to protect the environment, a precautionary approach should be widely applied, meaning that where there are threats of serious or irreversible damage to the environment, lack of full scientific certainty should not be used as a reason for postponing cost-effective measures to prevent environmental degradation' (Rio Declaration on Environment and Development).

6 Of course, the principle could also be defined more generally with a view to harm as such, but this opens the principle to the objection of an irrational and extreme risk aversion. See also Gardiner (2011), who restricts the principle to 'unacceptable outcomes' drawing on John Rawls's theory of justice. UNESCO-COMEST (2005) refers to 'harm to humans or the environment that is threatening to human life or health, or serious and effectively irreversible, or inequitable to present or future generations, or imposed without adequate consideration of the human rights of those affected'.

7 According to UNESCO-COMEST (2005), 'the judgment of plausibility should be grounded in scientific analysis. Analysis should be ongoing so that chosen actions are subject to review. Uncertainty may apply to, but need not be limited to, causality or the bounds of the possible harm.'

8 UNESCO-COMEST (2005) specifies actions as 'interventions that are undertaken before harm occurs that seek to avoid or diminish the harm. Actions should be chosen that are proportional to the seriousness of the potential harm, with consideration of their positive and negative consequences, and with an assessment of the moral implications of both action and inaction. The choice of action should be the result of a participatory process.'

9 The threshold specification is also important for reasons of intergenerational justice. The focus with a view to future generations cannot be harm to specific, future individuals (non-identity problem). Rather, the notion of harm is set with a view to general conditions that we expect would harm future generations.

10 For a detailed discussion of this classificatory scheme, see Gorke (2010).

11 While the extension of the moral community for sentient beings is a relatively well-known topic due to the animal rights movement, the same cannot be said about our relation to plants. For a pioneering study, see Kallhoff (2002).

12 This push is typical for environmental ethics, which is not only driven in the direction of application of frameworks to specific problems, but also in the direction of further reflection on how to think about specific principles, etc. (Ott 2008).

13 There is an extended debate in ethics about basic needs, needs and wants, central capabilities and capabilities, basic interests, etc. (O'Neill 2011). For the sake of not further complicating an already complex discussion, we bracket these distinctions here and only assume that such a threshold distinction is possible and plausible (even if there are different ways of drawing the distinction).

14 With a view to climate change, see Caney (2010) for one idea of how such a process might look.

15 This kcal estimate includes assumptions about current food waste and loss, while actually consumed kcal levels are lower (around 2200–2300 km^3/yr).

16 This is a simplified presentation that merges Taylor's first and fourth priority rule. In his original discussion of the distributive justice principle, Taylor notes that 'fair share' does not require sacrifices in terms of basic interests (Taylor 1986, pp. 294–5). This suggests the simpler statement of the priority rules presented here: his original principle of self-defence can be expanded to a principle of self-preservation that entitles moral agents (just as moral patients) to meet and protect their basic interests.

17 This biocentric possibility might be too easily dismissed if priority rules are not fully discussed. For example, Düwell in his discussion of human dignity and future generations, claims that biocentrism must assume that the lexical priority of human rights is overruled and by implication that there is no viable discussion that takes dignity seriously in a more-than-human world (see Düwell et al. 2014, p. 552).

18 In his discussion of a hierarchy of needs for water governance, Feitelson (2012) makes a similar suggestion when he clusters human basic freshwater needs and basic freshwater needs of ecosystems at the ethical bottom of his hierarchy before politics and markets (also see the chapter by Feitelson in this volume for further elaboration).

19 We bracket here the philosophical discussion whether there are also non-human moral agents. This move seems justified as humans are the central agent forcing the putative exit from the Holocene conditions.

20 So as not to increase complexity even further, we bracket here legitimate non-basic interests, which in Taylor's theory require an account of culturally important water uses, as well as the further challenges of compensation and reparation in accordance with the principle of restitutive justice where needs have been violated. Feitelson (2012, p. 56) discusses spiritual needs of humans in relation to water as basic needs.

21 From a biocentric perspective, holistic methods that combine hydrological, hydraulic, habitat simulation methods and expert knowledge for an in-depth understanding of aquatic ecosystems are clearly the preferred approach (Pastor et al. 2014, p. 5044). The current obstacle with respect to this goal is mostly pragmatic, i.e. a lack of global eco-hydrological data (Richter et al. 2006; Poff & Zimmermann 2010). In the absence of such data, hydrological estimates are the 'second best' (for an overview of methods, see Pastor et al. 2014) as a rough indicator for the global level to be complemented as much as possible with more fine-grained accounts on the regional level (see also Dearing et al. 2014).

22 For an interesting suggestion, see Feitelson's discussion of a hierarchy of needs in relation to ethical, political and economic governance mechanisms in this book.

23 An important question resulting from this concerns aggregation rules and judgments about sufficient space in the real world: if in some river basins 'good to very good' river status is not achieved, are we still in a sufficient space in the world as a whole? If yes, what are permissible reasons?

24 See http://therightsofnature.org/universal-declaration/, last accessed 31 January 2016.

25 Article 318 of the new 2008 constitution of Ecuador includes this priority list (within a constitution that recognizes rights of nature): 'The State, through the sole authority for water, shall be directly responsible for planning and managing water resources for human consumption, irrigation to guarantee food sovereignty, ecological wealth and productive activities, in this order of priority' (http://pdba.georgetown.edu/Constitutions/Ecuador/english08.html, last accessed 17 December 2015).

26 Likewise, the conceptual framework for the Intergovernmental Platform on Biodiversity and Ecosystem Services (IPBES) that has been developed over the last five years suggests that anthropocentric and biocentric/holistic perspectives can and need complement each other in global sustainability efforts. Driven by actors favouring the Mother Earth perspective led by the delegates from Bolivia, the final framework version includes alternative anthropocentric (e.g. 'ecosystem goods and services') and physiocentric (e.g. 'nature's gifts') terms (Borie & Hulme 2015).

References

Borie, M & Hulme, M 2015, 'Framing global biodiversity: IPBES between Mother Earth and ecosystem services', *Environmental Science and Policy*, vol. 55, pp. 487–96.

Caney, S 2010, 'Climate change, human rights, and moral thresholds', in SM Gardiner, S Caney, D Jamieson & H Shue (eds), *Climate ethics: Essential readings*, Oxford University Press, Oxford, pp. 163–80.

Chenoweth, J 2008, 'Minimum water requirement for social and economic development', *Desalination*, vol. 229, nos 1–3, pp. 245–56.

Crépin, A & Folke, C 2014, 'The economy, the biosphere and planetary boundaries: towards biosphere economics', *International Review of Environmental and Resource Economics*, vol. 8, no. 1, pp. 57–100.

Dearing, JA, Wang, R, Zhang, K, Dyke, JG, Haberl, H, Hossain, MS, Langdon, PG, Lenton, TM, Raworth, K, Brown, S, Carstensen, J, Cole, MJ, Cornell, SE, Dawson, TP, Doncaster, CP, Eigenbrod, F, Flörke, M, Jeffers, E, Mackay, AW, Nykvist, B & Poppy, GM 2014, 'Safe and just operating spaces for regional social-ecological systems', *Global Environmental Change*, vol. 28, pp. 227–38.

Döll, P, Fiedler, K & Zhang, J 2009, 'Global-scale analysis of river flow alterations due to water withdrawals and reservoirs', *Hydrology and Earth System Sciences*, vol. 13, pp. 2413–32.

Döll, P, Müller Schmied, H, Schuh, C, Portmann, F & Eicker, A 2014, 'Global-scale assessment of groundwater depletion and related groundwater abstractions: combining hydrological modeling with information from well observations and GRACE satellites', *Water Resources Research*, vol. 50, pp. 5698–720, accessed 18 July 2016, doi: 10.1002/2014WR015595.

Düwell, M 2014, 'Human dignity and future generations', in M Düwell, J Braarvig, R Brownsword & D Mieth (eds), *The Cambridge Handbook of Human Dignity*, Cambridge University Press, Cambridge, pp. 551–8.

Falkenmark, M 1986, 'Fresh water: Time for a modified approach', *Ambio*, vol. 15, no. 4, pp. 192–200.

Falkenmark, M & Rockström, J 2004, *Balancing Water for Humans and Nature: The New Approach in Ecohydrology*, Earthscan, London.

Feitelson, E 2012, 'What is water? A normative perspective', *Water Policy*, vol. 14, pp. 52–64.

Food and Agriculture Organization (FAO) 2004, Economic valuation of water resources in agriculture, accessed 18 December 2015, www.fao.org/docrep/007/y5582e/y5582e04.htm.

Gardiner, S 2011, *A Perfect Moral Storm: The Ethical Tragedy of Climate Change*, Oxford University Press, Oxford.

Gardiner, SM, Caney, S, Jamieson, D & Shue, H (eds) 2010, *Climate Ethics: Essential Readings*, Oxford University Press, Oxford.

Gerten, D, Heinke, J, Hoff, H, Biemans, H, Fader, M & Waha, K 2011, 'Global water availability and requirements for future food production', *Journal of Hydrometeorology*, no. 12, pp. 885–99, doi: http://dx.doi.org/10.1175/2011JHM1328.1.

Gerten, D, Hoff, H, Rockström J, Jägermeyr, J, Kummu, M & Pastor, AV 2013, 'Towards a revised planetary boundary for consumptive freshwater use: role of environmental flow requirements', *Current Opinion in Environmental Sustainability*, vol. 5, no. 6, pp. 551–8.

Gerten, D, Rockström, J, Heinke, J, Steffen, W, Richardson, K & Cornell, S 2015, 'Response to Comment on "Planetary boundaries: guiding human development on a changing planet"', *Science (New York, NY)*, vol. 348, no. 6240, p. 1217.

Gleick, PH 1996, 'Basic water requirements for human activities: meeting basic needs', *Water International*, vol. 21, no. 2, pp. 83–92.

Gorke, M 2010, *Eigenwert der Natur*, Hirzel, Stuttgart.

Griggs, D, Stafford Smith, M, Rockström, J, Öhman, MC, Gaffney, O, Glaser, G, Kanie, N, Noble, I, Steffen, W & Shyamsundar, P 2014, 'An integrated framework for sustainable development goals', *Ecology and Society*, vol. 19, no. 4, pp. 49–62.

Hedlund-de Witt, A 2014, 'Rethinking sustainable development: considering how different worldviews envision "development" and "quality of life"', *Sustainability*, vol. 6, no. 11, pp. 8310–28.

Kallhoff, A 2002, *Prinzipien der Pflanzenethik: Die Bewertung pflanzlichen Lebens in Biologie und Philosophie*, Campus Forschung, Bd. 840, Campus, Frankfurt.

Mauser, W 2007, *Wie lange reicht die Ressource Wasser?: Vom Umgang mit dem blauen Gold*, Forum für Verantwortung, 2nd edn, vol. 17273, 2nd edn, Fischer Taschenbuch, Frankfurt am Main.

Mehta, L, Veldwisch, GJ & Franco, J 2012, 'Introduction to the special issue: water grabbing? Focus on the (re) appropriation of finite water resources', *Water Alternatives*, vol. 5, no. 2, pp. 193–207.

Meisch, S 2014, 'The need for a value-reflexive governance of water in the Anthropocene', in A Bhaduri, J Bogardi, J Leentvaar & S Marx (eds), *The Global Water System in the Anthropocene: Challenges for Science and Governance*, Springer, Heidelberg, pp. 427–37.

Millennium Ecosystem Assessment (MEA) 2005, *Ecosystems and Human Well-being: Synthesis*, Island Press, Washington, DC.

Nolt, J 2014, *Environmental Ethics for the Long Term: An Introduction*, Routledge, London.

Nussbaum, M 2006, *Frontiers of Justice: Disability, Nationality, Species Membership*, Harvard University Press, Cambridge, MA.

O'Neill, J 2011, 'The overshadowing of needs', in F Rauschmayer, I Omann & J Frühmann (eds), *Sustainable Development: Capabilities, Needs, and Well-being*, Routledge Studies in Ecological Economics, Routledge, New York, pp. 45–64.

Ott, K 2008, 'Umweltethik zwischen Grundlagenreflexion und Politikberatung', in K Bruckmeier & WH Serbser (eds), *Ethik und Umweltpolitik – Humanökologische Positionen und Perspektiven*, oekom, München.

Paech, N 2012, *Befreiung vom Überfluss: Auf dem Weg in die Postwachstumsökonomie*, oekom, München.

Pastor, AV, Ludwig, F, Biemans, H, Hoff, H & Kabat, P 2014, 'Accounting for environmental flow requirements in global water assessments', *Hydrology and Earth System Sciences*, vol. 18, no. 12, pp. 5041–59.

Poff, NL & Zimmerman, JKH 2010, 'Ecological responses to altered flow regimes: a literature review to inform the science and management of environmental flows', *Freshwater Biology*, vol. 55, no. 1, pp. 194–205.

Raworth, K 2013, 'Defining a safe and just space for humanity', in E Assadourian, T Prugh & L Starke (eds), *Is Sustainability Still Possible?*, State of the world, vol. 2013, Island Press, Washington, DC, p. 28.

Richter, BD, Warner, AT, Meyer, JL & Lutz, K 2006, 'A collaborative and adaptive process for developing environmental flow recommendations', *River Research and Applications*, vol. 22, pp. 297–318.

Rockström J, Steffen W, Noone K, Persson Å, Chapin III FS, Lambin EF, Lenton TM,

Scheffer M, Folke C, Schellnhuber HJ, Nykvist B, de Wit CA, Hughes T, van der Leeuw S, Rodhe H, Sörlin S, Snyder PK, Costanza R, Svedin U, Falkenmark M, Karlberg L, Corell RW, Fabry VJ, Hansen J, Walker B, Liverman D, Richardson K, Crutzen P & Foley JA 2009; 'A safe operating space for humanity', *Nature*, vol. 461, pp. 472–5.

Shiklomanov, IA & Rodda, JC 2003, *World Water Resources at the Beginning of the Twenty-first Century,* Cambridge University Press, Cambridge.

Siebert, S & Dieb, P 2010, 'Quantifying blue and green water uses and virtual water contents in global crop production as well as potential production losses without irrigation', *Journal of Hydrology*, vol. 384, pp. 198–217.

Spangenberg, JH (ed.) 2003, *Vision 2020: Arbeit, Umwelt, Gerechtigkeit – Strategien für ein zukunftsfähiges Deutschland*, oekom, München.

Statistisches Bundesamt 2013, Die Wassernutzer, accessed 23 December 2015, www.umweltbundesamt.de/daten/wasser-als-ressource/wasserressourcen-ihre-nutzung.

Steffen, W, Richardson, K, Rockström J, Cornell, SE, Fetzer, I, Bennett, EM, Biggs, R, Carpenter, SR, Vries, W de, de Wit, Cynthia A, Folke, C, Gerten, D, Heinke, J, Mace, GM, Persson, LM, Ramanathan, V, Reyers, B & Sörlin, S 2015, 'Sustainability. Planetary boundaries: guiding human development on a changing planet' (eng), *Science (New York)*, vol. 347, no. 6223, p. 1259855.

Stern, N 2010, 'The economics of climate change', in S Gardiner, S Caney, J Jamieson & H Shue (eds), *Climate Ethics: Essential Readings*, Oxford University Press, Oxford, pp. 39–76.

Taylor, PW 1986, 'Studies in moral, political, and legal philosophy', in *Respect for Nature: A Theory of Environmental Ethics*, Princeton University Press, Princeton, NJ.
UNESCO-COMEST 2005, *The Precautionary Principle*, UNESCO, Paris.
White, L 1967, 'The historical roots of our ecologic crisis', *Science*, vol. 155, no. 3767, pp. 1203–7.

8 The relevance of ethical factors in the pursuit of integrated water resources management

Maite Martínez Aldaya, Pedro Martínez-Santos and M. Ramón Llamas

Introduction

During the last three decades, a good number of new concepts have emerged such as integrated water resources management (IWRM), water governance, life cycle analysis (from cradle to cradle instead of from cradle to grave), decoupling economic growth from natural resource use, green growth, circular economy and others. All these ideas aim to manage scarce water resources efficiently and avoid waste.

The concept of integrated water resources management (IWRM) has gradually gained recognition over the last decades (UNEP 2012). The concept was developed from the down-to-earth experience of practitioners and advocates for a coordinated approach in a way that balances social and economic needs with care for nature. Many parts of the concept have been around in the water resources management sector since the United Nations Conference on Water in Mar de Plata in 1977. But it was not until 1992 that the concept was incorporated in the 'Dublin Statement on Water and Sustainable Development' and the Agenda 21 of the United Nations Conference on Environment and Development (UNCED) (Martínez-Santos et al. 2014).

The Global Water Partnership (GWP 2000) provides a widely accepted definition of IWRM as 'a process which promotes the coordinated development and management of water, land, and related resources in order to maximize the resultant economic and social welfare in an equitable manner without compromising the sustainability of vital ecosystems'. The IWRM approach has been internationally accepted as the way forward for efficient, equitable and sustainable development and management of the world's limited water resources and for coping with conflicting demands (UNEP 2012). Target 6.5 of the United Nations Sustainable Development Goals has the goal to 'by 2030, implement integrated water resources management at all levels, including through transboundary cooperation as appropriate' (UN 2015). Nevertheless, IWRM is both controversial and elusive in practice (Martínez-Santos et al. 2014). Not the least of its shortcomings is the fact that many decisions affecting water are made outside the sphere of conventional water planning.

Tracking how and where water is used and analysing this in terms of monetary value is an important first step in understanding the potential for improving IWRM within a given context. However, reallocating water among users or from users to nature is an almost impossible task, similar to the challenge of trying to 'square the circle' (De Stefano & Llamas 2012). Initiatives portrayed as solutions to the water governance 'puzzle' – for example, water trade, improved water use efficiency, users' collective action, public participation – are not free of difficulties and shortcomings. There is a growing need for maintaining natural capital and for knowing better the human component of water governance – people's needs, wishes, (vested) interests, aspirations – that often determine the result of decisions and, sometimes, lead water management to a deadlock (ibid.). This dimension is becoming more and more relevant. For instance, the recent Encyclical issued by Pope Francis (2015) emphasizes the crucial role of ethical values to achieve sustainable development.

This chapter deals with aspects which often escape the debate of IWRM including water accounting, water footprint, virtual water (food) trade, green water (rainwater stored in the soil as soil moisture), blue water (surface water and groundwater), water pollution, environmental externalities and the 'intangible' values involved in the efficient water allocation, equitable distribution at an appropriate scale. Water governance goals are invariably underpinned by ethical values – which may be different from country to country – or among different social groups or interest groups. After analysing these elements, the chapter concludes that a better understanding of the ethical values behind the water policy and decisions could contribute to the overarching goal of water and water-related food security through better integrated water resources management.

The components of IWRM

According to UN-Water (2013), water security is 'the capacity of a population to safeguard sustainable access to adequate quantities of acceptable quality water for sustaining livelihoods, human wellbeing, and socio-economic development, for ensuring protection against water-borne pollution and water-related disasters, and for preserving ecosystems in a climate of peace and political stability'. This entails dealing with interconnected and multifaceted aspects – biophysical, infrastructural, institutional, trade, political, social and financial – and interdisciplinary collaboration across sectors, communities and political borders, for which integrated resources management is needed. There are ethical issues imbedded in all aspects of IWRM decision-making, including planning, regulating, operating, financing and investing, designing and implementing.

With the purpose of clarifying and simplifying the components of IWRM, they have been classified in two main groups (De Stefano & Llamas 2012; Martinez-Santos et al. 2014):

- Measurable utilitarian components, which can be objectively quantified. Examples include water footprint, virtual water trade, economic productivity and some social and ecological aspects.
- Intangible components, which are not easily computed: religious, spiritual, emotional and cultural values; value of nature; intergenerational equity.

Both measurable and intangible factors are closely intertwined. In fact, there are elements that are in between both components having parts belonging to both, such as 'natural capital' which has a quantifiable dimension (e.g. environmental water requirements), and an intangible component related to the intrinsic value of nature, that is, the ethical or philosophic value of nature 'in itself' or 'for its own sake'. In the same line, some social aspects are measurable (e.g. employment) while others are not (cultural values). As research and development advances, social scientists are able to quantify more and more of the intangible components. This classification is not static but dynamic and changeable in time.

The measurable components

The water footprint

The quantification of water uses and their monetary value is a first important step in understanding how reallocation of water among uses could help mitigate many of the current water problems and achieve a more IWRM (see Figure 8.1 and Table 8.1) (De Stefano & Llamas 2012). See for instance the Spanish case in Table 8.2. Here a transparent framework is provided integrating the water uses with the economic and social values of the different socioeconomic sectors. These measurable components, relatively easy to properly use, are still applied in a partial and confused way.

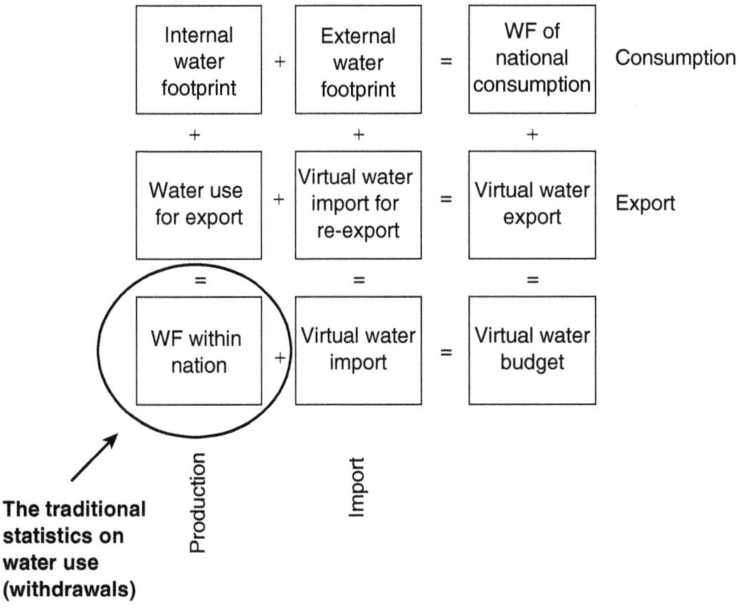

Figure 8.1 National water accounting framework.

Source: Hoekstra et al. (2011a).

Table 8.1 Global water footprint (WF) of production (1996–2005) (1 gm^3 = 1 km^3)

	Agricultural production			*Industrial production*	*Domestic water supply*	*Total*
	Crop production	*Pasture*	*Water supply in animal raising*			
Global WF of production (gm^3/yr)						
– Green	5771	913	–	–	–	6684
– Blue	899	–	46	38	42	1025
– Grey	733	–	–	363	282	1378
Total	7404	913	46	400	324	9087
WF for export (gm^3/yr)		1597		165	0	1762
WF for export compared to total (%)		19		41	0	19

Source: Mekonnen and Hoekstra (2011).

Total global water resources amount to about 110,000 km^3/yr, of which *green water* – that is, water stored in the soil – is calculated at 70,000 km^3/yr, whereas *blue water* – that is, fresh surface and groundwater – account for 40,000 km^3/yr (WWAP 2006). Of all available blue and green global water, human water use is around 8%, with most (70%) going to agriculture, while industry accounts for 20% and domestic water use 10% (CAWMA 2007; WWAP 2009). When considering consumptive use, the agricultural percentage is even higher, reaching 90% of total water consumption (Mekonnen & Hoekstra 2011).

The water footprint, calculated according to the WFN Manual (Hoekstra et al. 2011a), is a good method for the estimation of water consumption and pollution for each activity in a river basin and offers an interesting framework for water accounting. The water footprint concept was introduced by Arjen Hoekstra in 2002 as an indicator that could map the impact of human consumption on global freshwater resources (Hoekstra 2003). It refers to all forms of freshwater use (consumption and pollution) that influence the production of goods and services consumed by the inhabitants of a certain geographical region (Hoekstra et al. 2011a).

On one hand, one can talk about the water footprint (WF) of a product, which is the amount of water consumed directly (operations) or indirectly (supply chain) to produce the product and is synonymous to the virtual water of a product. On the other hand, one can talk about the water footprint of an individual, which refers to the total amount of freshwater used to produce the goods and services (directly and indirectly) consumed by this individual. In the first case, the water footprint of products, the basic information provided by water footprints can be used by the private sector to perform risk assessment, as a planning tool, to identify hotspots in their supply chains or to couple it with tools like Life Cycle Assessment (LCA) methods in order to perform benchmarking of products. For most products, the vast majority of water is consumed indirectly (Hoekstra 2013). For instance, more than 90% of a sugar-based beverage company's water footprint is in its supply chain (Ercin et al. 2010). In the second case, the WF of individuals or the inhabitants of a given region or country, the basic information provided by WFs can be used by governments, academia, NGOs or other organizations for awareness raising or for understanding changes and trends in consumption patterns as related to water resources. WF information can also be used by water managers at different spatial levels, for example, to compare the water footprints of different production sectors with the availability of water in the specific geographical area to inform decision-making (UNEP 2011).

A water footprint assessment has four distinct phases: (1) Setting goals and scope, (2) Water accounting, (3) Sustainability assessment, and (4) Response formulation (Hoekstra et al. 2011a). Currently, the ethical dimension is not considered when developing a water footprint assessment, yet there are implications and inherent difficulties in setting goals and acting ethically. For instance, selectively presenting information to promote the perception of a product or organization as being environmentally or eco-friendly could be a form of greenwashing. Considering ethical principles is essential.

The water footprint indicator provides information to policy-makers that can complement conventional measures of water withdrawals (see Figure 8.1). Typical national water use accounts are based on direct blue water withdrawal within the country, whereas the water footprint assessment provides information on green and blue water consumption and pollution (grey water) including data on direct and indirect water use (virtual water flows). By looking only at direct water use, most governments have a blind spot to the issue of sustainability of national consumption because they are effectively externalizing the portion of their water footprint, which takes place within the producing countries associated with the goods and services consumed. Data and knowledge of the dependency on water resources elsewhere is relevant for a national government, not only when evaluating its environmental policy but also when assessing national water and food security. Cooperation between communities and nations is required in terms of information sharing. In this globalized world, a global water ethics dimension is fundamental to achieve real sustainability. This makes the water footprint very different from other indicators and very useful for achieving a more integrated water resource management. Making national water footprint accounting a standard component in national water statistics would provide a stronger information basis to formulate a national water plan and river basin plans that are coherent with national policies with respect to the environment, agriculture, energy, trade, foreign affairs and development cooperation (Hoekstra et al. 2011a).

More and more studies are recognizing the usefulness of the water footprint as a green indicator for water resources management (the French Government in Mordant & Pasquier 2012; UNEP 2011). At a national level, Spain is the first country to include water footprint analysis into governmental policy-making in the context of the EU Water Framework Directive. In September 2008, the

Table 8.2 Water uses, related economic values and employment in Spain

Activity	*Consumptive use (10^9m^3)*	*GDP (10^9 Euro)*	*Workforce (%)*
Agriculture & livestock	15 (75%)	25 (3%)	4
Industry	1 (5%)	160 (16%)	17
Urban uses	3 (12%)	5 (0.5%)	1
Services (excl. tourism)	0.5 (4%)	600 (60%)	67
Tourism and golf courses	0.5 (4%)	110 (11%)	11
Bottled water	0.1 (–%)	3 (0.2%)	0.1
Total	20 (100%)	900 (100%)	100

Source: Garrido et al. (2010).

Spanish government approved a regulation that incorporates the analysis of the water footprint of the different socio-economic sectors as a technical criterion for the development of the River Basin Management Plans as part of the requirements of the Water Framework Directive (Official State Gazette 2008). Unfortunately, however, this Spanish regulation has been somewhat ineffective (Water Observatory 2014). There is a need for a systematic, standardized and comprehensive water accounting systems (UNSD 2012). Provision of partial information may lead to misleading conclusions. For instance, addressing local water use efficiency alone might unintentionally stress water bodies at the basin scale (WWAP 2016).

Virtual water trade

The concept of virtual water was introduced by Tony Allan in the 1990s (Allan 1993, 2011) when he studied the possibility of importing virtual water as a partial solution to problems of water scarcity in the Middle East. The virtual water content of a product refers to the volume of water consumed or polluted for producing the product, measured over its full production chain (Allan 2011). Both virtual water content and water footprint refer to the water used to make a product. The water footprint makes reference not only to the volume of water used, but also to the source and sort of water used (green, blue, grey) and to when and where the water was used. Furthermore, the water footprint concept has a wider application (e.g. water footprint of a consumer, producer, geographical location). The virtual water term is used when referring to trade. If a region exports/imports a product, it exports/imports water in virtual form. One can speak about virtual water export or import, or more generally about virtual water flows or trade (Hoekstra et al. 2011a).

Most of the global virtual water flows refer to agricultural commodity trade. It is widely recognized that international trade in agricultural products can contribute to addressing problems related to the unequal geographical distribution of water (Allan 2011; Hoekstra 2010; Hoekstra et al. 2011b; Jackson et al. 2014; UNEP 2011). In fact, current global virtual water trade has a moderating impact on the demand for irrigation water, as three of the four major exporters (United

States, Brazil and Argentina) mostly produce in highly productive rainfed conditions (Aldaya et al. 2010a; de Fraiture et al. 2004; Mekonnen & Hoekstra 2011) (Figure 8.2), while most importers rely (at least partially) on their blue water resources. The virtual water trade can efficiently redistribute global water and partially help address the impacts of production. Virtual water trade, thus, can play a role in ensuring water and water dependent food security in water-short countries.

On the other hand, at a smaller scale, there might be exceptions such as Thailand's water problems due to irrigation of rice for export, Uzbekistan's overuse of water to produce cotton for export or Kenya's water depletion around Lake Naivasha to produce flowers for export to the UK and the Netherlands (Hoekstra 2010). If effective national and international policies and incentives are not properly developed and implemented, virtual water trade can have side-effects. International trade in water-intensive commodities takes water in the exporting countries, which can no longer be used for other (domestic) purposes. Furthermore, the social and environmental costs associated with water use remain in the exporting countries; they are not included in the price paid for the products by the consumers in the importing countries. Essentially, the mechanism of virtual water trade is itself neither good nor bad but requires ethical evaluation under which condition it might be useful.

It is well known that water is seldom the dominant factor determining trade in water-intensive commodities. Many factors influence virtual water trade, such as direct or indirect subsidies, availability of land, labour and technology, level of socio-economic development, national food policies and international trade agreements (Aldaya et al. 2010a; Rogers & Ramirez-Vallejo 2011a, 2011b). Virtual water flows are subject to world trade rules (Hoekstra et al. 2011b). The European Single Market and World Trade Organization (WTO) frameworks have the potential to address the link between international trade and sustainable water use. Hoekstra (2010) identifies several mechanisms to better ensure that trade and sustainable water use go hand in hand, such as: 1) the principles of product transparency and non-discrimination (though the criteria to evaluate the similarities of goods are not clear), 2) international agreements on sustainable water use, 3) an international water pricing protocol, 4) an international water label for water-intensive products,[1] or 5) an international water footprint permit system. More recently, the WTO has started looking at WTO interventions that could influence water-related policies on either the production side (irrigation subsidies) or the consumption side (water footprint labelling) (Jackson et al. 2014). Ethical principles are at the core of these mechanisms, such as environmental protection and social fairness. More work is needed to clarify key concepts and to enhance transparency in order to have a more comprehensive understanding of the ways in which these rules alter water resource outcomes.

Water and food security is today much more related to economic capacity and trade than to physical water scarcity. Knowledge about the virtual water flows entering and leaving a country can cast a completely new light on the actual water scarcity of a country. This shift in perception means that it forces a re-consideration of what are the main problems of food security, away from

Figure 8.2 National virtual water balances. Arrows show gross virtual water flows >15 gm^3/yr (1 gm^3 = 1 km^3).

Source: Mekonnen and Hoekstra (2011).

pure physical scarcity and technological fixes. The main issues that have to be addressed globally in relation to food security are economic, socio-political and mainly ethical in nature: such as the hidden oligopolies that currently exist in the WTO, the potential threat of political embargoes and the need for domestic social reforms to be realized (Dalin et al. 2012).

Economic water productivity

Another key accounting component to consider is the economic dimension of water management (Figure 8.3). The impact of the different water uses on the national or local economy, measured as blue water productivity (US$ per m^3), is important information. A water productivity analysis can be useful to make strategic choices on whether water is allocated to high value uses such as urban water supply, industrial uses or cash crops (e.g. vegetables, fruit and flowers), or low value ones (like growing cereals or alfalfa) (WWAP 2006).

In Spain today, 4% of all the blue water used in irrigated agriculture accounts for 66% of total gross value added (Aldaya et al. 2010b). Conversely, about 60% of the water used in this sector produces just 5% of the total gross economic value added in agriculture (ibid.). This implies that Spain is mainly producing blue water-intensive, low-value crops. This is the legacy of the country's food self-sufficiency policies. Today this paradigm is becoming obsolete due to the spectacular increase of the international food–virtual water trade. Within the European Union there are no barriers to trade. It might be cheaper to buy maize from Germany or France than irrigating new crops in Spain. Currently, such inter- or transnational dynamics are agreed under the umbrella of international organizations, such as the European Union or the World Trade Organization. However, more transparency and ethical principles need to be brought to the negotiating table.

Research suggests that the water crisis both in Spain and in many other arid and semiarid countries is a result of poor water management, such as obsolete irrigation systems or excessive blue water use for growing low economic value crops (Llamas 2005; UNDP 2006; CAWMA 2007). Concepts such as the water footprint, virtual water trade, and water productivity analyses offer new data and perspectives that enable a more optimistic outlook on water security.

However, policies aimed at increasing local water use efficiency and productivity alone may unintentionally stress water supplies at a broader basin scale and result in a rebound or take-back effect (WWAP 2015). This occurs when the water savings obtained from improved water-use efficiency are reinvested in further production. Therefore, although the process becomes more efficient, total water use does not decrease (WWAP 2015). This is the case of the irrigation modernization process in some areas in Spain, which shows how efficiency improvements lead to greater water consumption (Dumont et al. 2013).

Simply increasing the economic benefit per unit of water is not enough. Policies for enhancing the value of water used in agriculture involve not only increasing yields, changing from low to high-value crops, reallocating water from low to higher value sectors or lowering the cost of inputs. They should also optimize the

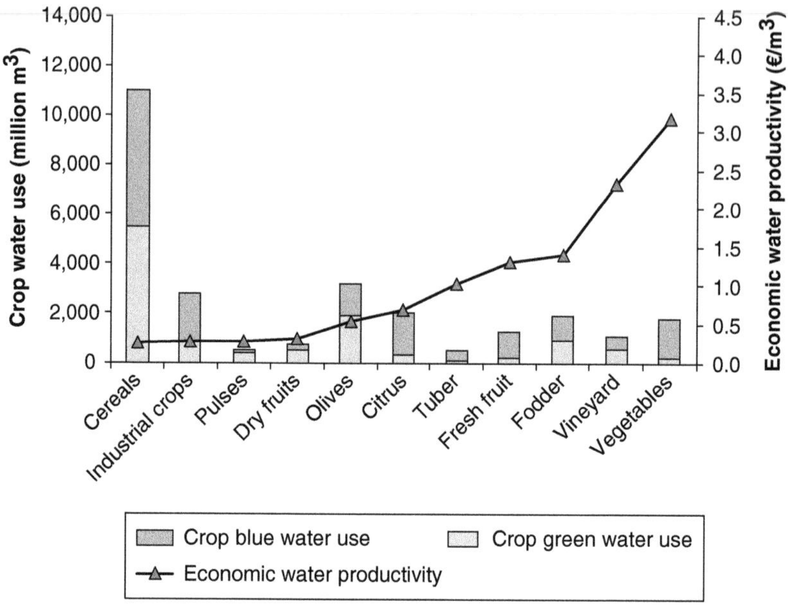

Figure 8.3 Water productivity and blue and green water footprint of crop production in Spanish agriculture (2002).

Source: Garrido et al. (2010).

creation of quality jobs (social ethic) and related environmental aspects (environmental ethic). For instance, policies should achieve more livelihood support, while increasing health benefits and the value of ecological services of agriculture (CAWMA 2007).

The concept of 'more crop per drop' should be adjusted to include 'more added value per drop' in agriculture, where added value refers to the socio-economic and environmental welfare. The transition, however, is not easy since the real-world situation is generally difficult to change. For example, some farmers cultivate certain low-value crops in areas where they cannot grow something else; others do not have an investment or entrepreneurial capacity to move to high-value crops, etc.

Social aspects

The social aspects related to water are the most difficult to quantify since, in many cases, the link to water is not straightforward (WWAP 2016). The social impacts refer to the effect of an activity on the well-being of individuals. This includes employment, conflicts, health problems and equity, among others.

It is widely recognized that the water uses in a catchment are socially unsustainable when basic human needs are not met for all people in that catchment or when basic rules of fairness are not met. Water-related basic human needs include a minimum amount of clean drinking water and sanitation and a minimum allocation of water to food production to secure a sufficient level of food supply. The human right to water and sanitation was explicitly recognized by the United Nations General Assembly on 28 July 2010, through Resolution 64/292. An analogous 'right to water for food' has not been formally acknowledged, but food itself as a human right has already been established in the Universal Declaration of Human Rights (1948).

Another relevant basic human need is employment and quality jobs. Policies for enhancing the social value of water need to consider the creation of quality jobs. For instance, policies should achieve more livelihood support per unit of water (more jobs, nutrition and income for the same amount of water) while increasing health benefits (CAWMA 2007).

Businesses have a relevant role in terms of decent job creation through extending their influence beyond the company gates to their supply chains. For instance, the WBCSD WASH Pledge is already securing appropriate access to safe water, sanitation and hygiene (WASH) for employees at the workplace (WBCSD 2015). By signing the pledge, companies commit to respect the right to WASH. The positive effects are numerous within and beyond facilities, including a healthier workforce and associated communities. For instance, behavioural changes related to hygiene promotion can be brought from the workplace to households and to the community as a whole.

Basic human needs and rules of fairness are difficult to quantify. The presence of conflict over water can be a proxy indicator of violations of formal or informal social values about rights to water or rules of water equity and fairness within a catchment (Gleick 2016; Oregon State University 2016).

Natural capital

Natural capital is the world's stock of natural resources, which includes water ecosystems. This perspective is complementary to the water footprint analysis, which focuses on the human appropriation of water resources. Natural capital not only contributes to producing consumption goods or producing utility through environmental amenities, but it also has an intrinsic value. The intrinsic value of nature, or the value that nature has in itself, is often ignored since it is not instrumental to human goals, and its quantification is unexplored.

The research on this topic is still in its infancy. Costanza et al. (1997) and the Millennium Ecosystem Assessment (2005) provided a preliminary study of ecosystem water needs and services. Environmental water requirements, mainly green, have rarely been taken into account in water resources management, even while considered the main water user (CAWMA 2007). Few studies include both aquatic and terrestrial ecosystem water requirements, though recent reports imply that this might be crucial (Aldaya et al. 2010b; Salmoral et al. 2011; Willaarts 2012) (Figure 8.4). Our understanding of intangible environmental values, such as the intrinsic value, is even worse; they are often overlooked since they are not easily quantifiable.

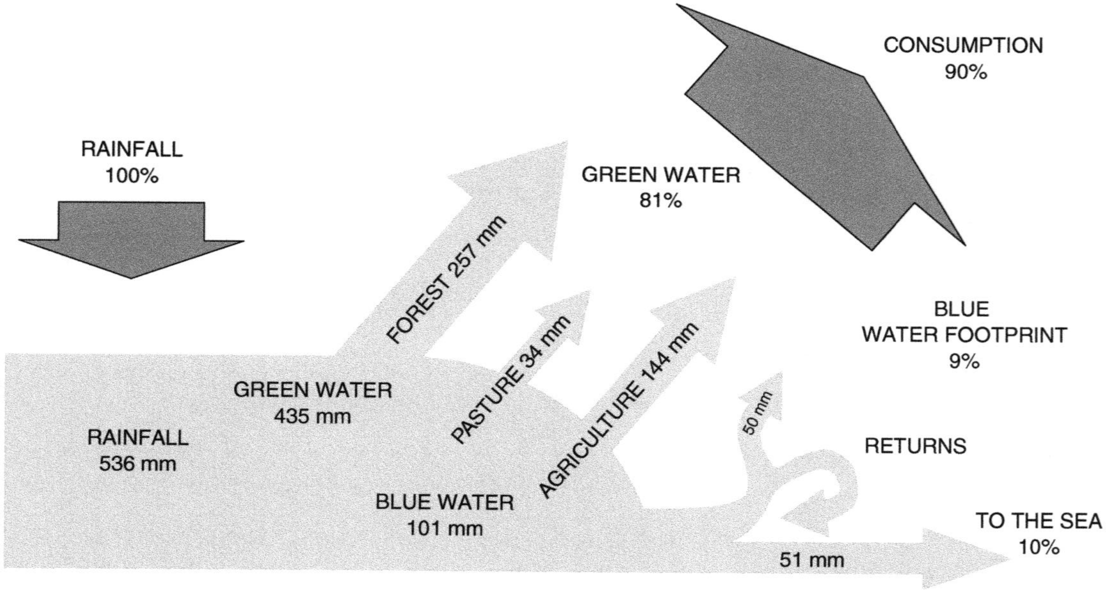

Figure 8.4 Green and blue water balance in the Guadalquivir River Basin, Spain.

Source: Salmoral et al. (2011).

The intangible components

Among the intangible components of water, cultural and spiritual values are of fundamental importance at two levels: (1) at the policy-making level, cultural values that influence the decisions of policy-makers, and (2) at the policy implementation level, local cultural values that influence the success or failure of those policies.

In almost all water policy decisions there are a number of cultural or religious factors or influences from the dominant value system, which have significant and often defining roles in shaping the policy decisions. This is common to all countries. Water policy is generally dictated by the interests and values of parties outside the formal water box. The final policy decision is often due to a mixture of vested interests, ignorance, arrogance, negligence, corruption and lack of transparency and accountability (Ragone & Llamas 2006).

At the local level, the great variation of cultural traditions matters. For instance, when implementing modern agricultural technologies, it is very important to know the customs of the inhabitants of the region. The establishment of modern technological advances and policy reforms quickly ignoring the cultural traditions may fail. Giordano and Shah (2014) describe the failure of implementing a water policy reform in Sri Lanka as the result of ignoring the customs of the farmers in that area. Sections of the press, non-governmental organizations, religious groups and some farmer organizations claimed that the process was top-down and closed despite the apparent efforts for inclusion. Other criticisms included the failure to draft the policy document in the local language, insensitivity to the cultural aspects of water and that the whole exercise was to satisfy donor interest rather than cater to national needs.

Conclusions

This chapter illustrates how a better understanding of the ethical implications of water policies and practices can contribute to a more integrated water resources management and ultimately to achieving water security for both humans and the environment. These intangible social, cultural and spiritual factors, even if they are becoming more significant every year, are easily overlooked.

The application of the IWRM, with its pros and cons, can certainly foster good decision-making. There are already many technologies and tools that are helping to achieve a more integrated approach and forestall the often-cited water crisis. These include accounting tools (e.g. the water footprint, virtual water trade, water productivity, quantification of socio-economic aspects and natural capital) as well as technological innovations (like desalination and the use of GIS and remote sensing). These recent advances are helping to ensure water supply and protection, and water-dependent food security and have demonstrated that usually there is no physical water scarcity but rather deficient water governance or 'mental aridity'. These scientific and technological advances, however, should not be applied in isolation from environmental and social initiatives. Both technological as well as social and environmental innovations are needed.

It seems clear that the development of concepts such as trade in virtual water and water footprints have generated new ways of looking at old problems of water and food security, and concern that humanity will soon be 'water stressed'. Trade in virtual water constitutes a key element in helping to eliminate or at least soften the global water crisis. However, there is no easy panacea and methods have to be further developed, data have to be improved and social, political and ecological side-effects or unintended consequences need to be further studied.

In order to solve water problems, social and political perspectives provided by social sciences, psychology and humanities should be combined with the physical and natural sciences. Today humanity is a technological giant but almost as a child in environmental ethics. As Pope Francis emphasizes, the technological solutions may be useful but not sufficient (Pope Francis 2015). Fortunately, awareness about this gap seems to be increasing.

Science and technological innovation already offer cheap, accessible options to quantify and assess water uses in areas experiencing so-called 'water stress', but these assessments need to be grounded in the sustainable, equitable and efficient allocation of water. There is enough fresh water to ensure water security and produce food for the global population now and in the future, as long as we focus on needs and not on wants. Improving water resources management and achieving the Sustainable Development Goal on water calls for a problem-solving mindset based on ethics, and not only on science and technology.

Note

1 See Sym (in this volume) for a discussion of the Water Stewardship Standard, which could evolve into a water label.

References

Aldaya, MM, Allan, JA & Hoekstra, AY 2010a, 'Strategic importance of green water in international crop trade', *Ecological Economics*, vol. 69, no. 4, pp. 887–94.

Aldaya, MM, Garrido, A, Llamas, MR, Varela-Ortega, C, Novo, P & Casado, RR 2010b, 'Water footprint and virtual water trade in Spain', in A Garrido & MR Llamas (eds), *Water Policy in Spain*, CRC Press, Leiden, The Netherlands, pp. 49–59.

Allan, T 1993, 'Fortunately there are substitutes for water otherwise our hydro-political futures would be impossible', in *Priorities for Water Resources Allocation and Management*, ODA, London, pp. 13–26.

Allan, T 2011, *Virtual Water: Tackling the Threat to Our Planet's Most Precious Resource*, I.B. Tauris, London, p. 384.

CAWMA 2007, 'Comprehensive assessment of water management in agriculture', in *Water for Food, Water for Life: A Comprehensive Assessment of Water Management in Agriculture*, Earthscan, London and International Water Management Institute, Colombo.

Costanza, R, Dostan, R, de Groot, R, Farber, S, Grasso, M, Hannon, B, Limburg, K, Naeem, S, O S, Oot, R, Paruelo, J, Raskin, RG, Sutton, P & van den Belt, M 1997, 'The value of the world's ecosystem services and natural capital', *Nature*, vol. 387, p. 387.

Dalin, C, Konar, M, Hanasaki, N, Rinaldo, A & Rodriguez-Iturbe, I 2012, 'Evolution of the global virtual water trade network', in *Proceedings of the National Academy of Sciences, USA*, vol. 109, pp. 5989–94.

De Fraiture, C, Cai, X, Amarasinghe, U, Rosegrant, M & Molden, D 2004, 'Does international cereal trade save water? The impact of virtual water trade on global water use', Comprehensive assessment research report, vol. 4, Comprehensive Assessment Secretariat, Colombo, Sri Lanka.

De Stefano, L & Llamas, MR (eds) 2012, *Water, Agriculture and the Environment in Spain: Can We Square the Circle?*, Taylor & Francis, London.

Dumont, A, Mayor, B & Lopez-Gunn, E 2013, 'Is the rebound effect or Jevons Paradox a useful concept for better management of water resources? Insights from the irrigation modernisation process in Spain', *Aquatic procedia*, vol. 1, pp. 64–76.

Ercin, AE, Aldaya, MM & Hoekstra, AY 2011, 'Corporate water footprint accounting and impact assessment: the case of the water footprint of a sugar-containing carbonated beverage', *Water Resources Management*, vol. 25, no. 2, pp. 721–41.

Garrido, A, Llamas, MR, Varela-Ortega, C, Novo P, Rodríguez-Casado, R & Aldaya, MM 2010, *Water Footprint and Virtual Water Trade in Spain: Policy Implications*, Springer, New York, 153 pp.

Giordano, M & Shah, T 2014, 'Non-integrated water resources management', in P Martínez-Santos, MM Aldaya & R Llamas (eds), *Integrated Water Resources Management in the 21st Century: Revisiting the Paradigm*, Taylor & Francis, London, pp. 37–46.

Gleick, PH 2016, Water conflict chronology, accessed 15 July 2016, www2.worldwater.org/conflict/.

GWP (Global Water Partnership) 2000, 'Integrated Water Resources Management', Background Papers, no. 4, Technical Advisory Committee (TAC), 69 pp.

Hoekstra, AY (ed.) 2003, 'Virtual water trade: Proceedings of the International Expert Meeting on Virtual Water Trade, 12–13 December 2002', *Value of Water Research Report Series* no. 12, UNESCO-IHE, Delft, The Netherlands.

Hoekstra, AY 2010, 'The relation between international trade and freshwater scarcity', Working Paper, ERSD-2010-05 (January 2010), World Trade Organization, Geneva, Switzerland, 25 pp.

Hoekstra, AY 2013, *The Water Footprint of Modern Consumer Society*, Earthscan/ Routledge, London, 204 pp.

Hoekstra, AY, Chapagain, AK, Aldaya, MM & Mekonnen, MM 2011a, *The Water Footprint Assessment Manual: Setting the Global Standard*, Earthscan, London, 206 pp.

Hoekstra, AY, Aldaya, MM & Avril, B (eds) 2011b, 'Proceedings of the ESF strategic workshop on accounting for water scarcity and pollution in the rules of international trade (Amsterdam, 25–26 November 2010)', *Value of Water Research Report Series*, no. 54, UNESCO-IHE, Delft, The Netherlands.

Jackson, LA, Pene, C, Martinez-Hommel, MB, Hofmann, C & Tamiotti, L 2014, 'Water policy, agricultural trade and WTO rules', in P Martinez-Santos, MM Aldaya & MR Llamas (eds), *Integrated Water Resources Management in the 21st Century: Revisiting the Paradigm*, Taylor & Francis, London, pp. 59–75.

Llamas, MR 2005, 'Los colores del agua, el agua virtual y los conflictos hídricos. Discurso inaugural del año 2005–06' [Water colours, virtual water and water conflicts. Opening address of the 2005–06 Course]', *Revista de la Real Academia de Ciencias Exactas, Físicas y Naturales* (Madrid), vol. 99, no. 2, pp. 369–89.

Martínez-Santos, P, Aldaya, MM & Llamas, R (eds) 2014, *Integrated Water Resources Management in the 21st Century: Revisiting the Paradigm*, Taylor & Francis, London.

Mekonnen, MM & Hoekstra, AY 2011, 'National water footprint accounts: the green, blue and grey water footprint of production and consumption', *Value of Water Research Report Series*, no. 50, UNESCO-IHE, Delft, The Netherlands.

Millennium Ecosystem Assessment (MEA) 2005, Millennium Ecosystem Assessment, accessed 15 July 2016, www.millenniumassessment.org/en/index.html.

Mordant, G & Pasquier, JL 2012, 'An estimate of the water footprint for France', Ministère de l'écologie, du développement durable et de l'énergie – Service de l'observation et des statistiques, Government of France.

Official State Gazette 2008, 'Approval of the water planning instruction', Ministry of the Environment and Rural and Marine Affairs, Official State Gazette, no. 229, accessed 15 July 2016, www.boe.es/boe/dias/2008/09/22/pdfs/A38472-38582.pdf.

Oregon State University 2016, 'The transboundary freshwater dispute database', Oregon State University, Department of Geosciences, Corvallis, accessed 15 July 2016, www.transboundarywaters.orst.edu/database/.

Pope Francis 2015, 'Laudato si (Praise be to you – On Care for Our Common Home) – Encyclical', Vatican Press, Vatican.

Ragone, SE & Llamas, MR 2006, 'The Alicante Declaration: steps along the pathway to a sustainable future', *Ground Water*, vol. 44, no. 4, pp. 500–03.

Rogers, P & Ramirez-Vallejo, J 2011a, 'Failure of the virtual water argument', in AY Hoekstra, MM Aldaya & B Avril (eds), *Value of Water Research Report Series*, no. 54, UNESCO-IHE, Delft, The Netherlands.

Rogers, P & Ramirez-Vallejo, J 2011b, 'Proceedings of the ESF Strategic Workshop on accounting for water scarcity and pollution in the rules of international trade (Amsterdam, 25–26 November 2010)', *Value of Water Research Report Series*, no. 54, UNESCO-IHE, Delft, The Netherlands.

Salmoral, G, Dumont, A, Aldaya, MM, Rodríguez-Casado, R, Garrido, A & Llamas, MR 2011, 'Análisis de la huella hídrica extendida de la cuenca del Guadalquivir', Botin Foundation, Santander.

UN 2015, *Sustainable Development Goals. Sustainable Development Knowledge Platform*, United Nations, New York.

UNDP 2006, 'Beyond scarcity: power, poverty and the global water crisis', *Human Development Report* (2006), United Nations Development Programme, New York.

UNEP 2011, *Water Footprint and Corporate Water Accounting for Resource Efficiency*, United Nations Environment Programme, Nairobi.

UNEP 2012, *Status Report on the Application of Integrated Approaches to Water Resources Management*, UN Water Report, United Nations Environment Programme, Nairobi.

UNSD 2012, 'SEEA-Water: system of environmental-economic accounting for water', United Nations Statistics Division, Department of Economic and Social Affairs Statistics Division. New York.

UN-Water 2013, 'Water security & the global water agenda', *A UN-Water Analytical Brief*, United Nations University Institute for Water, Environment & Health (UNU-INWEH), Hamilton, Canada.

Water Observatory 2014, Conclusions of the 13th national seminar 'Water footprint as a tool for water planning and conflict reduction', Water Observatory, Botin Foundation, accessed 15 July 2016, www.fundacionbotin.org/observatory-area/13th-national-seminar-la-huella-hidrica-como-instrumento-para-la-planificacion-hidrologica-y-reduccion-de-conflictos.html.

WBCSD (World Business Council for Sustainable Development) 2015, WASH Pledge at the workplace, accessed 15 July 2016, www.wbcsd.org/washatworkplace.aspx.

Willaarts, B 2012, 'Linking land management to water planning: estimating the water consumption of Spanish forests', in L De Stefano & MR Llamas (eds), *Water, Agriculture and the Environment in Spain: Can We Square the Circle?*, CRC Press, Taylor & Francis, New York, pp. 139–52.

WWAP (The United Nations World Water Assessment Programme) 2006, 'Water – a shared responsibility', *The United Nations World Water Development Report*, no. 2, Paris, UNESCO.

WWAP (The United Nations World Water Assessment Programme) 2009, 'Water in a changing world', *The United Nations World Water Development Report* no. 3, Paris, UNESCO.

WWAP (The United Nations World Water Assessment Programme) 2015, 'Water for a sustainable world', *The United Nations World Water Development Report 2015*, Paris, UNESCO.

WWAP (The United Nations World Water Assessment Programme) 2016, 'Water and jobs', *The United Nations World Water Development Report 2016*, Paris, UNESCO.

9 A hierarchy of water needs and their implications for allocation mechanisms

Eran Feitelson

Introduction

The professional discourse usually relates to water as a unitary substance with many uses: domestic, agriculture, industrial, and more recently also for nature (i.e. the maintenance of ecosystems). This view underlies the water balances that are calculated on the basis of the hydrological cycle, which as Linton (2010) has argued, is the outcome of the modernization of water. Prior to the modern era, water was often viewed in the plural term 'waters', constituting a variety of cool wet substances (Hamlin 2000), only to be replaced by the unitary view in the last couple of centuries. In recent years, however, this modernist view of water as a single 'thing' is being questioned from a social-normative perspective (Feitelson 2012).

On one hand, there is increasing recognition that the supply of water is not really limited to the natural freshwater resources, which are the focus of the unitary view, as increasing quantities of water are provided by desalination or through recycling of wastewater (Miller 2006; Elimelech & Phillips 2011; Feitelson & Jones 2014). Moreover, there is a possibility to reuse water also within households (greywater) (Friedler et al. 2005). In addition, freshwater is increasingly differentiated into 'blue water', which is the focus of most hydrological studies and discourse, and 'green water' that produces much of the food supply, which is the primary use of water for human consumption (Falkenmark & Rockstrom 2006).

On the other hand, we are witnessing widening discourse regarding the ethical dimensions of the use of water. In essence, there is increasing recognition that some water should be viewed as an essential human right (Gleick 1998; Gupta et al. 2010), and some is essential for the maintenance of other forms of life (Postel 1997; Breckenbridge 2005), thereby arguably having a priority over the economic use of water. Furthermore, the human right to water has now been incorporated into recent progressive water laws, such as those in South Africa and Ecuador (also see Rodina in this volume). As a result the argument has been made that rather than viewing water as a unitary substance, it should be viewed from an ethical perspective in the plural term 'waters', though in a different way than it was viewed historically (Feitelson 2012).

The main insight of this new view of water is that it should be differentiated on two lines. One, according to the differentiation between 'needs' and desires; and two, between natural water and produced water (Feitelson 2012). In this chapter I ask what are the implications of this new view of water for the way in which water should be allocated and managed.

There are three basic views regarding the ways water should be allocated and managed, which are usually presented as alternatives. The first is that water is a resource that should be allocated according to ethical principles, which put justice considerations at the centre (Sultana & Loftus 2012). These should be reified in laws and legal structures, both within countries and internationally (Leb 2012). Examples include the recognition of the human right to water in the Ecuadorian and South African water laws noted above, as well as in the UN Water Courses Convention (Article 10). The second is that water is contested and hence allocations are ultimately political decisions (Feitelson 2005; Mollinga 2008). Therefore, decisions should be made in the political domain. As laws are also promulgated in the political arena, politically structured decision-making institutions are the place to make the allocation and management decisions. The third basic approach is that water should be allocated through market processes. The argument for this approach is that water is essentially a scarce resource for which there are different values, and which can be produced at a price (Jaeger et al. 2013). Hence, water is not really different from other resources and commodities that are bought and sold in the market, and that market mechanisms assure that water will be put to the best use (Winpenny 1994; Fisher et al. 2005). These approaches are usually thought of as mutually exclusive. Yet they do not need to be so, if water is viewed in the plural rather than as a unitary resource.

My argument in this chapter is that all three approaches to the allocation of water are pertinent. But each pertains to a different type of water. Hence, I argue there exists a hierarchy of water, in terms of use, and different allocation and management principles pertain to different levels in this hierarchy.

To make this point I begin by outlining the different types of water, differentiated according to their use. These uses are then ordered in a hierarchical manner, and the different allocation approaches are related to the different levels of the hierarchy. Finally, I take this approach one step further by relating the hierarchy and allocation/management principles to the sources of water, and then to the policy approaches that should be used for each combination of use and source.

Water needs/rights[1]

The view that water should be considered an inalienable right stems from the fact that water is necessary for survival. At the most basic level clean potable water is needed for drinking, food preparation and hygiene. Consequently, there is an emerging consensus that all humans should be provided with the amount of water that is necessary to supply such basic needs, regardless of their ability to pay. Peter Gleick (1996) estimated that 50 l/c/d can supply these basic needs.

However, humans require more than their mere survival. Humans can rightfully expect to enjoy a dignified life (Hiskes 2009). Human dignity underlies, for example, the Millennium Development Goals (MDG), whose premise is that extreme poverty is unacceptable, as it does not allow for dignified living, and hence includes access to clean water and sanitation as part of these goals. Yet, the MDGs do not specify how much water is sufficient to assure human dignity, or what is meant by 'access' (Marin et al. 2009).

Chenoweth (2008) identifies two approaches to the specification of basic water needs. The first, which is the one used by Gleick (1996, 1998) as well as by many others, tries to quantify the amount of water that is needed for basic domestic functions. However, there is wide variance in the amounts needed to carry out various domestic functions, both between individuals and across societies. Climatic, land use and societal differences lead to variations in the amount of water used (Gleick 2003), and hence it is difficult if not impossible to come up with a practical answer to the question how much water should be allocated to meet the basic needs. Moreover, this approach does not have a clear analytical answer to the question which domestic functions are critical for a dignified living.

The second approach, termed by Chenoweth (2008) the development-efficiency approach, begins by identifying societies that offer a dignified level of living and searching for the one that does so with the minimal amount of water per capita. Using this approach Chenoweth (2008) finds that a country can meet its domestic water requirement and maintain a non-agricultural economy that is capable of sustaining a high level of development with as little as 120 l/c/d. Adding low (but not excessively so) losses in the water system he concludes that 135 l/c/d (which is akin to 50 m^3/c/yr) should be viewed as the minimal threshold required for social and economic development that permits a dignified level of living. However, these numbers are based on assumptions regarding what constitute reasonable losses in the supply system, and the extent to which they reflect water consumption patterns in all parts of the world. Allowing for somewhat higher water loss levels, and greater aridity in the Israeli-Palestinian case, Feitelson et al. (2011) suggest that the amount of water needed to allow a dignified level of living should be 60 m^3/c/yr (which is equivalent to 165 l/c/d).[2] Regardless, of the exact number, the important point for the purpose of this chapter is that there is a (relatively low) quantity of water that can be considered as the basic human need, and which allows for a dignified level of living (see Ziegler, Gerten and Döll in this volume).

So far the discussion of human needs has alluded to humans as individuals with physical needs. However, human existence is more than merely physical. Humans live in societies and have spiritual needs too. In many societies, water has symbolic properties (Linton 2010). Water is viewed as having spiritually cleansing properties (beyond hygiene), and hence is widely used in religious and cultural ceremonies. In some cases specific water bodies, such as the Jordan River or the Ganges, are viewed as holy or as deities. The quantity of water used in religious and cultural ceremonies is usually small. But in some cases other attributes of water (such as clarity, or the continuous natural flow) has spiritual meaning for the believers. To date this aspect of water has not been incorporated in water policies, as the sorry state of the Jordan River attests to (Courcier et al. 2005), but from a normative perspective it can be argued that this aspect should be seen as an additional need, or right, of humans.

However, humans are not the only ones who are dependent on freshwater for survival. The same is true for most terrestrial flora and fauna. The recognition that as humans we have an ethical responsibility toward nature has now been well recognized, though the nature of this responsibility is debated. In the case of water, this ethic has been couched in the language of stewardship (Postel 1997); ecohydrosolidarity (Falkenmark & Folke 2010) or biocentric ethic whereby humans are only the first among equals (Merchant 1997; Breckenbridge 2005).

Once the ethical responsibility of humans toward nature, regardless of is exact definition, has been recognized the question shifts to the amount of water and flow characteristics that are needed for the survival of ecosystems and the species within them. Yet, identifying the quantity and flow attributes which are needed for the survival of ecosystems is more complex than the case of basic human needs, as they vary widely across space and time as a function of environmental, ecological and climatic variables (Covich 1993). Moreover, as the flow patterns and quality attributes of water have implications for ecosystem dynamic and feedback mechanisms, which produce non-linear results, it is impossible to identify in advance the thresholds that should constitute minimal water needs (Anderson et al. 2006). Therefore, it is impossible to come up with definite globally applicable thresholds (Poff & Zimmerman 2010), as was attempted with regard to human needs.

Hence, for the purpose of this chapter it is sufficient to note that the water needed for ecosystems to survive has to be accounted for, though the quantities and flow requirements have to be determined locally. It is also important to note that the survival of ecosystems, for which minimal thresholds should be determined, is not the same as maintenance of a healthy 'ideal' ecosystem. Moreover, as essentially any use of water by humans that involves abstraction from natural water courses or aquifers affects the flow and quantity available for ecosystems the water retained for nature is not the water that was available in a pristine state. Hence, in all cases it is necessary to strike a balance between the water retained for nature, some of which can be determined to be a minimal need, and that used by humans.

Water for agriculture/food

The main use of water by humans is to produce food. It can be argued thus that water is essential for food security (Falkenmark & Lannerstadt 2010). Yet, much of the food, and particularly water-intensive foodstuffs, is increasingly supplied through the global market (Allan 2011). The water that is used to produce food that is traded globally is thus disassociated from the consumption of food. As a result, the main flows of water can be argued to be the flows of virtual water embedded in foodstuff (Hoekstra & Hung 2005). These flows balance out the spatial discrepancies between the distribution of people and the distribution of irrigable land and water, thereby re-scaling the question of food security to the global level. Hence, food in this globalized market is a commodity, and the water used to produce it is an input in a market process.[3] This is a clear break from the past when much of the food was locally produced, particularly in developing countries. However, much food is still locally produced in many parts of the world. Moreover, the sustenance of many communities is based on agriculture, mainly in the developing world but also in developed countries, and states are still held accountable for higher food prices and for the welfare of rural inhabitants. Thus, despite the re-scaling of much of the supply of foodstuffs, agriculture-related policies are still important at the national scale.

Villholth (2009) suggests that from an ethical perspective food production should be differentiated according to the producers. Essentially, she differentiates between food that is produced for subsistence and food that is produced for the market. Water is an input in the production of food in large commercial farms. But in smaller farms water for agriculture has wider implications.

Agriculture is not only an economic sector. Farming, also in developed countries, is largely seen as a way of life that has many positive externalities, both in terms of landscape and in terms of culture and meaning (Maguire et al. 1997). It is for this reason that agriculture is widely supported through subsidies in most developed countries. Yet, the agriculture that is supported is in many places mainly small-scale agriculture. Such agriculture often is the backbone of communities. This insight suggests an additional ethical factor. People do not live only as individuals. Rather people, and societies, see a value in communities as such. In some cases, such as in Switzerland, the community is seen as the central unit of society. This is also the main tenet of some of the anarchist movements and more recently of the communitarian movement (Etzioni 1998). From this perspective the failure to supply adequate water to sustain agricultural activity in the cases where without an agricultural base communities lose their economic base has widespread negative implications.

The implications of the differentiation of agricultural water use is that different policies should be directed toward commercial farming vis-á-vis small-scale farming. Obviously a major question is how to differentiate between the two, as much of the small-scale farming is also geared toward the market. One possibility, raised in the Israeli-Palestinian case, is to differentiate between small-scale agriculture that is within the commute-shed of urban centres, and hence have alternative options for employment, and peripheral areas where there is no alternative employment within a reasonable commute time from the village (Feitelson et al. 2011). In these latter cases lack of water can lead to the abandonment of villages and destruction of communities, and hence sufficient water has to be assured, from an ethical perspective, to allow farming to continue. This water should be supplied regardless of the ability of farmers to pay the full cost of supply. Therefore, there should be two pricing schemes for water used in agriculture. In the case of small-scale, peripheral and subsistence agriculture there is a clear case for supply and subsidization of the minimal amount of water that is needed to sustain the agricultural base of the community when water is efficiently utilized.[4] In the case of water used for large-scale commercial farming the case of subsidies is not clear, and it can be argued that full cost pricing is warranted. However, in practice full cost pricing of water for agriculture is not practised almost anywhere (Molle & Berkoff 2007), suggesting that the normative view of agriculture as a special sector has continuous widespread political resonance worldwide.

What about other uses of water?

The discussion of water for agriculture has bridged the gap between water that is supplied on a normative basis and water that should be viewed as a factor of production. This latter view of water is not limited however to agriculture. Water is also a factor of production in many other sectors, such as industry and tourism (Gossling et al. 2012; Florke et al. 2013). These sectors, however, do not have the same normative and political standing as agriculture. Therefore, water supply to commercial, industrial, tourism and service uses should be priced as all other inputs. However, as water use has negative external effects on natural systems, and in many cases (such as in aquifer abstraction) on other water users, efficient pricing should include the external costs of water supply. While water supply that has normative underpinning can justifiably be subsidized, that is not the case for water used as an input in economic sectors other than agriculture. Moreover, the failure to impose full-cost pricing in these cases may lead to excessive water use and lack of incentives for water conservation, and hence market failure.

In the domestic sector we have argued that there is a certain quantity that should be viewed as a basic need that allows for a dignified level of living. However, there are additional activities, such as water fountains, jacuzzis and similar consumptive uses, that people desire, but which cannot be viewed as necessary for a dignified level of living. This does not mean that such consumptive uses should be forbidden. On the contrary, domestic use beyond the basic need is legitimate. But it has to be paid for. In other words, different drops of water that come to the same household from the same tap should be treated differently according to their use. If the water is used for hygiene, direct consumption, food preparation and so forth, it can be viewed as supplying a basic need. But if it is used in a jacuzzi or to grow tropical plants in a garden, such water should be viewed as a luxury good. This differentiation suggests that block-rate pricing should be used with regard to domestic water, whereby water that is viewed as a basic need is moderately priced (or is provided free of charge in extreme cases), while water use beyond this quantity should be fully priced. That is, consumptive use beyond the quantity determined as necessary to allow a dignified level of living should be priced to include the externality cost of providing this water.

This approach, differentiating water that is used for the supply of market goods and luxury goods from needs applies also to the bottling of water. Bottled water is a commodity that is sold. Therefore, despite the fact that it serves for drinking, it should be viewed as a commodity, and hence be fully priced. This statement can be modified, however, where municipal water supply is of low quality, and may incur health issues. In this particular circumstance, there may be a justification for not pricing bottled water at its full social cost, due to their positive health implications, at least until the municipal water supply is improved. The same can be said for the supply of bottled water in quality crisis situations, such as the crisis that occurred in Flint, Michigan. In such a situation, the subsidization of bottled water is justified.

The water needs hierarchy

Up to this point we have outlined the normative differences between the different types of water. In Figure 9.1 these different types are presented in a hierarchical manner, reflecting Maslow's (1943) hierarchy of needs. At the bottom of the pyramid, reflecting Maslow's physiological and safety needs, are the basic needs of humans (those needed for a dignified level of living). Above them I place the water needed for ecosystem survival and for fulfilling spiritual and cultural needs. These three types of needs should be supplied universally regardless of cost. Hence the allocation of these needs is based on normative grounds, and is universally applicable.

The next set of waters is those that are widely contested. The contested issues include the question to what extent social needs, such as the maintenance of communities in peripheral areas, should be subsidized; how much water should we allocate to nature, beyond the minimal amounts needed for ecosystem survival; how much water should be allocated to small-scale farming and at what cost, and

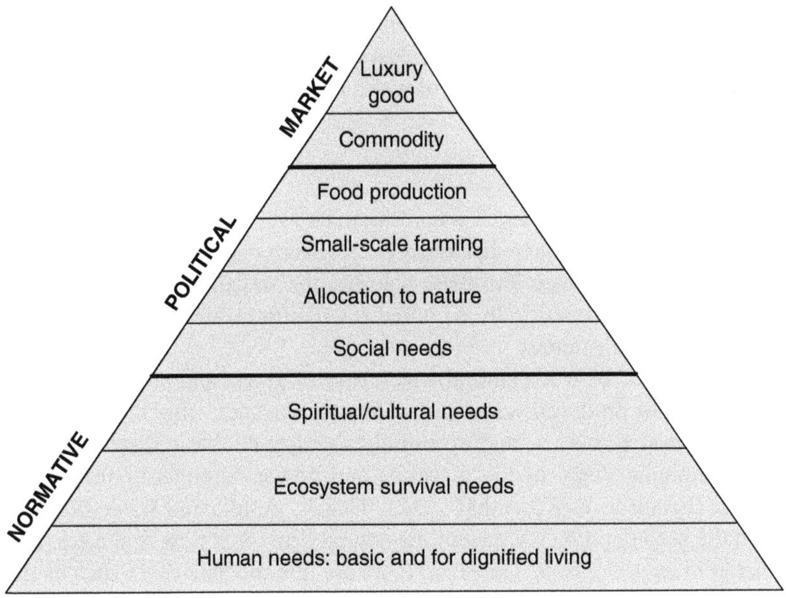

Figure 9.1 Relating the needs hierarchy to allocation mechanisms.

Source: own.

the allocation and pricing of water for commercial agriculture. Different societies in different time periods are likely to come to different answers, depending on the balance of power between different interest groups (including civil society), and the shifts in the dominant discourse. In the United States, for example, the political power structures led to widespread building of dams and provision of cheap water to agriculture in much of the twentieth century (Reisner 1986). In Israel water was provided at below-cost rates to all forms of agriculture for many decades due to the internal power structures and the ideological domination of the ruralistic ideal (Feitelson 2005). However, the shift from a welfare state to a neo-liberal state in Israel has resulted in a shift in the dominant discourse, as the economic story-line has become the dominant story-line, and in the empowerment of the Treasury vis-à-vis other ministries. Moreover, with the decline of agriculture as an economic sector and the subsequent decline in the power of the agricultural lobby, pricing policies in Israel shifted and subsidies for water to agriculture were greatly reduced.

These issues are fought over and deliberated in the political arena, and rightly so. Hence, the allocation and management of these types of water should be determined through political processes within the political arena, as this is the arena where contentious value-laden issues are resolved in democratic societies.

Finally, at the top of the pyramid are the uses for which subsidies cannot be justified on normative grounds. These are the use of water as an input to production, whether in industry or services (tourism included) and the provision of luxury benefits to households. In both cases the use of water for attracting tourists, producing goods or providing enjoyment in the domestic sector[5] should be allowed. However, these uses should be fully priced. Hence the allocation of water to these uses should be through the market mechanisms. That is, anybody who is willing to pay the full social price (including the cost of negative externalities in supply) should be able to purchase the amount of water he is willing to pay for.

Are all water sources equal?

So far water has been differentiated according to its use. The implied assumption behind this is that all the water discussed comes from natural sources, whether surface water or groundwater. But these sources, the so-called 'blue' water, are only a small part of the water cycle. As noted at the outset water is supplied today from several additional sources.

The main divide is between natural water, blue or green, and produced water. The clearest case of produced water is desalinated seawater. This source has no relation to freshwater sources, and essentially changes the basic flow direction, thereby undermining many of the premises and power relations of freshwater management (Feitelson & Rosenthal 2012). That is, desalinated water flows in pipes from the sea inland, in contrast to the natural flows that flow to the sea, and is not related to natural basins. However, there are intermediate cases such as the recycling of wastewater and the desalination of brackish water. These are based on natural sources, but are made usable only the massive application of capital and technology.

From a normative perspective, the important difference is that produced water is an outcome of an investment decision in an industrial process. It can be argued that such waters are inherently a commodity. They belong to whoever produced them.[6] Of course, a somewhat similar argument can be made also in the case of blue water, as the diversion and pumping of such water are also the outcome of investments. However, in the case of produced water the extent to which they are an outcome of human intervention and ingenuity is much greater and the source of the water supplied is not freshwater. As a result, produced water is conveyed in pipes, often from the sea inland, utterly disregarding the natural water geography (Feitelson & Rosenthal 2012). Its management is thus disassociated from that of natural water, and by necessity is manipulated by supply and demand considerations, rather than being a function of natural watercourses.

If water is produced it can be expected that it will be bought and sold as any other commodity. Thus, the default pricing policy is that produced water will be fully priced. The case of recycled wastewater is more complex, as the treatment of wastewater is warranted from an environmental perspective. That is, the polluter should pay for treatment of wastewater. The question is thus to whom the water that has been treated belongs, and to what level should the wastewater be treated. Does treated wastewater belong to the party that treated it, or does it return to the natural hydrological cycle, and thus should be treated in a similar manner to blue water. It is likely that the answer to this question will be partly an outcome of climate. In humid areas treated wastewater is often returned to the river from which the water was abstracted. In arid areas wastewater is more often used directly for irrigation. Yet, the array of crops to which it can be used is a function of politically determined health standards, and the level to which wastewater has been treated.[7] It is likely therefore that the normative decision regarding the utilization of recycled wastewater will also be determined in the political arena, with the answer differing according to the different circumstances in different regions and countries.

An additional question arises when the normative demands, as outlined in the hierarchy presented above cannot be met from natural resources alone, and thus merit subsidies from a needs perspective, yet supply at least partially has to be augmented by produced water. This is the topic of the next section.

The policy implications of a normative approach

Table 9.1 summarizes the relations between uses, sources, allocation mechanisms and policies. As can be seen in the table, for every type of water (except green water) different pricing and allocation principles may apply. Pluses signify the cases where green water is used, but as green water cannot be controlled, there are no policies, which are pertinent for these combinations. Minuses signify infeasible or highly unlikely combinations.

In the case of blue water for domestic consumption, block-rate pricing is warranted. The quantity that is determined normatively to be the basic need merits a subsidy (left-hand column). In the case of poor households, this subsidy may reach 100% (i.e. the basic quantity will be free of charge). For any additional use beyond this basic need (i.e. luxury goods in Table 9.1) the household should pay the full cost, including the negative externality costs associated with supply. These may vary by location. Such block-rate pricing systems are well known, and are widely used. They do require, however, that all water use be metered. In countries where there is universal metering block-rate pricing can be applied nationally (which is the case in Israel).

Subsidies are warranted also for the use of blue water for environmental and spiritual needs. These needs have to be assessed locally, according to the ecosystems that are affected, and the local cultural, religious and spiritual customs. As the spiritual needs are usually minute the supply and subsidization of these needs should not be problematic. However, where the religious or spiritual needs require that water flow freely in rivers or streams (such as in the case of the Jordan River), the ability to allow the necessary flow and its pricing is essentially similar to the case of environmental needs, which also require water to flow freely in nature.

In the case of flows to nature, beyond the question how much water should be left to flow freely to allow the ecosystems to survive, it is necessary to determine whether anybody has to pay for this free flow of water. In Israel, for example, the Nature and Parks Authority is required to pay for water released to national parks or nature reserves, but it is reimbursed by the government. However, it can be argued that ecosystems should not be charged for water, on normative grounds, as the water was naturally there. This can be taken a step further and an argument can be made that nature should be compensated for the water that is abstracted from it. While this logic may not have been fully implemented, programmes whereby water rights were purchased in order to return water for nature have been put in place in Oregon and several other places in the USA (Landry 1998). In many cases these purchases were made by environmental NGOs.

The allocation of water and the pricing of water for small-scale agriculture, peripheral farming, as well as for commercial farming will be determined politically. While there is a compelling argument for charging the full cost from commercial agriculture, it is unlikely that this will be implemented in practice, given the special role of agriculture in the political arena. In contrast, a stronger argument can be made for subsidizing water that is needed to maintain communities that have no alternative sources of employment except agriculture (peripheral farming) and for small-scale agriculture. So in practice, it is likely that all these uses will be subsidized to some extent, though the extent to which they will be subsidized will be subject to the political discourse and power dynamics in each locale at each point in time.

The allocation of water to nature beyond the mere minimum that is needed for ecosystem survival is also a political decision. The same is true for the pricing of such water, and the regime of flows – a regime that may be crucial for the ecosystem, as ecosystem dynamics are a function of flow attributes, and not only total quantities. In this case the proponents of increasing water allocations for nature, and maintaining an appropriate flow regime, are likely to come from civil society. The allocation of blue water to both farming and nature should also be a

Table 9.1 Allocation mechanisms, policies and water sources

Water source	*Allocation mechanism*							
	Normative			*Political*			*Economic/market*	
	Basic human	*Ecosystem survival*	*Spiritual*	*Small-scale & peripheral farming*	*Nature*	*Commercial farming*	*Economic production*	*Luxury goods*
Blue water	Subsidized	Subsidized/ no cost	subsidized	Subsidies	Allocation, no cost?	Priced, but not fully	Full social cost	Full social cost
Green water	–	+	–	+	+	+	+	+ (gardening)
Recycled	–	For stream rehabilitation	–	If available, subsidized	Subsidized	Priced, but not fully	Priced, but not fully	Priced, but not fully
Desalination	Subsidized	–	–	–	–	–	Full social cost	Full social cost

Source: own.

function of the availability of green water (rain water). In essence, the supply of blue water is somewhat of an inverse function of the availability of green water, as in drought situations more water has to be suppled artificially than in normal or rainy years. It is for this reason that the cases where green water supplies some of the needed water are noted in Table 9.1.

Recycled wastewater can be used for irrigation, to supplant environmental flows, for watering gardens and lawns in municipal parks or recreation areas and in industry. It is not used for domestic use (except grey water recycling within the household) or for spiritual needs. As the use of recycled wastewater is limited to these uses, and the extent to which it can be utilized is also restricted by the level of treatment, the competition for recycled wastewater is less acute than for freshwater. Moreover, wastewater recycling is desirable in order to prevent the pollution of blue water sources and for health reasons. Hence, it is unlikely that recycled wastewater will be charged at freshwater rates. It is thus likely that it will be cheaper than freshwater for all the pertinent uses. In the case of environmental flows it is also most likely that the reuse of wastewater within a river will incur no cost at all for environmental agencies, even in water-scarce regions. Quite obviously this ideal may not seem practical in most developing countries, in many of which there is not sanitation at the household level. But from an ethical perspective wastewater recycling should be encouraged, and hence due to its positive externality value it is most likely that the recycling outlays (in contrast to the wastewater treatment cost) may be subsidized – that is, the full social cost will not be charged. Yet, the decisions regarding wastewater treatment standards, recycling options and charges will be determined in the political arena. It should also be noted that in some industries and resorts it is possible that wastewater will be recycled internally, in which cases there will not be a formal pricing policy. In these cases, the only policies that are pertinent are the regulation of recycling standards and practices.

Desalinated seawater is produced through industrial processes. Hence, it should be viewed as an industrial product that is owned by whoever produced it. For this reason desalinated seawater, which is used mainly for domestic consumption, should be sold at a market price. However, this price has to reflect the full social cost of desalination, as desalination has negative environmental effects.[8] An exception to this principle is the case where desalinated seawater is needed to supply basic human needs. This is, for example, the case in Gaza, where due to the extent of over-pumping and pollution of the local aquifer there is not alternative supply of freshwater, or in the arid Gulf States. In these cases seawater desalination may merit a subsidy in order to allow for the block-rate pricing discussed above.

Conclusions

In this chapter I argue that the three different allocation mechanisms, normative rule-based, political power-based and economic money-based, are not mutually exclusive. Rather, each of them pertains to different types of water, differentiated according to use and source. The differentiation of waters is presented in a hierarchical form, reminiscent of Maslow's needs hierarchy. In our case needs should be supplied on the basis of ethical principles, which may serve as a basis for legislation. These include basic human needs (broadly defined to include the water necessary for a dignified level of living), spiritual-cultural-religious needs and environmental flows needed for ecosystem survival.

The fact that there is wide agreement over these normative needs allows them to be potentially useful also in transboundary settings. In a joint Israeli-Palestinian study these needs were used to assess the degree to which they can be supplied from the shared resources. It was shown that all three needs can be fully supplied with relatively moderate changes from current water allocations between the parties, and hence can serve as a basis for an equitable division of water, for the case of the West Bank (Feitelson et al. 2011). This was not the case for Gaza, which therefore was found to merit subsidized seawater desalination.

Beyond these needs there is an additional set of uses, which have normative facets. These include small-scale farming, peripheral farming (farming which is crucial for the maintenance of communities where there is not alternative employment within a reasonable commute) and environmental flows, above those needed for the survival of ecosystems. However, the importance of these normative factors and their policy implications are often contested within societies. Hence the allocations of water and the pricing of these allocations should be determined politically. The scales at which these decisions will be made are a function of the issue at hand and the institutional structures that govern water allocations. In humid regions these may be at the basin scale, while in small arid country cases they may be taken at the national scale. In the case of trans-boundary water, they may be taken partially at the international level, in bodies that were established to deal with the trans-boundary issues, and partially at the intra-national level, as ultimately representatives at the international level are always constrained by the decisions made intra-nationally (Putnam 1988).

Where water is used as an input for economic activities, or for the provision of services beyond the basic needs, there is no justification for subsidizing water. Hence, in these cases water should be fully priced, where full cost is defined as the full social cost (i.e. including a surcharge to account for the negative externalities of water supply). This is true also for produced water; as such water is also essentially an industrial product. There are, however, a few exceptions to these principles. One pertains to commercial agriculture, which is an economic activity, but is also imbued with normative values. Hence, in practice, pricing is determined politically – not least due to the centrality of food prices in assuring stability in many low-income societies. A second exception pertains to the situation where seawater desalination is necessary to supply basic human needs, and hence some desalinates seawater should, ethically, be subsidized.

There are, however, several potential ethical pitfalls in the discussion of the policies that are derived from the discussion so far. The first is that basic human needs apply to all humans, regardless of citizenship. Statistics usually pertain only to citizens or permanent residents. However, in situations where there is a large number of refugees within a country, or a region, such as is the case today (2017) in the Kingdom of Jordan, Turkey, Lebanon and the Greek island of Lesbos, there is a danger that these refugees, many of which may be unregistered, will not be taken into account in water allocations. In such cases there is a danger that the most vulnerable people, the refugees that are not accounted for, will be forced to use expensive unmonitored water vending services.

A second potential pitfall is the possibility that those not represented in the political arena at a certain point in time will be overlooked and disregarded in making decisions over the water uses whose allocations are made in the political arena. These may include in addition to marginalized communities or social groups also future generations, non-citizens and nature.

The outcome of political decision-making is a function of power. As there are discrepancies in power between different social groups and stakeholders, it is likely that political decisions will reflect the power of the stakeholders involved. Hence, there is a need to assure that minorities and groups without due representation will not be overridden, and that their interests will also be addressed. Involving civil society organizations in decision-making may help partially address this concern.

To implement the full-cost pricing principle for the cases where such pricing is warranted it is necessary to identify and quantify the negative externalities. This is far from simple (Birol et al. 2006). In many cases, such as assessing the monetized detrimental effects of water supply on various ecosystems, it is impossible to get theoretically valid, agreed-upon values. As a result, in most cases second- or third-best estimates are used, if at all. To address this difficulty it is advisable that agreed-upon procedures for assessing negative externalities will be put in place. Even if the values derived are not the first-best 'true' values, such procedures can reduce the chance that the negative externalities will be overlooked.

In conclusion, this chapter has outlined an ethics-based approach to allocate and price water. Yet, this approach is not free from pitfalls. Hence, it is mandatory that the potential pitfalls identified here will be taken into account when formulating the ethics-based policies, so that water allocations and pricing will indeed be fair to the extent possible.

Notes

1 This section is largely based on Feitelson (2012).
2 Higher minimal basic needs were estimated earlier by Shuval (1992) and by Falkenmark (1986) based on the Israeli experience. However, Chenoweth (2008) found lower figures based on the Maltese experience. Hence these are presented here.
3 This re-scaling though comes at the price of a wider water footprint (see Aldaya et al. in this volume). Still, the estimation of the water footprint is fraught with difficulties (Chenoweth et al. 2014). This implies that internalizing the externalities associated with the wider footprint is at present technically problematic, if not infeasible.
4 This qualification implies that improved management and education are also needed in such cases to prevent the supply and subsidization of inefficient practices.
5 The domestic sector includes municipal use.

6 This holds true also in the case where a public entity owns the desalination plant. In this case, the water produced belongs to that entity (for example, a municipal or state water company). These entities can decide to provide them at-cost or to charge a block-rate price. But this is their purview.
7 In essence, the higher the level of treatment, the wider the array of crops to which the treated wastewater can be applied.
8 The environmental effects of desalination include the energy intensiveness of these processes, and resulting GHG emissions, as well as the effects of the brine outflow on marine life (Lattemann & Höpner 2008).

References

Allan, T 2011, *Virtual Water: Tackling the Threat to Our Planet's Most Precious Resource*, Tauris, London and New York.

Anderson, K, Paul, AJ, McCauley, E, Jackson, LJ, Post, JR & Nisbet, RM 2006, 'Instream flow needs in streams and rivers: the importance of understanding ecological dynamics', *Frontiers in Ecology and the Environment*, vol. 4, pp. 309–18.

Birol, E, Karousakis, K & Doundouri, P 2006, 'Using economic valuation techniques to inform water resource management: a survey and critical appraisal of available techniques and an application', *Science of the Total Environment*, vol. 365, pp. 105–22.

Breckenbridge, LP 2005, 'Can fish own water? Envisioning non human property in ecosystems', *Journal of Land Use and Environmental Law*, vol. 20, pp. 293–335.

Chenoweth, J 2008, 'Minimum water requirements for social and economic development', *Desalination*, vol. 229, pp. 245–56.

Chenoweth, J, Hadjikakou, M & Zoumides C 2014, 'Quantifying the human impact on water resources: a critical review of the water footprint concept', *Hydrology and Earth System Sciences*, vol. 18, pp. 2325–42.

Courcier R, Venot J-P & Molle F 2005, 'Historical transformations of the Lower Jordan River basin (in Jordan): changes in water use and projections (1950–2025)', International Water Management Institute, Colombo, Sri Lanka.

Covich, AP 1993, 'Water and ecosystems', in P Glieck (ed.), *Water in Crisis: A Guide to the World's Freshwater Resources*, Oxford University Press, Oxford.

Elimelech, M & Phillip, WA 2011, 'The future of seawater desalination: energy, technology, and the environment', *Science*, vol. 333, no. 6043, pp. 712–17.

Etzioni, A (ed.) 1998, *The Essential Communitarian Reader*, Rowman and Littlefield, Lanham, MD.

Falkenmark, M 1986, 'Fresh water – time for a modified approach', *Ambio*, vol. 15, pp. 192–200.

Falkenmark, M & Folke, C 2010, 'Ecohydrosolidarity: a new ethic for stewardship of value-adding rainfall', in PG Brown & JJ Schmidt (eds), *Water Ethics: Foundational Readings for Students and Professionals*, Island Press, Washington, DC, pp. 247–64.

Falkenmark, M & Lannerstadt, M 2010, 'Food security in water-short countries – coping with carrying capacity overshoot', in L Martinez-Cortina, A Garrido & E Lopez-Gunn (eds), *Re-thinking Water and Food Security*, CRC Press/Balkema, Leiden.

Falkenmark, M & Rockstrom, J 2006, 'The new blue and green water paradigm: breaking new ground for water resources planning and management', *Journal of Water Resources Planning and Management*, vol. 132, no. 3, pp. 129–32.

Feitelson, E 2005, 'The political economy of groundwater exploitation: the Israeli case', *International Journal of Water Resources Development*, vol. 21, pp. 413–23.

Feitelson E 2012, 'What is water: a normative perspective', *Water Policy*, vol. 15, pp. 52–64.

Feitelson, E & Jones, A 2014, 'Global diffusion of XL-capacity seawater desalination', *Water Policy*, vol. 16, pp. 1031–53.

Feitelson, E & Rosenthal, G 2012, 'Desalination space and power: the ramifications of Israel's changing water geography', *Geoforum*, vol. 43, pp. 272–84.

Feitelson, E, Tamimi, AR, Bein, A, Laster, R, Marei, A, Rosenthal, G & Salhout, S 2011, 'Defining Water Needs for Fully Exploited Resources: A Necessary Step for Israeli-Palestinian Reconciliation', report submitted to Green Cross International, the Peres Center for Peace and the Jerusalem Institute for Israel Studies.

Fisher, F et al. 2005, *Liquid Assets: An Economic Approach for Water Management & Conflict Resolution in the Middle East & Beyond*, Resources for the Future Press, Washington, DC.

Florke, M, Kynast, E, Barlund, I, Eisner, S, Wimmer, S & Alcamo, J 2013, 'Domestic and industrial water uses in the past 60 years as a mirror of socio-economic development: a global simulation study', *Global Environmental Change*, vol. 23, pp. 144–56.

Friedler, E, Kovalio, R & Galil, NI 2005, 'On-site greywater treatment and reuse in multi-storey buildings', *Water Science and Technology*, vol. 51, pp. 187–94.

Gleick, PH 1996, 'Basic water requirements for human activities: meeting basic needs', *Water International*, vol. 21, pp. 83–92.

Gleick, PH 1998, 'The human right to water', *Water Policy*, vol. 1, pp. 487–503.

Gleick, PH 2003, 'Water use', *Annual Review of Environment Resources*, vol. 28, pp. 275–314.

Gossling, S, Peeters, P, Hall, CN, Ceron, JP, Dubois, G, Lehmann, LV & Scott, D 2012, 'Tourism water use: supply demand and security. An international review', *Tourism Management*, vol. 33, pp. 1–15.

Gupta, J, Ahlers, R & Ahmed, L 2010, 'The human right to water: moving towards consensus in a fragmented world', *Review of European Community & International Environmental Law*, vol. 19, pp. 294–305.

Hamlin, C 2000, '"Waters" or "water"? – master narratives in water history and their implications for contemporary water policy', *Water Policy*, vol. 2, pp. 313–25.

Hiskes, RP 2009, *The Human Right to a Green Future*, Cambridge University Press, Cambridge.

Hoekstra, AY & Hung, PQ 2005, 'Globalisation of water resources: international water flows in relation to crop trade', *Global Environmental Change*, vol. 15, pp. 45–56.

Jaeger, WK, Plantinga, AJ, Change, H, Dello, K, Grant, G, Hulse, D, McDonnell, J, Lancaster, S, Moradkhani, H, Morzillo, AT, Mote, P, Nolin, A, Santelmann, M & Wu, J 2013, 'Toward a formal definition of water scarcity in natural-human systems', *Water Resources Research*, vol. 49, pp. 4506–17.

Landry, C 1998, 'Market transfers of water for environmental protection in the western United States, *Water Policy*, vol. 1, pp. 457–69.

Lattemann, S & H S &, T 2008, 'Environmental impact and impact assessment of seawater desalination', *Desalination*, vol. 220, pp. 1–15.

Leb, C 2012, 'The right to water in a transboundary context: emergence of seminal trends', *Water International*, vol. 37, pp. 640–53.

Linton, J 2010, *What is Water? A History of a Modern Abstraction*, University of British Columbia Press, Vancouver.

Maguire, M, Foote, R & Vespe, F 1997, 'Beauty as well as bread', *Journal of the American Planning Association*, vol. 63, pp. 317–28.

Marin, LE, Sandoval, R, Tagle, F, Sanchez, E & Martinez, VH 2009, 'Water as a human right and as an economic resource: an example from Mexico', in MR Llamas, L Martinez-Cortina & A Mukherji (eds), *Water Ethics*, CRC Press/Balkema, Leiden, pp. 115–25.

Maslow, AH 1943, 'A theory of human motivation', *Psychological Review*, vol. 50, pp. 370–96.

Merchant, C 1997, 'Fish first! The changing ethics of ecosystem management', *Human Ecology Review*, vol. 4, pp. 25–30.

Miller, GW 2006, 'Integrated concepts in water reuse: managing global water needs', *Desalination*, vol. 187, pp. 65–75.

Molle, F & Berkoff, J 2007, 'Water pricing in irrigation, mapping the debate in light of experience', in F Molle & J Berkoff (eds), *Irrigation Water Pricing: The Gap between Theory and Practice*, CABI Publication, Wallingford, accessed 18 July 2016, www.cgiar.org.

Mollinga, P 2008, 'Water politics and development: framing a political sociology of water resource management', *Water Policy*, vol. 1, pp. 7–23.

Poff, NL & Zimmerman, JK 2010, 'Ecological responses to altered flow regimes: a literature review to inform the science and management of environmental flows', *Freshwater Biology*, vol. 55, pp. 194–205.

Postel, S 1997, *The Last Oasis*, 2nd edn, Norton, New York.

Putnam, RD 1988, 'Diplomacy and domestic politics: the logic of the two-level games', *International Organization*, vol. 42, pp. 427–60.

Reisner, M 1986, *Cadillac Desert: The American West and Its Disappearing Water*, Penguin Books, New York.

Shuval, H 1992, 'Approaches to resolving the water conflicts between Israel and her neighbors – a regional water for peace plan', *Water International*, vol. 17, pp. 133–43, 155–65.

Sultana, F & Loftus, A 2012, 'The right to water: prospects and possibilities', in F Sultana & A Loftus (eds), *The Right to Water: Politics, Governance and Social Struggles*, Earthscan, London.

Villholth, KG 2009, 'Water and ethics in food production and provision – how to ensure water and food security and equity in the 21st century?', in MR Llamas, L Martinez-Cortina & A Mukherji (eds), *Water Ethics*, CRC Press, Boca Raton, FL.

Winpenny, J 1994, *Managing Water as an Economic Resource*, Routledge, London and New York.

10 Reflections on water ethics and the human right to water in Khayelitsha, South Africa

Lucy Rodina

Introduction

Water is life and all humans need water to survive and live an adequate and dignified life. And yet, in 2015 2.4 billion people in the world lacked access to improved sanitation and 663 million continued to use unimproved drinking water sources[1] (WHO/UNICEF 2015). Over the past few decades this has served as justification for arguments in support for a human right to water – a right that today 'entitles everyone to sufficient, safe, acceptable, physically accessible and affordable water for personal and domestic uses' (UN CESR 2003). In 2010 the United Nations General Assembly recognized 'the right to safe and clean drinking water and sanitation[2] as a human right that is essential for the full enjoyment of life and all human rights', marking an important milestone for water justice advocates throughout the world. Of course, even before 2010 interpretations of this right already served as a powerful tool for marginalized and vulnerable groups to claim access to improved sources of water (and sanitation) and demand accountability from authorities (Mirosa & Harris 2011). Today many challenges remain as the definition of the right to water and its implementation measures remain highly contested in different contexts. The successful implementation of the human right to water continues to pose significant challenges in an increasingly changing water governance environment, where water security, climate change and ongoing environmental degradation gain prominence.

And while many have seen much promise in the formal Human Right to Water (HRW) to help achieve universal access to water, especially for poor and marginalized populations (Harris et al. 2015; Sultana & Loftus 2015), many others have raised a series of issues, including its Western origins and bias (Parmar 2008), its inability to preclude water privatization (Bakker 2012), and its lack of environmental considerations (Feris 2014; Harris et al. 2015). Furthermore, access to water remains deeply unequal around the globe, with millions of poor and marginalized people, especially in urban and peri-urban slum settlements, lacking access to safe and adequate water (Mehta 2014). As such, in order to achieve universal access to sufficient and safe water for the full enjoyment of life and other rights, the implementation of the HRW should engage with on-the-ground realities of unequal access to resources and the processes that produce these inequalities.

This chapter aims to provide practical examples and insights from critical scholarship in political ecology and related fields on the uneven effects of the implementation of the right to water through the case of impoverished, peri-urban and informal settlements in Cape Town, South Africa. South Africa is particularly interesting because it demonstrates the multiple contestations that emerge in the context of progressive water legislation,[3] high social inequality (McDonald 2007) and strong civic and political commitment to redressing past inequalities through providing equitable access to services (Tissington 2010). This case demonstrates that questions of equity and justice in access to water are obfuscated by the narrow focus on basic needs, thus limiting the broader justice agenda behind the HRW discourse.

As will be discussed in more detail below, the implementation of the HRW in Cape Town is focused predominantly on providing basic amounts of water and is effectively folded into service delivery programs. While this approach appears to be effective in extending access to basic services to more people, it also serves to maintain historical inequalities in water access by effectively limiting access for the poor to only the free basic minimum amounts. This narrow implementation of HRW limits its potential for achieving justice in access to water in that it hinders the opportunities for improved and more diverse livelihoods. In addition, this approach tends to lock in already poor populations in using water only for mere survival, and thus perpetuates historical inequalities. This chapter aims to engage with scholarship on water ethics in questioning the ethics of the basic needs approach to implementing HRW. Aligning with Parmar (2008), this chapter argues that the fair and just implementation of the human right to water requires moving beyond a short-sighted basic needs framing towards broader conceptions of livelihoods needs, and social justice.

Human right to water in the making

Human rights became a powerful international political tool to protect human life and promote fair development after the end of the Second World War, through the United Nations Economic and Social Council's 'Declaration on the Essential Rights of Man', which later led to the legally binding International Covenant on Civil and Political Rights as well as the International Covenant on Economic, Social and Cultural Rights (Obani & Gupta 2014). Neither of these Covenants explicitly included the right to water and sanitation, however many scholars have argued that access to water is a prerequisite for the realization of the right to life and other rights, such as the right to food, clothing and housing.

In 2010 the United Nations General Assembly adopted resolution A/64/292 which explicitly recognizes that every human being has a right to safe and clean *drinking water and sanitation* as they are essential for the full enjoyment of life and all human rights. In the same year the UN Human Rights Council adopted the 'Human Right on Access to Safe Drinking Water and Sanitation' resolution, affirming that HRW 'is derived from the right to an adequate standard of living and inextricably related to the right to the highest attainable standard of physical and mental health, as well the right to life and human dignity' (Obani & Gupta 2014). General Comment 15 of the Committee on Economic, Social and Culture Rights, which provides clarification on the different aspects of this right, is often cited in guiding HRW implementation. According to the General Comment 15, HRW 'entitles everyone to sufficient, safe, acceptable, physically accessible and affordable water for personal and domestic uses'. While this definition has been widely adopted internationally, General Comment 15 also stipulates, that 'water is required for a range of different purposes, besides personal and domestic uses, to realize many of the Covenant rights', including water to produce food (right to adequate food) and ensure environmental hygiene (right to health) (UN CESR 2003, section 6). Further, water is recognized as 'essential for securing livelihoods (right to gain a living by work) and enjoying certain cultural practices (right to take part in cultural life)' (ibid.). However, priority is given to personal and domestic uses to prevent disease and starvation due to lack of water.

The volume of water considered sufficient for basic needs, according to the World Health Organization and Peter Gleick's research, is a minimum of 50 litres per person a day – a number that includes minimum drinking water for survival, water for sanitation, water for bathing, and water for food preparation (Gleick 1996). Water also needs to be safe, meaning free of harmful micro-organisms, chemical substances and radiological hazards, and it should be of acceptable colour, odour and taste (General Comment 15, section 12b). Water and water facilities should be accessible to everyone without discrimination, within 'safe physical reach for all sections of the population', where 'physical security should not be threatened during access to water facilities' (section 12c). Lastly, water should also be economically affordable to all sections of the populations (section 12c).

By implication the UN's formal recognition of the HRW created immediate obligations on states to take steps toward the full realization of the right and to guarantee that it is exercised without discrimination (Obani & Gupta 2014). The UN Resolution A/RES/64/292 mandates states and international organizations 'to provide financial resources, capacity-building and technology transfer' through international cooperation and assistance, particularly to developing countries, who are often under-resourced to be able to provide water for all. To date 15 national constitutions contain explicit recognition of the right to water, including Mexico, Uruguay, Uganda, South Africa, Nicaragua, Ecuador, DRC, Maldives, Bolivia, Kenya, Niger, Morocco, Somalia, Zimbabwe and Tunisia; however, only four of them have an explicit recognition of the right to sanitation – Mexico, Uruguay, Maldives and Kenya (Obani & Gupta 2014).

With the passing of the 2010 UN resolution, as well as its adoption in a range of national constitutions and legislations, the human right to water is here to stay. Of course, to fully understand how the human right to water is (or should be) realized, how claims to this right are made, and what water access actually means in people's daily lives, we need to critically engage with the on-the-ground realities, politics and struggles around water access (Kowarsch in this volume; Sultana &

Loftus 2012). Scholars and policy-makers continue to debate the successes and failures of the implementation of HRW, as the ways this right is negotiated, experienced and struggled for in different contexts still remain key challenges (Mirosa & Harris 2011; Parmar 2008; Sultana & Loftus 2012). From a social justice perspective, HRW is potentially a highly effective mechanism to make progress towards realizing the goal of safe and affordable water for all, regardless of ability to pay (Barlow 2007; Mirosa & Harris 2011). HRW has also been successfully used to challenge the state and to regain access to traditional lands and livelihoods even in countries that do not have constitutional guarantees of the HRW, as in the case of the San and Bakgalagadi peoples of Botswana (Morinville & Rodina 2013).

A number of critiques against the HRW have also been put forward. Among the biggest controversy around the HRW leading up to the 2010 UN resolutions was the question of private sector participation in the water sector. Resistance against water privatization did indeed fuel the global social activism in support of the human right to water (Barlow 2007). And yet, as others have argued, HRW is still compatible with the notion of private property rights and does not preclude privatization (Bakker 2012). Another critique of the way HRW has been interpreted and implemented is the unsatisfactory progress on improving sanitation, especially in lesser-developed contexts. The right to sanitation is implicitly part of the right to water; however, as we have seen in the progress on the Millennium Development Goals, access to improved sources of drinking water has significantly exceeded progress on sanitation (WHO/UNICEF 2015). According to some estimates, 1.8 million people around the world die every year from diarrhoeal diseases, the majority of whom are children under the age of 5. Further, close to 1.1 billion people globally have no access to any form of toilet (Feris 2014).

The focus of HRW on the rights of individuals, rather than communities, is also highly problematic in some contexts. Schmidt and Mitchell, for instance, take issue with the fact that the Universal Declaration on Human Rights, which informs the legal framing of the HRW, has no mention of collective rights, and the notion of rights is very much consistent with the liberal tradition that prioritized the individual above the community (Schmidt & Mitchell 2014). In a similar vein, Parmar (2008) argued that the HRW can be particularly hostile to the needs and goals of historically marginalized communities, including indigenous populations, particularly given the focus on individual rights. However, because of the nature of the resource and its fundamental role in supporting life, the right to water cannot be separated from the water commons (Perera 2015; Schmidt & Mitchell 2014). At the core of this debate is the idea that water is a collective resource (Schmidt & Mitchell 2014), often entrusted in the hands of government to manage the common good.

Further, Mehta (2006) critiques the focus of HRW on water for domestic needs, rather than livelihood needs – a narrow approach that can leave marginalized communities with substandard access. The focus on water for individual domestic needs, albeit insufficient, as noted in the General Comment 15 (see above), lends itself to rather narrow technical implementation approaches, which tend to focus only on infrastructure extension and basic service provision. While basic infrastructure is indeed necessary for the realization of the HRW, it does not allow much room for social mediation as it typically does not involve community consultation or other social considerations. Further, this focus on individual water access in isolation from community capacity-building is likely unsustainable because it does not enable communities to manage water issues in the long run.

Water is variable across time and space and its state and availability depend on biophysical conditions, technologies (taps, pipes, dams), institutions and different allocation regimes, shaped by norms, politics, discourses and everyday realities of water access. The way HRW is actualized depends not only on the way water resources are accessed, distributed and governed, but also on the ways water is valued and understood ontologically, culturally and spiritually. Articulating and implementing a human right to water should be more attentive to material conditions of water access and water use, as well as water's meanings and value in people's daily lives. The HRW emerged from decades of debates and activism around the just, fair and equitable access to water of good quality to support life and livelihoods. It is thus an important theme within global water ethics, being at the intersection of debates on fairness and justice for all humans and moral considerations of the environment. The normative dimensions of the human right to water, including established principles and contested norms, effectively bring together environmental ethics with the social principles of justice and fairness. This is partly why some have argued that HRW has a strong, but largely unmet, potential to serve environmental justice agendas (Harris et al. 2015).

A number of scholars have already brought attention to the ethical dimensions of water governance, arguing that the special role water plays in creating and sustaining life on the planet has particular philosophical and moral implications (Armstrong 2009; Schmidt & Peppard 2014). Over the last three decades, a growing body of work on water ethics (Groenfeldt 2012; Groenfeldt & Schmidt 2013; Brown & Schmidt 2010) has provided insights into how to incorporate ethical values into decision-making and water policy, specifically in contexts of conflict resolution, incorporating diverse cultural and spiritual values and uses of water, and moving away from technocratic approaches to water management (Groenfeldt 2013). Others have pointed out the necessity to engage in a pragmatic reflection of actual outcomes of water policies and management practices, in order to evaluate the philosophical and ethical principles underlying water governance (Kowarsch in this volume). Bringing in insights from other disciplinary and interdisciplinary fields can thus inform water ethics, both empirically and conceptually.

Speaking to this growing body of work, this chapter argues that debating the merits and the ethical foundations of a right to water as defined by the global water governance discourse will not achieve much in terms of actualizing the ethical claim of water justice for all. It is equally important to critically engage with the on-the-ground realities in diverse socio-political contexts where this right is continuously negotiated and contested in order to evaluate how successful HRW actually is (or can be). As we will see in the case of Khayelitsha, the realization of the human right to water has been effectively folded into service delivery agendas, bound by the economic and technocratic limitations of infrastructure and service delivery models. The narrow focus on basic needs and service provision limits the transformational and justice potential of the HRW by sidelining the realities of social inequality and ongoing social struggles.

Methodology

This chapter is based on a Master's research project conducted in 2012 and 2013 in one of the largest townships in Cape Town – Khayelitsha, located on the sandy Cape Flats to the southeast of the Cape Town city centre. Khayelitsha carries historical significance as one of the key sites of resistance against the apartheid state. Over the past two decades it has grown significantly, although the current population size is debatable, with official statistics stating roughly 400,000, and researchers and NGOs citing numbers between 500,000 and 1,000,000 (Thompson & Nleya 2008). With the help of a research team consisting of community members from the study site and university students, we conducted 39 interviews and four focus groups in Site C, a subsection of Khayelitsha. This area was selected through connections with a local NGO – Iliso Care Society – and researchers at the University of Western Cape.[4] Due to security concerns related to high levels of criminal activity and increased incidence of social protests, field visits were scheduled only in the mornings during the working week. One downside of this approach was the possible exclusion of members of the community with jobs during the regular work hours of the week and residents who commute to the city either for work or in search of employment opportunities.

The final sample of interview participants was almost evenly split between men and women, with roughly half residing in shacks and half in government subsidized brick or cement homes (see more details in Rodina 2016). The interviews and focus groups consisted of conversations around the diverse aspects of water services – conditions, experiences with access, sensibilities around payment for water services, etc. The qualitative data were transcribed and later analysed for content, key themes and similarities and differences between different respondent groups (men, women, home residents, shack dwellers) using NVIVO qualitative software. Two journal articles have been produced from this Master's work: Rodina (2016) and Rodina and Harris (2016). This book chapter summarizes some of the key findings of this work, framed in relation to debates in global water ethics. Specifically, the goal is to inform the growing water ethics scholarships through a specific case of the uneven and highly contested roll-out of water services in Cape Town, South Africa.

Site C, Khayelitsha in Cape Town, South Africa

South Africa is often highlighted as an example of progressive realization of the human right to water for its constitutional guarantee of the right of all citizens to access water. South Africa has national guidelines that pose limitations on discontinuation of services for non-payment (Water Services Act no. 108 of 1997) and a policy that sets minimum amounts of water for basic needs free of charge (Free Basic Water policy, hereafter FBW). The FBW policy was implemented in 2001 to ensure that the constitutional right to water is realized for everyone in South Africa, regardless of ability to pay. At present, this policy mandates municipalities to provide at least 25 litres per person per day of free water for basic needs within 200 metres from home, with the expectation to provide larger amounts when possible, as 25 l/p/d is below the international standards of 50 l/p/d set by the WHO (Howard & Bartram 2003).

Equitable, fair and universal access to water in South Africa remains politically contested as levels and quality of drinking water and sanitation services continue to vary significantly across different communities throughout urban and rural areas of the country (Dugard 2013; Mirosa & Harris 2011). For instance, the roll-out of pre-paid meters from the early 2000s onwards has led to a number of public health concerns, including cholera outbreaks (Mirosa & Harris 2011), fire damage (Dugard 2009), as well as high levels of indebtedness (Smith & Hanson 2003). The technologies and economic mechanisms put in place to actualize HRW in South Africa have in fact resulted in a disproportionate burden on already marginalized populations (Bond 2012; Loftus 2006; Schnitzler 2008; Smith & Hanson 2003). This is particularly problematic in the context of the post-apartheid state's political commitment to redressing deeply rooted historical inequalities through improvements of quality of life for impoverished residents of former apartheid townships. South Africa's civil society has been at the forefront of social movements and struggles, both through the court system (Dugard 2009) and through citizen protests (Dugard & Tissington 2013; Nleya 2011), to directly challenge and redefine aspects of the way this right has been implemented (i.e., challenge the pre-paid meters and the practice of cut-offs after the basic amount has been used up; see more in Dugard 2009; Loftus 2006; Schnitzler 2008).

The City of Cape Town stands out among other municipalities in South Africa as it has been relatively successful in extending basic access to improved sources of water to the many impoverished areas, including informal and semi-informal black and coloured townships, thus complying with the Free Basic Water (FBW) policy and the constitutional guarantee of access to sufficient water. The population of the City of Cape Town is just over 3.5 million (CCT 2012b) with over 20% of all households living in informal settlements (CCT 2012a). According to some estimates, Cape Town has more than 220 informal settlements throughout the city, with an estimated total population of 900,000 people (Mels et al. 2009). From the perspective of the municipality, service provision is a not an easy task, as the city faces a series of challenges, including aging or failing infrastructure, and concerns related to sea level rise, floods and droughts. For instance, townships such as Khayelitsha, already face serious flooding and household damage during the heavy winter storms (Ziervogel et al. 2014). These challenges are particularly acute due to persistently high levels of social and economic inequality (McDonald 2007) and widespread problems with basic service delivery (Dugard 2013).

In 2011, the national Census reported that an estimated 96.6% of households in Cape Town had access to piped water within 200 meters from home while

87% of households had access to piped water inside their dwelling or yard (CCT 2012a). These numbers can be misleading, however, as they vary significantly within semi-informal and squatter settlements in townships such as Khayelitsha, where only 62% of households have access to piped water in their dwelling or inside their yard (CCT 2013), while the rest use less adequate communal services. Khayelitsha has seen a considerable increase in the supply of communal taps since 2001 in compliance with the Free Basic Water policy. In addition, as part of national efforts to supply housing and eradicate slums (Huchzermeyer 2003), Khayelitsha is undergoing a formalization process whereby shacks are gradually being replaced with government-subsidized homes, or RDP homes, named after the Reconstruction and Development Program of 1994 (Tissington, 2010).

In addition to built homes, the process of formalization includes provision of in-house water taps as well as in-house sanitation facilities. The in-house services come with installation of water meters, or water demand management devices, often installed without the consent of the homeowner and resulting in high water bills due to leaks in the pipe system (Pereira 2009; Wilson 2011). At the time of this research, in Site C the meters were not yet being read and households were not be billed for water use until the formalization process was to be complete. However, the installation of water meters and debt due to high water bills have been key issues in other townships in Cape Town (Wilson 2011). In anticipation of payment of services, the residents of Site C had already begun adjusting water consumption accordingly to avoid the possibility of high water bills (Rodina 2016).

In Site C, those living in shacks generally use communal taps, commonly located on the sides of streets, and communal toilets, typically located in designated areas in the community (Rodina 2016). Interviews with community members focused on access to drinking water revealed that communal water taps are often 'filthy, unsafe, and unacceptable', posing health risks as well as exposure to already high levels of criminal activity (Rodina 2016).[5] With respect to drinking water taps, logics of water use also change with the shift from communal to in-house taps. Specifically, those using in-house taps feel pressured to use water conservatively to avoid accumulating high water bills, and are therefore often unwilling to share water with shack dwellers (see also Tapela 2012). This tendency signals a shift away from customary communal uses of water and also has the potential to deprive them of alternative forms of water access when communal taps are unavailable or out of commission (Rodina 2016). For instance, when asked how difficult it is to ask for water from people with in-house connections, many residents would admit to being uncomfortable, or outright denied water. One interview participant shared having had conflicts in the past when she tried to use water from a neighbour's yard tap (interviews with community member 2012). As many people with in-house taps have water meters, they are worried about giving water away to shack dwellers as it costs money. They are worried about getting high water bills later on (see full discussion in Rodina 2016).

Further, water services in Khayelitsha mean much more in people's daily lives than mere access to water. This transition from communal to in-house 'upgraded' taps also carries particular notions of citizenship, and even a right to live in these communities. Receiving in-house services also comes with responsibilities of service consumers, such as payment for water, and water conservation to avoid high bills. The formalization process in this sense validates the idea of citizenship for people who are typically seen by the state as 'squatters' – temporary or illegal residents. The state is also seen by residents of Site C as imposing certain regulations and restrictions on water use through installation of water meters, and the expectation for future payment for services (see Rodina & Harris 2016). In other words, the formalization of housing and service provision in part serves to 'formalize' particular visions of citizenship with greater definition of rights and responsibilities by both the state and citizens. In contrast, as Rodina and Harris (2016) argue, shack dwellers are excluded from this form of citizenship, and in turn reject and delegitimize the state as not living up to its promises to provide better services. In this sense, while the HRW is typically interpreted simply as access to water infrastructure, water provision is a much larger societal project that goes beyond access to basic services. This is particularly the case in Cape Town which has seen a growing social mobilization around demands for improved services in the townships (Alexander 2010; Dugard & Tissington 2013).

To summarize, the reality of the implementation of the human right to water in South Africa and Cape Town specifically, albeit successful in some ways, has been uneven in that it has contributed to inequalities between shack dwellers and residents of formalized housing both in terms of access to improved forms of water services and in terms of sense of belonging in the democratic process (Rodina & Harris 2016). Users of both types of services still deal with a range of challenges that question the ethics of service delivery. Specifically, users of communal taps are exposed to health risks due to the unclean state of the communal taps, and safety risks due to high levels of crime in the community. At the same time, RDP residents are faced with affordability concerns, as many of them cannot afford to pay for water beyond the free allocation. In addition, while most residents do indeed have basic access to drinking water, communal sanitation services in Site C lag behind.

While HRW is framed as access to 'safe and sufficient water for personal and domestic uses', this framing essentially omits a discussion on inequality in terms of the broader benefits HRW can bring. The evidence presented here shows that the narrow and technical discourses around HRW that focus on service coverage alone do not address the inequalities in access to sufficient, safe and affordable amounts of water for different types of water services. This case also demonstrates multiple tensions around the different material dimensions of water access, such as cleanliness, safety and location of water access points, that would remain invisible if we only consider service coverage statistics and water policies in place (Rodina 2016).

Discussion

The human right to water in this context is materially and symbolically realized through service delivery, and its potential is limited by the politics of the post-apartheid service delivery crisis (Alexander 2010; Dugard 2013; Thompson & Nleya 2010). Water and sanitation services are brought to the urban townships by the state through the housing formalization process, which creates two tiers of service consumers – those with private in-house connections and those using communal services. Communal services, as reported by local residents, are often inadequate and pose health and security concerns for many residents, which questions the safety and accessibility of these services. Users of in-house water services, on the other hand, will be charged for water use beyond their free basic minimum once all the houses in Site C have been provided with water meters, according to interviews with city officials and community members. Due to high levels of unemployment in Khayelitsha, estimated at 38% in 2011 (CCT 2013), this effectively limits water access only to the basic minimum, which is not enough for a series of livelihood options, such as running small businesses, growing food, etc. In addition, these communities are likely to become more vulnerable due to projected climate change impacts of annual flooding and increased storm surges (Ziervogel et al. 2010), which are likely to cause service disruptions. Stormwater drainage – including sewerage and grey water as well as stormwater – are already among the highest recurrent risks in these areas, posing serious health and environmental risks in already degraded lands (Jiusto & Kenney 2015). All of these challenges point to the multiple limitations of the basic needs approaches to actualizing the universal human right to water.

In the context of Cape Town and Site C specifically, the state is the sole provider of water services. Water is delivered through centralized infrastructure with a strong focus on engendering responsible individual water consumers. While the municipality of Cape Town has indeed been successful in extending access to basic services, the lack of focus on building community capacities around safe water access points, or around other key concerns in these neighbourhoods, is proving problematic. Achieving water justice for residents of Site C and similar areas will indeed require a broader agenda that looks beyond basic service provision and employs a variety of alternative solutions to both water supply and water demand management. These could involve more capacity-building around enabling residents of Site C to access enough water to pursue a variety of livelihoods, to create safe access to water, to enable growing food safely without being penalized with high water bills, among other possible solutions.

Overall, HRW has indeed contributed to enabling access to basic water services for a large number of the formerly underserved populations, mostly in developing countries (WHO/UNICEF 2015). However, there is still a need to (re)define how the right to water is framed, interpreted and implemented in different contexts (Kowarsch in this volume) to enable it to achieve a broader social (and environmental) justice agenda (see Harris et al. 2015). This process of redefining is actively ongoing through court cases, new legal interpretations of the right and through political struggles for better and more secure access to water (e.g. Harris & Roa-Garcia 2013). For example, consider the debate around the need to articulate the right to sanitation as separate from the right to drinking water (Ellis & Feris 2014). Thus far, water and sanitation have been conceptualized together because of the many ways in which they are indeed linked. The downside of this 'coupling' is that sanitation remains 'in the shadow' of drinking water – or what some have called an 'institutional orphan' because of the consistent lack of interest and commitment on the part of international development agencies (Ellis & Feris 2014). In 2015, the UNGA adopted resolution 70/169 that distinguished between the two rights, albeit still keeping them under the umbrella of a HRW. This is a recent example of the continued re-defining of the right, even as many other challenges remain to be addressed.

While the broad definition of the right to water as 'necessary for the enjoyment of all other rights and a more dignified life' holds much transformational and aspirational meaning, in practice the implementation of HRW tends to strip these aspects by reducing HRW to access to water for basic needs. While there are many practical reasons for why that is the case (limited resources, infrastructure challenges), HRW will not bring the promise of water justice for all if it does not open up opportunities to consider broader water justice goals, such as enabling diverse livelihoods and participation in determining how this right should be realized (see Clark 2012). The solution to such a challenge is not easy, particularly in light of concerns with the sustainable use of water resources in largely overburdened watersheds throughout the globe. Simply increasing the amount of free basic water or increasing overall water supply to feed growing demands may not be possible or even desirable from a water sustainability point of view. However, alternative solutions and innovations in water use, such as grey water recycling and others, may provide hope for achieving a more ambitious water justice agenda. As such, HRW should adopt a broader global water ethic of respect of both humans – and the environment (see Barlow 2007) – rooted in a more holistic and nuanced understanding of the social relations around water.

Water is needed for diverse human and ecological functions. It is therefore conceptually and philosophically positioned at the intersection of social and environmental values – adding to the already complex trade-offs in implementing the HRW. Schmidt and Shrubsole (2013) illustrate the ethical tensions of the Western 'modernization' of water premised on nature/society dualism, which tends to reduce the meaning of water to a 'resource'. The authors specifically take issue with the normative philosophy behind 'water as a resource' for enabling a technocratic management paradigm, which has in fact contributed to contemporary water problems (Schmidt & Shrubsole 2013). This problematic framing is also evident in the case of Khayelitsha. Overly technical interpretations of HRW risk failing the moral imperative to ensure all humans have water not only to survive, but also to thrive and enjoy a dignified life. Further, ignoring both the aspirational potential of the HRW as well as the need to protect ecosystem health and integrity risks recreating inter- and intra-generational justice issues and inequality (Barlow 2015). The continuous ability to uphold the HRW requires more critical consideration of ongoing inequality and social struggles in different contexts. As suggested by Harris et al. (2015), there are yet other ways to push forward, and continually reimagine, what 'HRW' might mean, including going beyond anthropocentric biases and potentially including environmental rights of water. The HRW is thus continuously contested and reimagined, in light of a suite of complex political, cultural, social, hydrological and climate factors that come into play.

Conclusions

Recognizing access to safe and sufficient water as a human right marks a milestone in addressing global inequalities in water access. It has been an important development that puts pressure on governments to ensure access to sufficient water for all. This chapter argues that despite significant achievements in both international and national policy discourses around water provision, more critical investigation of the policy and practical implementation of the right to water is still needed. Through the case study from Khayelitsha, South Africa, this chapter has tried to unpack the ethical implications of the uneven implementation of the human right to water by focusing on actual water access in an impoverished urban area in Cape Town. The evidence suggests that predominantly technical understandings of the right to water in terms of basic needs sideline important justice and ethical concerns for both current and future generations. While the foundation of the HRW is a broad justice agenda for universal human access to water to enjoy a dignified life, its legal interpretation and implementation tend to focus on a much narrower set of goals: access to water for basic needs. While there is no easily implementable alternative framing of the HRW, this chapter highlights the need for critical engagement with on-the-ground realities of unequal access and the processes that produce these inequalities in order to fully achieve the justice potential of the right to water. A deeper engagement with the different material and socio-political processes that shape inequality can help open up spaces for alternative and more innovative approaches to ensuring water justice for all in ways that do not excessively burden already stressed watersheds. This will require innovation in both supply-side and demand-side solutions that enable the possibility of millions of impoverished people to enjoy safe and clean water for improved and more dignified livelihoods.

Acknowledgements

This research would not have been possible without the support of Iliso Care Society in Site C, Khayelitsha, the Environmental Monitoring Group in Cape Town, Dr Jacqueline Goldin and her research group, Anthropology of Water, as well as the African Centre for Citizenship and Democracy at the University of the Western Cape. I am also very grateful for the support, insights and feedback from Dr Leila Harris, the EDGES research collaborative and the Water Ethics Network.

Notes

1 Unimproved sanitation facilities do not ensure hygienic separation of human excreta from human contact. They include pit latrines without a slab or platform, hanging

latrines and bucket latrines. Unimproved drinking-water sources include unprotected dug wells, unprotected springs, carts with small tank/drum, surface water and bottled water (WHO/UNICEF 2015).

2 This research focused primarily on drinking water and not sanitation. However, in the research process it became immediately obvious that sanitation services in the townships of South Africa are lagging behind provision of drinking water – a problem that is often cited as one of the key reasons for widespread service delivery protests (Dugard 2013).

3 South Africa has a constitutional guarantee of universal access to water.

4 A 250-household survey of residents in Khayelitsha and Philippi served to provide additional background (Harris et al. 2012).

5 The same concerns apply to toilets as well, as they are often poorly maintained, and can be unsafe for similar reasons.

References

Armstrong, A 2009, 'Viewpoint - further ideas towards a water ethic', *Water Alternatives*, vol. 2, no. 1, pp. 138–147.

Alexander, P 2010, 'Rebellion of the poor: South Africa's service delivery protests – a preliminary analysis', *Review of African Political Economy*, vol. 37, no. 123, pp. 25–40.

Bakker, K 2012, 'Commons versus commodities: debating the human right to water', in F Sultana & A Loftus (eds), *The Human Right to Water: Politics, Governance and Social Struggles*, Earthscan, New York.

Barlow, M 2007, *Blue Covenant: The Global Water Crisis and the Coming Battle for the Right to Water*, The New Press, New York.

Barlow, M 2015, *Our Right to Water: A People's Guide to Implementing the United Nations' Recognition of the Right to Water and Sanitation*, The Council of Canadians, Ottawa, Canada.

Bond, P, 2012, 'The right to the city and the eco-social commoning of water: discursive and political lessons from South Africa', in F Sultana & A Loftus (eds), *The Human Right to Water: Politics, Governance and Social Struggles*, Earthscan, New York.

Brown, P & Schmidt, J 2010, *Water Ethics: Foundational Readings for Students and Professionals*, Island Press, Washington, DC.

CCT 2012a, *Cape Town Overview – 2011 Census (December 2012)*, City of Cape Town, South Africa.

CCT 2012b, *Statistics for the City of Cape Town – 2012*, City of Cape Town, South Africa.

CCT 2013, *City of Cape Town – 2011 Census Suburb Khayelitsha*, City of Cape Town, South Africa.

Clark, C 2012, 'The centrality of community participation to the realization of the right to water: the illustrative case of South Africa', in F Sultana & A Loftus (eds), *The Human Right to Water: Politics, Governance and Social Struggles*, Earthscan, New York.

Dugard, J 2009, 'Rights, regulation and resistance: the Phiri water campaign', *South African Journal on Human Rights*, vol. 24, no. 3, pp. 593–611.

Dugard, J 2013, 'Urban basic services: rights, reality and resistance', in M Langford, B Cousins, J Dugard, & T Madlingozi (eds), *Symbols or Substance: The Role and Impact of Socio-economic Rights Strategies in South Africa*, Cambridge University Press, Cambridge.

Dugard, J & Tissington, K 2013, 'Civil society and protest in South Africa: a view from 2012', in *State of Civil Society 2013 – Creating an Enabling Environment*, pp. 264–77, CIVICUS World Alliance for Citizen Participation, accessed 5 July 2016, http://socs.civicus.org/wp-content/uploads/2013/04/2013StateofCivilSocietyReport_full.pdf.

Ellis, K & Feris, L 2014, 'The right to sanitation: time to delink from the right to water', *Human Rights Quarterly*, vol. 36, no. 3, pp. 607–29.

Feris, L 2014, 'The human right to sanitation: a critique on the absence of environmental considerations', *Review of European, Comparative & International Environmental Law*, vol. 24, no. 1, pp. 16–26.

Gleick, P 1996, 'Basic water requirements for human activities: meeting basic needs', *Water International*, vol. 21, pp. 83–93.

Groenfeldt, D 2012, 'The Practical Importance of Water Ethics', Water-Culture Institute, Santa Fe, NM, accessed 30 May 2016, www.waterculture.org/uploads/WEC_Background_Paper_-_16Aug2013.pdf

Groenfeldt, D 2013, *Water Ethics: A Values Approach to Solving the Water Crisis*, Earthscan, New York.

Groenfeldt, D & Schmidt, JJ 2013, 'Ethics and water governance', *Ecology and Society*, vol. 18, art. 14, accessed 6 July 2016, http://doi.org/10.5751/ES-04629-180114.

Harris, LM & Roa-Garcia, MC 2013, 'Recent waves of water governance: constitutional reform and resistance to neoliberalization in Latin America (1990–2012)', *Geoforum*, vol. 50, pp. 20–30.

Harris, LM, Goldin, JA, Darwah, A & UBC EDGES Research Collaborative 2012, 'Water access and participatory governance survey of households in informal settlements in Accra, Ghana and Cape Town, South Africa.'

Harris, LM, Rodina, L & Morinville, C 2015, 'Revisiting the human right to water from an environmental justice lens', *Politics, Groups, and Identities*, vol. 3, no. 4, pp. 660–5.

Howard, G & Bartram, J 2003, *Domestic Water Quantity, Service Level and Health*, World Health Organization, Geneva.

Huchzermeyer, M 2003, 'Housing rights in South Africa: invasions, evictions, the media, and the courts in the cases of Grootboom, Alexandra, and Bredell', *Urban Forum*, vol. 14, no. 1, pp. 80–107.

Jiusto, S & Kenney, M 2015, 'Hard rain gonna fall: strategies for sustainable urban drainage in informal settlements', *Urban Water Journal*, vol. 13, no. 3, pp. 253–69.

Loftus, A 2006, 'Reification and the dictatorship of the water meter', *Antipode*, vol. 38, no. 5, pp. 1023–45.

McDonald, DA 2007, 'World class inequality: Cape Town, South Africa', *Queen's Quarterly*, vol. 114, no. 3, pp. 395–405.

Mehta, L 2006, 'Do human rights make a difference to poor and vulnerable people? Accountability for the right to water in South Africa', in PJ Newell & J Wheeler (eds), *Rights, Resources and the Politics of Accountability*, Zed Books, London.

Mehta, L 2014, 'Water and human development', *World Development*, vol. 59, pp. 59–69, http://doi.org/10.1016/j.worlddev.2013.12.018.

Mels, A, Castellano, D, Braadbaart, O, Veenstra, S, Dijkstra, I, Meulman, B et al. 2009, 'Sanitation services for the informal settlements of Cape Town, South Africa', *Desalination*, vol. 248, pp. 330–7.

Mirosa, O, & Harris, LM 2011, 'Human right to water: contemporary challenges and contours of a global debate', *Antipode*, pp. 1–21. http://doi.org/10.1111/j.1467-8330.2011.00929.x.

Morinville, C & Rodina, L 2013, 'Rethinking the human right to water: water access and dispossession in Botswana's Central Kalahari Game Reserve', *Geoforum*, vol. 49, pp. 150–9, http://doi.org/10.1016/j.geoforum.2013.06.012.

Nleya, N 2011, 'Linking service delivery and protest in South Africa: an exploration of evidence from Khayelitsha', *Africanus*, vol. 41, no. 1, pp. 3–14.

Obani, P & Gupta, J 2014, 'The evolution of the right to water and sanitation: differentiating the implications', *Review of European, Comparative & International Environmental Law*, vol. 24, no. 1, pp. 27–39, http://doi.org/10.1111/reel.12095.

Parmar, P 2008, 'Revisiting the human right to water', *The Australian Feminist Law Journal*, vol. 28, pp. 77–96.

Pereira, T 2009, 'Household impacts of water management devices', presented at the EMG Water and Climate Change Seminar, 26 November.

Perera, V 2015, 'Engaged universals and community economies: the (human) right to water in Colombia', *Antipode*, vol. 47, no. 1, pp. 197–215, http://doi.org/10.1111/anti.12097.

Rodina, L 2016, 'Human right to water in Khayelitsha, South Africa – lessons from a "lived experiences" perspective', *Geoforum*, vol. 72, pp. 58–66.

Rodina, L & Harris, LM 2016, 'Water services, lived citizenship, and notions of the state in marginalised urban spaces: the case of Khayelitsha, South Africa', *Water Alternatives*, vol. 9, no. 2, pp. 336–55.

Schmidt, JJ & Mitchell, KR 2014, 'Property and the right to water: toward a non-liberal commons', *Review of Radical Political Economics*, vol. 46, no. 1, pp. 54–69.

Schmidt, JJ & Peppard, CZ 2014, 'Water ethics on a human-dominated planet: rationality, context and values in global governance', *Wiley Interdisciplinary Reviews: Water*, vol. 1, no. 6, pp. 533–47, http://doi.org/10.1002/wat2.1043.

Schmidt, JJ & Shrubsole, D 2013, 'Modern water ethics: implications for shared governance', *Environmental Values*, vol. 22, pp. 359–79.

Schnitzler, von, A 2008, 'Citizenship prepaid: water, calculability, and techno-politics in South Africa', *Journal of Southern African Studies*, vol. 34, no. 4, pp. 899–917, http://doi.org/10.1080/03057070802456821.

Smith, L & Hanson, S 2003, 'Access to water for the urban poor in Cape Town: where equity meets cost recovery', *Urban Studies*, vol. 40, no. 8, pp. 1517–48, http://doi.org/10.1080/0042098032000094414.

Sultana, F & Loftus, A 2012, 'The right to water: prospects and possibilities', in F Sultana & A Loftus (eds), *The Right to Water: Politics, Governance and Social Struggles*, Earthscan, New York.

Sultana, F & Loftus, A 2015, 'The human right to water: critiques and condition of possibility', *Wiley Interdisciplinary Reviews: Water*, vol. 2, no. 2, pp. 97–105, http://doi.org/10.1002/wat2.1067.

Tapela, BN 2012, 'Social water scarcity and water use' (No. 1940/1/11), Water Research Commission, Cape Town, South Africa, 175 pp.

Thompson, L & Nleya, N 2008, 'Citizen reportback in Khayelitsha, Cape Town', African Centre for Citizenship and Democracy, University of the Western Cape, Cape Town, South Africa.

Thompson, L & Nleya, N 2010, 'Community activism and protest in Khayelitsha, Cape Town', African Centre for Citizenship and Democracy Publication, University of the Western Cape, Working Paper, No. 4, 1–26.

Tissington, K 2010, 'A review of housing policy and development in South Africa since 1994', Socio-Economic Rights Institute of South Africa.

United Nations Committee on Economic, Social and Cultural Rights (UN CESR) 2003, *General Comment No 15: The Right to Water (Arts 11 and 12 of the Covenant)* E/C. 12/2002/11.

Wilson, J (ed.) 2011, 'Water and climate change: an exploration for the concerned and curious', Environmental Monitoring Group, Cape Town, South Africa.

WHO/UNICEF 2015, *Progress on Sanitation and Drinking Water: 2015 Update and MDG Assessment*, WHO/UNICEF.

Ziervogel, G, Shale, M & Du, M 2010, 'Climate change adaptation in a developing country context: the case of urban water supply in Cape Town', *Climate and Development*, vol. 2, no. 2, pp. 94–110, http://doi.org/10.3763/cdev.2010.0036.

Ziervogel, G, Waddell, J, Smit, W & Taylor, A 2014, 'Flooding in Cape Town's informal settlements: barriers to collaborative urban risk governance', *South African Geographical Journal*, 1–20, http://doi.org/10.1080/03736245.2014.924867.

11 An ecocentric water allocation across competing demands in an arid inland river basin of northwest China

Jie Liu and Xiang Huang

Introduction

Water resources allocation is set according to rules, whether explicit or implicit, which assign priority to particular users or give equal weight proportional to demand. Natural conditions, economic conditions and social factors are usually considered in the process of water allocation (www.waterculture.org). Water ethics, as a 'soft pathway' for water management, defines a set of moral values that guide human interactions with water resources and are reflected in designing the rules of allocating water. However, water ethics cannot be very helpful on a merely general and abstract level. This study from the Heihe River Basin (HRB) in northwest China explores the practical role of ethics in water allocation of this area.

The HRB is the second largest endorheic (closed) basin in the arid area of northwest China, with an area of approximately 130,000 km^2. It is also fairly representative of many other arid regions in East Asia with typical arid ecosystems (Cheng et al. 2014). The main channel of the Heihe River has a length of 821 km, originating from the Qilian Mountains in the south, flowing northward through the Hexi corridor of Gansu Province and entering into the Gobi Desert, which is in the western part of the Inner Mongolia Plateau (Figure 11.1). The Hexi corridor is considered the 'green ribbon' of northwestern China, and is one of the major grain producing regions of the country. Water originating in the mountains from snow and melting glaciers turns this region into oases, preventing desertification and sustaining agriculture. The struggle for water between desert and oasis has been going on for decades, but has recently accelerated due to climate change and ever increasing demands for water resources from irrigation development in the middle HRB, driven by drastic population growth (Zheng et al. 2009).

Over the past 2000 years the evolution of population, areas of cultivated and natural oasis, human water consumption and the stream flow into the downstream terminal lakes in the HRB have experienced dramatic changes (Lu et al. 2015). The rapid population increases especially after the foundation of new China in 1949 and the corresponding increase of cultivated oasis area have caused continuous increase of water consumption by human beings. The discharge in the lower reach of the Heihe River has decreased significantly and more than 30 tributaries as well as the terminal lakes have dried up. A marked degradation of the environment, secondary salinization and desertification of land in the entire HRB has raised great concerns about water management issues in the region, especially water allocation, since people's livelihoods in the HRB strongly depend on reliable access to the river. The HRB offers an ideal context to explore the ethical considerations of water resources allocation, because of its acute water shortage, competitive water demands and complicated natural and social conditions.

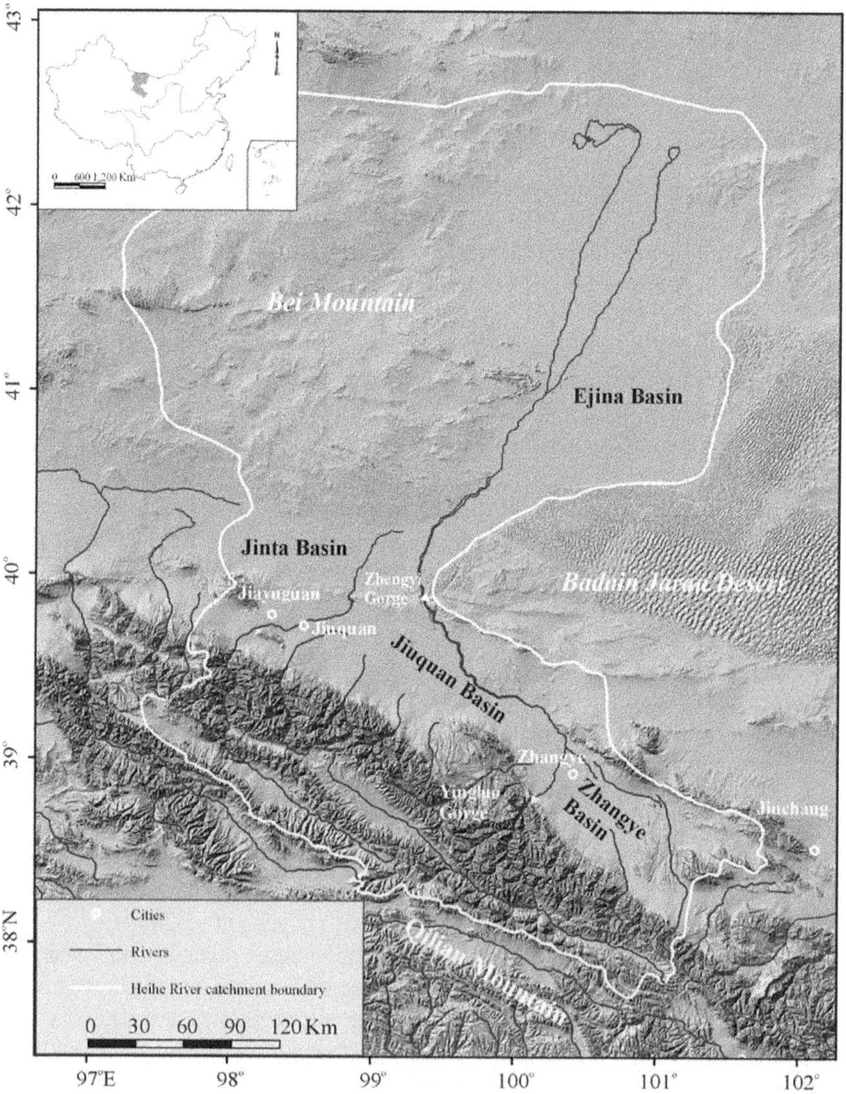

Figure 11.1 Location of the Heihe River Basin, China.

Source: based on Cao et al. (2016).

Major hydrological and water utilization characteristics of the Heihe River Basin

Interaction of surface water and groundwater

In the HRB, the exchange of surface water and groundwater is continuous and complex. The conjunctive use and management of surface water and groundwater resources is a key feature of water management of the basin. Figure 11.2 shows the hydrologic cycle in the HRB. Most water originates in the mountains from snow and glacial melt as well as permafrost thawing. The majority of groundwater discharges to streams in the mountainous upstream region. In the middle basin stream runoff recharges groundwater and sustains oases and agriculture. Groundwater also exchanges with river water and such exchanges vary temporally and spatially. In the desert downstream region, surface water gradually infiltrates downward and discharges to groundwater, which is consumed by evapotranspiration. This dynamic interaction between surface water and groundwater resources creates favourable conditions for people to utilize water resources, as surface water exploited by human activities may be replenished by groundwater and vice versa (Tang & Zhang 2001). But this also makes it difficult to accurately investigate the total available water resources, and creates challenges for ethical water allocation. The interaction mechanisms between surface water and groundwater need to be clearly characterized to avoid double counting of water resources and to provide accurate baseline data for water allocation decisions.

Water allocation among the upper, middle and lower Heihe River Basin

Agriculture in the middle HRB has for many decades consumed the majority of water, causing serious environmental deterioration in the lower basin. The Water Allocation Regulation, enacted in 1992 and then revised in 1997 and 2000, regulates the water allocation of the Heihe River based on the historical average annual runoff compared with the actual runoff of the present year. Water distribution is controlled through gauging stations between the upper and middle basins, and between the middle and lower basins. The upper-middle gauging station is located where the Heihe River flows out of the mountainous region. The flow at this point has increased since the 1940s, presumably due to warming temperatures leading to more snow, permafrost and glacial melting in the upper basin. Further downstream at the middle-lower gauging station, which is at the transition from middle reach agriculture areas to the downstream desert areas, there has been a significant decrease in runoff, indicating substantial increase in water consumption by the growing population plus increased urban, industrial and agricultural demands.

Water utilization of major water using sectors

A majority of available water is allocated to the middle HRB, where agriculture is the main water using sector. In 2008 (the latest year of available data), for

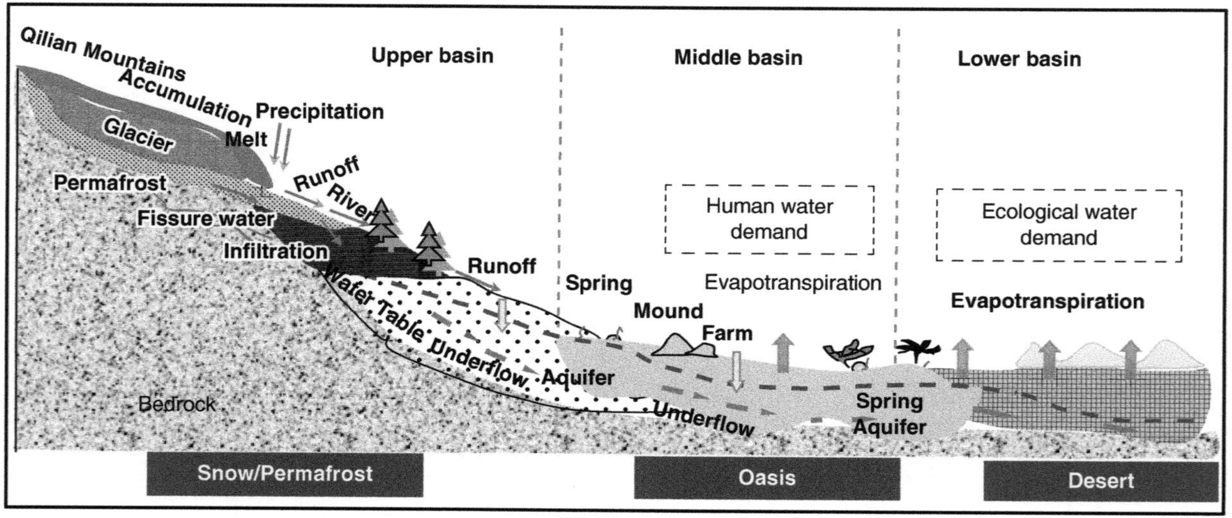

Figure 11.2 Interaction between surface water and groundwater in the Heihe River Basin.

Source: adapted from Kang et al. (2004).

example, 85% of water was used for agriculture in the middle basin. During the intensive irrigation period of May to June, competition for water among counties, townships and villages even resulted in illegal activities such as destruction of water infrastructures. In the region around Zhangye City, the major city in the middle basin, water-consuming crops like maize and wheat are still the major crop types. Allocating water for developing large-scale irrigated agriculture in such an arid region further exacerbates the water stress.

Water requirement of the environment

In addition to the three basic water use categories or 'sectors' (agriculture, urban/domestic and industrial water uses), the environment, which is not a 'using' sector, also has a requirement of water. Previous ignoring of this requirement in the HRB has caused deterioration to the riparian ecosystem and the wider environment. The annual drying-up of the lower reach of the Heihe has extended from 100 days in the 1950s to around 200 days in the 1990s, the west terminal lake dried up completely in 1961 and the east terminal lake dried up in 1992. Associated with these trends, the groundwater table has declined and the mineral content of the water rose considerably. In the lower basin, many springs and wetlands disappeared, and the area of desert plants, such as *Populus diversifolia* (HuYang) and *Elaeagnus angustifolia* (ShaZao), decreased by thousands of hectares annually. This has become the major driving factor for adopting ecocentric water allocation in the HRB. However, with the implementation of the Water Allocation Regulation, which deals only with surface water, the exploitation of groundwater in the middle basin increased from 312 million m^3 to 505 million m^3 during 2000 to 2011 (according to the first National Water Resources Survey for 2011). The net result of less diversion of surface water but more pumping of groundwater, is that the overall health of the middle-lower HRB ecosystem continues to decline.

Water ethics reflection in the Heihe River Basin

Evolution of water management in the Heihe River Basin

Before the foundation of new China in 1949, the water management system in the HRB was a hybrid of formal national regulations and informal village rules. As the population dramatically increased and more water disputes emerged during the Qing Dynasty (1616–1912), the government introduced a Water Averaging System ('JunShuiZhi' in Chinese) on the HRB, specifying the time and locations to irrigate croplands during the growing season. There were also local rules, documented in the form of inscriptions to indicate the water quota of each dam and the duties of managers for each county. The principle was to allocate water according to the cropped areas (Zhang & Zhao 2009).

After the foundation of new China, the HRB still followed the old Water Averaging System by allocating water twice a year. However, with the rapid growth of population and fast expansion of croplands, water resources in the HRB became over-exploited, which caused the terminal lakes to dry up. In 1997, the National Council approved a new Water Regulation for the Heihe Main Stream, and in 1999, the Heihe Water Administrative Bureau, under the supervision of the Yellow River Commission, Ministry of Water Resources, was formed to take responsibility of implementation. Then in 2010, a Water Permit System was introduced in the HRB. The Ministry of Water Resources takes charge of organizing, coordinating, supervising and instructing the water allocation of the entire Heihe Main Stream. The Yellow River Commission, through its Heihe River Administrative Bureau, takes charge of the implementation and inspection of water allocation. The local administrative departments in the three provinces (Gansu, Qinghai and Inner Mongolia) take charge of the implementation and investigation of water allocation within their jurisdictional limits.

The major driving factor leading to the water allocation regulations in the HRB is the continuous degradation of the lower basin ecosystem, reflected in the drying up of the terminal lakes, shrinking of natural oases and more frequent occurrence of sandstorms. Intensive water use for irrigation at the middle basin was blamed for this and therefore a series of ecocentric water allocation regulations have been implemented with the purpose to restore and protect the fragile ecosystem of the lower basin in particular. The new targets for restoring the lower basin ecosystem have been expanded from the basic aim of just bringing water back to the lake to more specifically define the allocation plans for the whole year and for specific climatic conditions (dry season and wet season).

With the intention to restore and protect the ecosystem of the lower HRB, the water allocation regulations have played an important role. Groundwater levels in the lower HRB have shown a slightly increasing trend and the eastern terminal lake has maintained a water surface area of 30 km^2, which effectively reduced the frequency and strength of sandstorms. However, new problems also emerged after implementing the water allocation plans. Groundwater levels in some areas of the middle basin started declining because of more groundwater pumping. On the other hand, more water released to the lower basin has led to expansion of watermelon planting which is counter-productive to the aim of ecosystem restoration. The optimal lake area that should be maintained has also become a point of controversy. A larger lake area means more evaporation loss while a smaller area will increase the frequency and intensity of sandstorms. Systematic consideration is needed of the total water flow and its dynamic variations, irrigation water demands, dynamic groundwater level changes and rational ratios of using surface water and groundwater.

Ethics underlying the current water allocation in the Heihe River Basin

Through the evolution of water management regulation, rules and practices in the HRB, ethical considerations are reflected although the term ‘water ethics’ is not mentioned explicitly. In the Regulation of Water Allocation of the Heihe Main Stream and the Rules for Implementation of Water Permit System in the HRB, the following ethical principles are evident:

- Water resources in the Heihe Main Stream are under unified management by the government and should be aligned with the overall planning of the basin and the water function zones.
- Domestic water use should have first priority, and the agricultural, industrial and eco-environmental water use should be reasonably secured.
- Total approved water permits should not exceed the total exploitable water and the withdrawn water may not be used for expanding irrigation areas.
- Surface water and groundwater should be conjunctively utilized and groundwater abstraction prohibited in groundwater overdraft or development-forbidden areas.
- Water allocation plans should be adjusted according to monthly and yearly changes, and corresponding penalties will be implemented if the agreed-upon water allocation scheme has not been met.

Through the evolution of water management and the related regulations and rules, it can be seen that the water allocation in the HRB is experiencing a transfer from an economic-centric water allocation to an ecocentric one. From a spatial perspective, the earlier pattern of the middle basin using disproportionately more water, has been replaced by a new pattern of unified water allocation among the upper, middle and lower basin. From the perspective of water using sectors, the former pattern of allocating water exclusively to the domestic, industrial and agricultural sectors has been changed, with a new priority on environmental and ecological water use under the current water allocation scheme. Expansion of irrigation areas is explicitly forbidden in the Regulation, and water permits will not be approved in cases that could threaten the environment, such as groundwater development in forbidden areas.

Missing pieces in the current water allocation of the Heihe River Basin

Although water allocation in the HRB is becoming more unified and ethical with regard to the spatial and sectoral perspectives, integrated and systematic water planning and distribution needs to be further strengthened. The ecocentric water allocation scheme in the HRB was initially motivated by ecosystem degradation in the lower basin. Drying up of the terminal lakes and desertification in the lower basin had drawn great attention from the government and the public. Ecological recovery in the lower basin was the major objective and driving factor of the current water management and allocation practices. However, without a systematic and integrated consideration of the economic development and environmental protection of the entire basin, it is impossible to make ethical decisions on water allocation.

In the HRB, although a certain amount of water is required to be released to the lower basin by the Regulation, the economic development strategy favouring agricultural activities in the middle basin has never been fundamentally changed. Therefore, a direct outcome after implementing the water allocation scheme after 2000 has been the rapid decline of groundwater storage in the middle basin.

Because the water allocation scheme applies only to the surface water in the Heihe Main Stream, the constraints on water withdrawals from the Heihe caused increased use of groundwater in the middle basin. The land area irrigated by wells has increased from less than 4000 hectares in the 1980s to nearly 19,000 hectares at the present. If both the irrigation demand and surface water allocation supplies remain unchanged, the decline of groundwater storage will continue (Wu et al. 2015). At the same time, the middle HRB exports virtual water to other regions through exporting agricultural products, which accounts for 25% of the total water use in Gansu Province and exacerbates water stress in the HRB (Cai et al. 2012; Zhao et al. 2015). Is it ethical to sacrifice the ecosystem of one region (the middle basin) for another (the lower basin), or to develop a local economy with the cost of environmental degradation and vice versa? Balancing economic development and environmental protection requires an overall consideration of economic ethics and environmental ethics.

The implementation of the Water Permit System and the Water Allocation Regulation in the HRB is proving more difficult than establishing the regulations in the first place. The de facto governance ethics in the HRB are resulting in the relentless deterioration of the HRB ecosystem. In Items 29 and 30 of the Water Allocation Regulation of the Heihe Main Stream, and Items 42, 43, 44 of the Rules for Implementation of Water Permit System, penalties are listed for behaviours that violate the regulation and rules. However, those items are too vague to be actually implemented in reality. If the planned water allocation is not able to meet demand in the present year, the deficit is simply accumulated without causing any change to the water withdrawal quota for the next year.

Informal regulations such as using cultural and moral constraints to help with the water allocation and utilization are also lacking. To some extent, this situation is a legacy of the past economic dominated development strategy. With the heavy price of environmental degradation and with the overall increase of economic living standards, people are realizing the importance of environment protection. Education and awareness play a critical role in the establishment of informal regulations, as well as in forming ethical frameworks of water allocation.

Ethical framework of water allocation

The value of water resources is determined both by its intrinsic attributes, and the way it is utilized by people (Llamas et al. 2009; Liu et al. 2011). Water resources allocation is actually the redistribution of the value of water resources, which will inevitably involve conflict of interests from various groups and stakeholders because of the characteristics of water as public goods (Hefny 2006; Liu et al. 2011). Based on the analysis of the ethical reflection of water allocation in the HRB case, an ethical framework of water allocation in a river basin is presented in Figure 11.3. The core purpose of the ethical framework is to evaluate the total water supply and water demand impartially and objectively. The prerequisite of rational water allocation is to accurately investigate the total amount of available water resource. This requires a full understanding of the hydrological cycle in the basin: where the water originates, how it flows, where it ends up, and how it is impacted by climate change and human activities. Hydrologists play an essential role at this stage by providing scientific and objective knowledge of the natural hydrological conditions. In addition, abstracting water from nature and conveying it to the end users will have impacts on the river ecosystem, and will relate to economic considerations. Whether the water supply project is environmentally friendly and economically efficient involves the concerns of both environmental and economic ethics.

From the water demand side, human demand is always given the first priority. Three basic water use categories or 'sectors' – agriculture, urban/domestic and industrial water uses – are usually defined as the human water demand. To evaluate the water demand of those sectors involves the evaluation of the social and economic development levels, for example, population, GDP, water-using efficiency and even living habit. Economic ethics, social and cultural ethics need to be considered, and social scientists play an important role at this stage. In addition to the three basic water use categories, the environment has requirements of water too. The important role that water plays in the ecosystem indicates the ecological

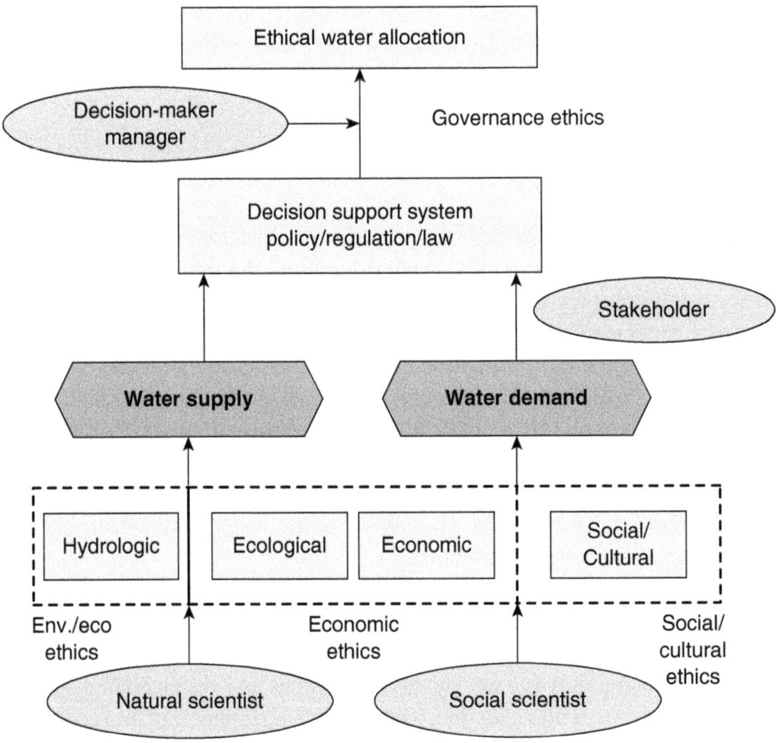

Figure 11.3 Ethical framework of water resources allocation.

Source: own.

value of water resources, which not only support the basic physiological consumption for living things, but also preserve the hydrological cycle and ecosystem stability. According to the basic environmental water requirements (Gleick 1998), a certain quality and quantity of water should be specially guaranteed for both natural ecosystems and highly managed ecosystems for the sake of maintaining ecological biodiversity and integrity. Stakeholders should be involved to ensure their water interests are fully represented.

With the water supply and water demand clearly and fairly defined, water can be allocated based on the water laws and regulations, which set the 'rules of the game'; without it, there is no fair play. Laws and regulations can greatly enhance the power of environmental/water ethics. When such laws and regulations are enforced, it results in better recognition of the value of water resources, and therefore is very effective at promoting better environmental ethics (Hennet 2007). Regulatory agencies must take the lead by adopting sound regulations and enforcing these regulations. This is an example of governance ethics.

Ethical water allocation involves a series of ethics including environmental ethics, economic ethics, social and cultural ethics, as well as governance ethics (Figure 11.3). It is an integrated system, which should involve natural scientists, social scientists, stakeholders, decision-makers and managers to ensure that all parties can work closely. With explicit attention to ethics in this integrated system, a better water allocation division is implied, which incorporates science, technology and social, economic considerations to provide enough and equal water for human beings and for nature.

Discussion and conclusions

As public goods, water resources have the features of externality and non-excludability. In the early stages of water development, the scale of development is relatively small compared to the carrying capacity of water resources; therefore, the use of water resources is non-competitive. However, with the continuous expansion of development, competition for water resources gradually appears (Yang et al. 2007). To manage the competing demands of various water using sectors, it is important to explicitly identify both the tangible and intangible values of water resources, which cannot be done only by engineering or economic criteria. Water ethics is an important element and its role in water resources management is gradually being noticed. There is no doubt that the water sectors will benefit from a more 'conscious' ethical reflection.

From the Heihe case, it is evident that the transfer from economic-centric to eco-centric water allocation has distinct effects on the ecosystem recovery in the lower basin. Even though the term 'water ethics' is not mentioned explicitly, ethical considerations are reflected in water management practices and water-related studies either unconsciously or with other motivations. An abundance of management measures and research projects have been launched which reflect a tacitly ethical approach. Integrated Heihe River Basin management has been proposed and is being implemented with the following objectives: (1) conjunctive development of surface water and groundwater resources; (2) coordination among domestic, industrial and ecological water uses; (3) overall water allocation planning for the upper, middle and the lower basins; (4) utilizing legal, administrative, economic, technical and educational means for comprehensive management. Since 2010, a new major research program of the National Natural Science Foundation of China, 'towards sustainable water resources management in arid ecosystems' has been investigating the ecology and hydrology of the HRB (Cheng et al. 2014). These and other efforts are indicative of ethical considerations regarding 'equity and sustainability' of water development and allocation in the HRB.

Knowledge is the basis for making ethical decisions and judgments. Natural scientists play an important role by providing objective information about the causes and effects of environmental processes. Long-term monitoring of water resources to explore regional water dynamics and total quantity of available water resources, development of basin-scale groundwater/surface water models to link with ecological and economic models, using isotope samples to date the water age and to quantify groundwater recharge rates are all examples of obtaining basic information that can support ethical water management.

To be effective, water ethics need to be incorporated into laws/regulations. 'Ethical guidelines that are agreeable and enforceable lead to laws and regulations' (Heuvelmans et al. 2010). For example, in the United States, various laws regulate the discharge of pollutants to water and establish goals for water protection. Enforcement of such regulations results in better recognition of the intrinsic value of natural resources, and therefore can be very effective at promoting better environmental ethics (Liu et al. 2010). The challenge facing environmental (water) ethics is finding consensus. Many factors contribute to the difficulties in reaching consensus on environmental ethics. Besides the lack of 'knowledge', the scope of environmental ethics is also a major factor. Some believe that environmental ethics should maximize the welfare of the human and therefore should be 'human-centred'; some extend the list of beneficiaries to 'all living things that can have interests'; others believe that this list should be even longer.

According to Leopold (Anderson 2007), the ethical community should be expanded from a human dimension to include all living, and possibly also inanimate, components of the environment. In addition, the external characteristics of the environment also affect the agreeability of environmental ethics. For example, an artificial canal was made from the Langxin Mountain (a water control station between the middle and lower HRB) to directly transport water to the east terminal lake for developing tourism. This has caused reduction in the water that is normally available to the middle basin and also increased the loss of water through evaporation over the open canal. An occasional visitor might think that it is ethical to save the attraction of the east terminal lake, whereas local residents would argue that satisfying their own water demands and protecting the local eco-environment is more ethical. Reaching agreement between the visitors and the local residents is difficult because the environmental cost is not being considered (Heuvelmans et al. 2010).

How to rationally determine the water use quota, how to ethically choose the principles of allocating water, and whether it is fair and ethical to trade water in the market, are all still largely unanswered. A lack of technical understanding and scientific knowledge are certainly important causes, but non-technical lack of awareness is also responsible for the current unsatisfactory situation in water management. Incorporating ethical considerations into water resources allocation is essential and indispensable. Moreover, the need for a more integrated approach to economically sound and environmentally sustainable management and allocation of water resources is evident. Changes in ethics of motivation, behaviour, morals and attitudes are a slow process; therefore, long-term campaigns to boost education and communication are needed to increase awareness and adoption of water ethics. In addition, subsidies, tax systems and other kinds of formal regulations should be considered as integral steps towards a more environment-friendly (and in this sense, more ethical) governance structure.

Finally, the subject of intergenerational equity is a fundamental ethical issue and closely linked to sustainability. As defined by the World Commission on Environment and Development (1987), sustainability is 'the use of resources to meet the needs of today's generation without inhibiting the ability of future generations to meet their needs'. To realize sustainable water development requires a balance of water requirements between the current generation and the future generation. Compared to contemporary people, future generations should enjoy an equal right to water resources. If current water development does not exceed the carrying capacity of water resources, then water requirements of future generations can be guaranteed. They will have the equal opportunity to access to water resources, and intergenerational equity is realized. In the HRB case, the current scope of technical measures (e.g. water saving measures and reservoir construction) and non-technical measures (e.g. demand management and integrated river basin management) focus on the water security of the present generation only. The issue of intergenerational equity has not been explicitly considered, because this would further increase ethical dilemmas about the already acute water scarcity and water allocating disputes. As population continues to grow, these dilemmas and disputes are likely to become more frequent and serious. New approaches to ethical water allocation between generations that incorporate principles of sustainability and equality are particularly urgent and important.

Acknowledgements

This study is funded by the National Natural Science Foundation of China (Grant nos 41271032 and 91225301). The authors are grateful to David Groenfeldt who provided insightful comments and led to significant improvement of the manuscript quality.

References

Anderson, M 2007, 'Ground water ethics', *Ground Water*, vol. 45, no. 4, p. 389.

Cai, ZH, Shen, LX, Liu, JG & Zhao X 2012, 'Applying input-output analysis method for calculation of water footprint and virtual water trade in Gansu Province', *Acta Ecologica Sinica*, vol. 32, no. 20, pp. 6481–8.

Cao, G, Zheng, CM & Craig, S 2016, 'Groundwater recharge and mixing in arid and semiarid regions: Heihe River Basin, northwest China', *Acta Geologica Sinica* (English edn), vol. 90, pp. 971–87.

Cheng, G, Li, X, Zhao, W, Xu, Zh, Feng, Q, Xiao, Sh & Xiao, H 2014, 'Integrated study of the water-ecosystem-economy in the Heihe River Basin', *National Science Review*, vol. 1, pp. 413–28, doi: 10.1093/nsr/nwu017.

Gleick, P 1998, 'Water in crisis: paths to sustainable water use', *Ecological Applications*, vol. 8, no. 3, pp. 571–8.

Hefny, M 2006, 'Actualizing water ethics in the regional context of the Arab world', *Management of Intentional and Accidental Water Pollution*, NATO Security through Science Series, pp. 277–310, doi: 10.1007/1-4020-4800-9_25.

Hennet, R 2007, 'Environmental ethics and the private sector consultant', presented at the 1st International Symposium of Water Ethics and Management, Beijing, 5 November.

Heuvelmans, G, Louwyck, A, Liu, J, Qi, Y & Zheng, C 2010, 'Comment on guest editorial: role of ethics in groundwater management', *Ground Water*, vol. 48, no. 4, pp. 480–1.

Kang, E, Li, X, Zhang, J & Hu, X 2004, 'Water resource relating to desertification in the Heixi area of Gansu Province, China', *Journal of Glaciology and Geocryology*, vol. 26, no. 6, pp. 657–67 (in Chinese).

Liu, J, Dorjderem, A, Lei, X, Liu, H, Macer, D, Qiao, Q, Sun, A, Tachiyama, K, Yu, L & Zheng, Y 2011, *Water Ethics and Water Resource Management*, UNESCO Ethics and Climate Change in Asia and the Pacific (ECCAP) Project Working Group 14 Report, UNESCO, Bangkok.

Liu, J, Zhang, M & Zheng, C 2010, 'Role of ethics in groundwater management', *Ground Water*, vol. 48, no. 1, p. 1.

Llamas, R, Cortina, L & Mukherji, A 2009, *Water Ethics*, CRC Press, Leiden, The Netherlands.

Lu, Z, Wei, Y, Xiao, H, Zou, S, Xie, J, Ren, J & Western, A, 2015, 'Evolution of the human–water relationships in the Heihe River basin in the past 2000 years', *Hydrology and Earth System Sciences*, vol. 19, pp. 2261–73, doi: 10.5194/hess-19-2261-2015.

Tang, Q & Zhang, J 2001, 'Water resource and eco-environment protection in the arid regions in northwestern China', *Progress in Geography*, vol. 20, no. 3, pp. 227–33 (in Chinese).

World Commission on Environment and Development 1987, *Our Common Future*, Oxford University Press, New York.

Wu, B, Zheng, Y, Wu, X, Tian, Y, Han, F, Liu, J & Zheng, C 2015, 'Optimizing water resources management in large river basins with integrated surface water-groundwater modeling: a surrogate-based approach', *Water Resources Research*, vol. 51, pp. 2153–73, doi: 10.1002/2014WR016653.

Yang, X, Wang, R & He, A 2007, 'Primary economic and social analysis on the eco-environmental problems in the Heihe River Basin', *Hydrogeology&Engineering Geology*, vol. 34, no. 6, pp. 100–03 (in Chinese).

Zhang, Y & Zhao, J 2009, 'Historical evolution and enlightenment of water system', *Forward Position*, vol. 2, pp. 98–101 (in Chinese).

Zhao, X, Liu, J, Liu, Q, Tillotson, M, Guan, D & Hubacek, K 2015, 'Physical and virtual water transfers for regional water stress alleviation in China', *PNAS*, vol. 112, no. 4, pp. 1031–5.

Zheng, C, Cheng, G & Xiao, H 2009, 'Sustainability of water resources in arid ecosystems: a view from the Heihe River Basin, China', *American Geophysical Union*, fall meeting, abstract #H14D-02.

12 Water, virtue ethics and traditional ecological knowledge in Rajasthan

Anupam Mishra and the rediscovery of water traditions[1]

Ricki Levi and Daniel Mishori[2]

Introduction

Sustainability is an enigmatic concept, which evades strict definitions (Dobson 2009; Kuhlman & Farrington 2010). Several studies have looked for patterns in different definitions, grouping them according the objectives of what ought to be sustained, for example, natural habitats, resources, human opportunities, economic value, etc. (Hopwood et al. 2005). The concept of sustainability replaced the previous 'sustainable development', which seemed prone to interpretations favouring a growth economy and a neo-liberal mindset (Redclif 2005). Beyond such definitions, sustainability is an ongoing social project, which explores new ways of living, and new values, which could support ecologically minded communities and lifestyles. Environmentalists commonly believe that current modern society and its values (the 'dominant paradigm'), is inherently unsustainable, and in desperate need of new ideas (Devall & Sessions 1985; Capra 1987). Such 'new' ideas could either be completely new and original, or rediscovered and reinvented,[3] including traditional ecological knowledge (TEK) and indigenous values, which may prove useful and insightful, and could provide inspiration for a sustainable future. The work of Anupam Mishra has this potential.

As we will show, the water philosophy and practices of Rajasthan – as rediscovered, portrayed and interpreted by the Indian Gandhian activist and author Anupam Mishra (1948–2016) – is rooted in traditional indigenous knowledge and in an ethical-spiritual worldview. This water ethics and worldview supports highly successful water practices, which enabled Rajasthani society and culture to thrive in the arid Thar Desert. Mishra's philosophy is also rooted in Gandhian philosophy and values (Mishra 1995), including the pursuit of Truth as both personal and collective enterprise, having spiritual and religious connotations. Following Gandhi's praise of simple living (which were manifested in his personal dress code of wearing traditional clothing made of Khadi [homespun cotton cloth]), Mishra opts for simplicity. This is manifested in a preference for traditional practices and modesty, and in the small-scale water ventures he promotes, which also serve to promote self-reliance and empowerment of individuals and local communities. In this, he is a true disciple of Gandhi, and a living proof of the relevance of Gandhi's teaching in the twenty-first century.

Water ethics discourse evolved when scholars began to consider injustices, inequalities and inequities pertaining to water distribution and usage, as well as issues regarding water pollution and conservation. Water ethics challenges the common view of water as resource and of water infrastructure as having only 'technical' or 'professional' aspects. In many cases a seemingly technical decision is based on human values. For example, when Israel faced a so-called water crisis in the past decade, a decision was made to base domestic consumption on desalination. Alternatives that could have been developed such as education and promotion of water saving, or harvesting rainwater which together could have provided similar quantities of water, and with fewer negative impacts (such as carbon emissions, pollution of the sea environment and unhealthy water [missing valuable chemicals such as magnesium]) were not seriously considered. This policy failure has since been criticized by a national inquiry commission.[4]

Water ethics considerations could have supported decision processes resulting in more sustainable water usage, conservation and infrastructure development. As such, water ethics has the potential to serve as a practical tool en route to sustainability (Groenfeldt 2013, pp. 1–7). This insight has evolved since the late 1990s by a growing number of water professionals and particularly through initiatives of UNESCO through its Commission on the Ethics of Scientific Knowledge and Technology (COMEST[5]; see also Brown & Schmidt 2010, p. 5).

There is a prevalent neo-liberal (capitalist) water ethics, which sees water as a mere product. This was the dominant worldview in Israel when Israeli officials opted for making water an industrial commodity (via desalination) instead of promoting saving and conservation decentralized practices. In other places, this capitalist ethics regards water as something which can be bought and sold in local or global markets, including legitimizing takeovers of local water resources and or infrastructures by foreign corporations (e.g., the Cochabamba Water War in Bolivia, 1999–2000). Faced with such challenges, water ethicists began to discuss alternatives, seeing water resources as intimately linked to natural habitats, and as inherently local and community-based. With such insights, focus was given to the study of indigenous water ethics and research into traditional conceptions of water and of water–nature–society relationships. Associated with such studies was an interest in water-related traditional ecological knowledge (TEK), which underlies traditional water practices, for which the work of Anupam Mishra serves as a superb example.

Anupam Mishra's reputation comes from decades of study into time-tested systems of water harvesting and conservation in several Indian states, principally in Rajasthan, dominated by the arid Thar Desert. Despite its warm climate and water scarcity, the Thar Desert is the most populated desert in the world (Mishra 2001, p. 31). This seeming paradox, a populated desert, stems from the fact that water is not merely a physical resource; Rajasthani inhabitants skilfully and efficiently managed their limited quantities of water, resulting not in scarcity but in abundance, historically enabling Rajasthan to play a dominant role in trade, culture and politics. According to Mishra: 'The people of Rajasthan scaled the peaks of trade, culture, art and standard of living *because of the depth of their philosophy of life. This philosophy gave a special space to water*' (Mishra 2001, p. 24, italics added). This philosophy is composed of a unique water ethics and highly sophisticated and complex set of traditional water skills and practices.

In this chapter, we will argue that this philosophy, as interpreted by Mishra, is a unique example of a virtue water ethic, intimately linked with TEK. Mishra elaborates on the ethical foundations of these unique water practices, describing them as inseparable from knowledge and skills. These ethical foundations are embedded in cultural-religious settings, which Mishra describes in detail. He attributes spiritual significance to water practices and ethics, and promotes a spirituality oriented interpretation of Rajasthan's water traditions. Mishra is not a detached academic but an activist, working on rediscovering and revitalizing water practices in Rajasthan and in India as a whole. He does it not only as a water conservationist, dedicated to managing resources, but also as a way to empower communities and to promote a more sustainable and just society, a fact which enables us to consider ethics in a wider sense, as a worldview and as a spiritual commitment.

Since our argument relies on the close affinity between water ethics and TEK, we begin by discussing 'water TEK' and practices in Rajasthan, as described by Mishra. Next we consider Mishra as a virtue ethicist by reviewing the concept of virtue ethics, the role of virtue ethics in environmental ethics, and the notion of water ethics. The next section focuses on Rajasthani water ethics, including the religious context, moral and spiritual virtues and values, commons' values and respect for nature. The last section discusses Mishara as a virtue ethicist, reflecting Gandhian moral and spiritual values, which are manifested in both the individual and the collective (society) realms.

Rajasthani water TEK

Traditional ecological knowledge (TEK) is a term that refers to the knowledge of indigenous communities which is derived from their direct connection and experience with nature over numerous generations. TEK is a source of 'knowledge about species, ecosystems, or practices held by people whose lives are closely linked to their natural environment' (Rist et al. 2010, p. 1). Nature is commonly perceived in such traditions as a comprehensive whole, including humans, animals, plants and landscapes (Pierotti & Wildcat 2000), a holistic worldview, often non-anthropocentric, emphasizing that all things are connected and related, in which nature is conceived as a family or a community. (Pierotti & Wildcat believe that such notions originate from the clan system.)

The notion of TEK gained popularity in the early 1990s (e.g. Berkes 1993; Inglis 1993; Berkes et al. 1995), and is presented as a tool for adaptive management (Berkes et al. 2000), capable of creating community resilience (Gómez-Baggethun et al. 2012), to work synergistically with conventional science and modern conservation practices (e.g. Becker & Ghimire 2003; Gagnon & Berteaux 2009), inspire ecological engineering (Martin et al. 2010) and to help mitigate climate change (Vinyeta & Lynn 2013). Research on TEK has been conducted throughout the world, including North American Inuits (Wenzel 1999), Ecuador (Becker & Ghimire 2003), Spain (Gómez-Baggethun et al. 2012) and East Jerusalem (Hamad 2016).

Rajasthani TEK

Mishra details many varieties of water TEK, infrastructures and practices, which we have described elsewhere (Levi & Mishori 2015). For our discussion here, we will only summarize them, to illustrate the enormous diversity in types and scale, which Mishra has rediscovered in Rajasthan.

Mishra describes ponds and wells (*kunds*), small ponds (*kundis*), and large reservoirs (*tankas*). According to Mishra, contrary to the famous proverb 'Running water, pure water', in Rajasthan water bodies were kept sealed in order to prevent evaporation, and to keep them clean, pure and fresh (ibid., p. 53).

In Rajasthan, three types of water are differentiated: *palar pani*, rainwater, collected in water tanks and reservoirs; *patal pani*, groundwater, accessible from wells and deep ponds; and *rejani pani*, capillary water, trapped in the ground, below the surface, before it reaches the saline underground water table (Mishra 2001, p. 45). All three types of water required sophisticated craftsmanship and community efforts to capture and to maintain. The ability to extract drinking water from the *rejani pani* is a uniquely Rajasthani practice. In certain areas in the desert, at a depth of 20 to 30 meters, there is a gypsum layer, which prevents water from percolating underground, and forms above it a layer of sand absorbed with moisture. It cannot be easily spotted from the surface, but miners know how to locate it, according to special characteristics (e.g. the presence of stagnated rainwater or certain trees). *Chelvanjis* (miners) dig very narrow and deep holes in the ground known as *kuin*. Each *kuin* can convert moisture into water at a rate of about two or three buckets a day.

TEK versus modern technologies

Mishra sees the British colonial rule in nineteenth-century India as responsible for the partial disintegration of Rajasthan's water traditions and infrastructure. Colonization led to privatization of communal and public assets and to a growing state control of public infrastructure, while the rapid growth in population created a growing demand for water. British colonial rule was characterized by strengthening central authority, which was unable to understand the traditional water systems (Mishra 1993, p. 72). Efforts to centralize the water systems led to the founding of the Public Works Department (PWD) in India in 1863, dispossessing the authority of local communities and commons. Mishra describes this process as an act of enclosure,[6] dispossessing the Indian people of their commons' rights, causing local communities to abandon their traditional role as guardians of water infrastructure. As a result, water lost its social significance as commons, and the corresponding socio-religious practices, whose ethical dimensions will be described below, began to disappear (Mishra 1993, pp. 74–5).

In Mishra's view, the 'top-down' management by the PWD was inefficient, leading to poor maintenance of ponds, which reduced the quantity and quality of water resources available to local communities. Mishra's negative interpretation of the role of the British rule is one-sided, and some researchers see the water policies of the British in a more favourable way.[7] Mishra sees India's independence in 1948 as making little difference since the modern Indian state inherited the colonial infatuation with modern technologies, policies and practices ('development' and 'modernization'). The liberalization policies of the 1990s added a positive outlook of globalization and privatization, which reinforced the previous preferences.

The importance of TEK in India

Mishra's efforts to rediscover and to reintroduce water TEK in Rajasthan is particularly justified in India, where 70% of the population depend on 'natural capital' for their daily survival (Narain 2006, p. 277). Agriculture, traditional medicine, hunting and gathering in forests, fishing and sheltering, are still common examples of ecosystem services that nature provides to significant portions of the Indian population (Shiva 2012). Deterioration of the environment leads to *ecological poverty* through 'lack of natural resources, both in quantity and quality, needed to sustain a productive and sustainable biomass-based economy' (Narain 2006, p. 277). Non-sustainable economic development severely affects indigenous communities, highlighting the urgent need for sustainable development in India (ibid.).

According to Vandana Shiva (1993), the daily connection with nature (through ecosystem services) is supported by social mechanisms such as commons management, which conforms to principles of justice that reflects understanding of nature as a sustainable system. Shiva argues that unlike the Western reductionist conceptions of Nature, where humans are perceived as passive or isolated beings detached from the environment, in India humans are perceived as engaging in an active cooperation and as an integral part of Nature (ibid.). Many Indian scholars are now rediscovering strategies for maintaining the health of the interlinked ecological and social systems, by strengthening traditional knowledge and skills, including rituals and religious practices that preserve and manage the environment in sustainable ways (Agarwal 2000; Shiva 1993). Mishra has similar motivations: to rejuvenate and revive the water traditions of Rajasthan.

Virtue ethics, environmental ethics and water ethics

Virtue ethics is a moral approach that emphasizes the building of moral character and personal virtues, in the context of a concrete community, practices and traditions (Statman 1997). Adherents of such ethics emphasize the role of model figures, which are supposed to exemplify moral character and personal virtues (e.g. Jesus Christ in Christianity, or the Guru in certain Indian and New Age traditions). It is expected that by following the example of model figures we can elevate our own virtues. Discussions in this tradition tend to focus on issues such as moral constraints, the good life and the practicality of ethics.

This ethical tradition began with Aristotle, but became prevalent in modern Western moral discourse only after Alasdair Macintyre (1981) published his book *After Virtue*. According to Macintyre, the Enlightenment project failed ethically, since it did not succeed in establishing the autonomous rationality, which was supposed to guide ethical decision-making: 'There seems to be no rational way of securing moral agreement in our culture' (p. 6). Religion, the old paradigm of ethical deliberation, was also unable to provide a 'shared background and foundation for moral discourse and action' in a world of 'secular rationality' (p. 50). From an environmental point of view, the failure of the Enlightenment project has led to an ecological crisis, which threatens Nature as well as the future of Humanity.

Macintyre criticized different concepts of individualism that have emerged from the Enlightenment project: Emotivism, which is 'the doctrine that all evaluative *judgments* and more specifically all moral judgments are *nothing but* expressions of preference, expressions of attitude or feeling, insofar as they are moral or evaluative in character' (p. 12); and Liberalism, which justifies the individual's freedom and autonomy, regardless of any connection to society (p. 119). This critique of individualism enables Macintyre to argue that morality in the modern era was left to be guided by abstract ethical principles and rules: 'Rules become the primary concept of the moral life. Qualities of character then generally come to be praised only because they will lead us to follow the right set of rules' (p. 119).

Macintyre's alternative to modern morality is based on a framework, which sees humans as predominantly social beings, who take part in cooperative, communal practices which are not purely instrumental. Macintyre emphasizes the 'dwelling' of humans in communities and institutions as sources of moral meanings, derived from the common good (i.e. morality is concrete and particular in its essence), and the ability of humans to mould themselves and to become ethical beings by developing virtues, character and personality in a continuous process.

Macintyre's solution involves practices, narrative and tradition. Practices are familiar social situations, which have rules, costumes, notions of excellence and purposes. In addition, they evolve and develop, thus enriching the community that engage with these practices. Therefore, practices are essential components to elaborate and acquire *Virtues*. Macintyre defines *virtue* as: 'an acquired human quality' the possession and exercise of which tends to enable us to achieve those goods which are internal to practices and the lack of which effectively prevents us from achieving any such goods' (p. 191). Modernity, in contrast, views individualism without narrative or context, resulting in sense of fragmentation, detachment and isolation. Thus, fostering virtues requires realization that the good (*telos*) is an ongoing, never-ending project, rooted and embodied in concrete, particular reality. 'Tradition' is the outcome of such social practices (pp. 204–20).

Environmental virtue ethics

Macintyre's insights have not passed unnoticed in environmental ethics, and significant work on environmental virtue ethics was done in the past two decades. Although it is argued that at first, environmental ethicists suffered from methodological 'virtue blindness' (Wensveen 2005, p. 17), the environmental virtue discourse emerged as an intriguing fresh discussion of ecological ethics, usually focused on describing environmental 'vices' and 'virtues', and in discussing exemplary environmental role models or 'heroes' (e.g. van Wensveen 2000; Sandler & Cafaro 2005; Sandler 2009, 2013). Environmental virtue ethics begins with the question 'How should we live?', which cannot be answered with accounts of right actions. Instead of framing environmental dilemmas in legal, technical, scientific or technological terms, virtue ethics introduces the language of character-building and how to nurture attitudes of doing good for nature, and consideration of human excellence, since it is 'always people – with character traits, attitudes and dispositions – who perform actions, promote policies, and lobby for laws (Sandler 2005).

> . . . as our environmental heroes also remind us . . . – by example and by word – environmental virtue is not merely instrumentally valuable as the disposition to identify and then perform proper action; it is also valuable in itself. It is life affirming and life enhancing.
>
> (Ibid., p. 3)

Hence, questions of the 'good life' turn into questions regarding the building of character, capable of nurturing and caring for the natural environment. Such virtuous exemplary figures or ecological heroes include Aldo Leopold (Shaw 1997), Rachel Carson, Henry David Thoreau (Cafaro 2000, 2001, 2004), John Muir, Edward Abbey (Andersen 2010), or Julia Butterfly Hill (Sandler 2005), an activist who lived for two years atop a threatened redwood. All these heroes showed respect and love for nature, courage to stand for what they believe in, and are recurring sources of inspiration.

Courage is an old virtue, but according to Hursthouse (2007), there are also completely new environmental virtues which include 'being disposed to feel wonder [towards nature] in the right way' (p. 161) and to understand or feel 'respect for nature' (p. 162), a concept derived from Taylor (1981, 1986). Most virtues, however, are simply the old ones attributed to ecology. For instance, Martin et al. (2009) discuss courage, temperance, compassion and humility and vices such as gluttony, greed and apathy. Passmore (1974) argued that greed, self-indulgence and short-sightedness might bring about ecological disaster. Van Wensveen (2000, 2005) compiled a catalogue of 189 different ecological virtues and 174 environmental vices, mostly adaptations of previously discussed vices and virtues to new environmental circumstances.

Water ethics discourse

We argue here that virtue ethics should be an important part of water ethics, as a sub-topic of environmental ethics. Water scarcity, pollution and adequate sanitation are some of the most pressing problems in many parts of the world, involving complex social-ethical issues. Water ethics addresses these complexities, while rejecting the prevalent view of water as supposedly bearing only technical aspects, which only experts can handle. Often, considerations masquerading as technical issues actually involve hidden values (Groenfeldt 2013, p. 3). The discourse of water ethics emerged as an attempt to expose and to discuss ethical presumptions and tacit values, and to provide ethical guidelines for water conservation and development (ibid.). For example, building a dam raises various questions such as (ibid., p. 1):

> What will be the impact on the river's fish, and on the traditional communities and cultures that depend on fishing? What will the electricity from the dam be used for, and what are the alternative energy options? What will happen to the people who live in the proposed reservoir area?

Instead of viewing the dam only from a crude technical viewpoint (e.g. which technology is best suited for the project), water ethics considers the overall imparts of such projects, and their impact on the natural and the social ecosystems. Accordingly, water ethics aims to expose the ethical principles and judgements of value of our water actions, and to set 'a normative framework guiding actions that affect water' (Brown & Schmidt 2010, p. 4). All this, under the assumption that if the rights and interests of Nature and of human communities are considered, as an integral part of every decision affecting water, the resulting policies will eventually become more sustainable and just (Groenfeldt 2013, p. 3).

Water ethics discourse often opts for 'Do's' and 'don'ts', that is, rules and duties as guidelines for moral action. Arguments often rely on deontological (Kantian) or utilitarian (Consequentialist) theories. In this, water ethics follow the general preference in modern Western philosophy for action ethics, that is, discussions regarding the morality or desirability of a certain course of action, and whether certain acts are right or wrong. However, Mishra's philosophy discusses values and morality in a very different way: he speaks mostly of virtues: ethical, religious and spiritual, both at the individual and community levels.

Rajasthani water ethics

The Rajasthani philosophy of water is closely linked with its religious background, a fact which Mishra illustrates by stories from Hindu mythology (e.g. Mishra 2001, pp. 24–5). Mishra regards 'Sri Krishna' as 'the Lord of the Desert' (ibid., p. 33), and refers to Indra as Goddess of rain (ibid., p. 104), who has to be seduced in order to bring rainwater to the desert (ibid., p. 128). Religious ritual is described as a way to force Indra to rain water (ibid., p. 116). Mishra also claims that water ponds were built and regarded as temples: stone pillars were positioned at entrances to water ponds, wells or lakes, symbolizing their sacredness; these pillars were carved with religious symbols and texts. People who entered such areas adapted their behaviour accordingly, including taking one's shoes off, observing appropriate behaviour such as the prohibition against spitting inside, etc. In large ponds, idols were placed at niches carved in the walls of *Ghats* (steps leading down the water), revered as guarding deities. Mishra also identifies notions of piety and impiety, normally associated with religious praxis (ibid., p. 133):

> To have a *kathri* [a wooden container kept on the coping of a well] made and to keep it on the well is considered to be a pious act and inversely to steal or break a *kathri* is considered to be a highly impious one. These implicit definitions of piety and impiety are nowhere couched in a written text; yet they are deeply engraved in the very soul of the local society.

Moral and spiritual virtues and values

The mythological-religious context, which Mishra uses, is intimately linked with nomenclature loaded with spiritual and ritual connotations. Mishra speaks of *riti* (identified with the traditional Rajasthanian notion of *voj*; Levi & Mishori 2015), saying 'they elaborated a *riti*, a tradition of preserving the rain water in each nook and cranny of every village' (ibid, p. 25). Literally, *riti* means a religious ceremony or a religious way of life, as well as a tradition or a social custom. By describing the water TEK as *riti* Mishra frames them as sacred practices, rather than as secular earthly efforts to sustain vital 'infrastructure' or 'resources'.

Another concept which Mishra uses is *savai*, intended perfectionism: doing something wholeheartedly, with more than 100%, including with one's treatment of his fellows. According to Mishra, '*savai* is a term used in mathematics but also in life. The term *savai* means 25% more, as in "I'm 100% and you're 125%." It is a kind of honour giving, to have more love and affection for somebody else than for oneself.' So *savai* means 'greater than us' (Mishra, in interview conducted by R. Levi, 3 October 2014, Delhi, India). In harvesting water, *savai* means 'the effort of harvesting is not 100%. 100% is perfect! They are not saying perfect, but more than perfect!' (ibid.). Mishra (2001, p. 24) sees the application of this virtue as explaining the ability of ancient Rajasthani cities to survive and even thrive in the desert:

> Without understanding the '*savai*' nature of this princely people, we will utterly fail to understand how in the last millennium big towns like Jaisalmer, Jodhpur, Bikaner, and even Jaipur were established according to all the rules of town planning. The towns were moreover highly populated and yet, in spite of the scarcity of water, they were no less equipped than other cities of

> the county. In fact, each of these towns, at different periods of time and for long durations, was an important centre of power, trade and art.

Likewise, building ponds was seen as an act of sacredness, and the pond-makers were regarded as spiritually elevated, while funding *kund* and *kundis* was regarded as an act of *Punya,* usually an act of gratitude and thanksgiving to celebrate joyful events (Mishra 2001, p. 57). A maker of ponds was regarded as (Mishra 1993, p. 69),

> a pious soul, an enlightened being. He who protects the pond is equally great. Thus pond is a miniature pilgrimage. The people gather here in the form of fairs. Such a fair-loving society makes pond an integral part of its heart, its vision.

In his book *Talaab*, Mishra (1993, p. 21) tells about a group of *gajdhars* (well and pond excavators) from the village of Pukharan that became Brahmins (priests) as a reward for building a pond in Pushkar, a sacred city for Hindus. In the context of the caste system, the rigid hierarchical social order of traditional Indian society, this story (whether truth or legend) is a remarkable illustration of the social-religious significance of water and water craftsmanship in traditional India.

Mishra's own spiritual-religious interpretation

Mishra not only describes the ways in which water practices are embedded in semi-religious or moral and spiritual notions, he also interprets these practices in his own Gandhian moral-spiritual terms (Levi & Mishori 2015). For example, the tedious techniques of collecting drops of water, which requires dedication and patience, are regarded as *tapasya,* ascetic dedication. According to Mishra (2001, p. 70),

> for the people of the desert, in the 10 days of rainfall, they have seen millions of priceless drops and they have undertaken to collect them in each house, each village, even in each town. The result of this *tapasya* (ascetic dedication) is manifest.

He reframes *tapasya*; instead of understanding it in the context of the individual he interprets it as a virtue in the realm of the community, making it a 'collective tapas', mutual moral dedication and commitment that generates a better, sustainable society, intimately embedded in its local environment (ibid, p. 70, and Mishra, interview by R. Levi, 3 October 2014, Delhi, India). This is a community water ethics, based on social virtues.

Other moral-religious notions that Mishra uses are *swayamsiddha* and *samaysiddha. Swayamsiddha* means self-achieved empowerment or accomplishment (with spiritual resonance), while *samaysiddha* means having achieved realization or perfection which defies time. The concept of *swayamisiddha* is used to describe the ability to identify suitable local materials, and *samaysiddha* to signify the ability to build with materials resistant to desert conditions and to the challenges of time: 'These have seen old times and new times and thus have passed the test of Time and become *samaysidh*' (Mishra 2001, p. 58). The word *sva* (self), in this context, means 'local', emphasizing the use of local materials: '. . . for their construction no material from outside is required' (ibid., p. 58). Mishra, then, ascribes spiritual value to the perfection achieved by local communities, on their quest for survival in the harsh Thar Desert: 'In Rajasthan all the water work has been the result of the spiritual commitment of the whole society which also enjoyed its fruits' (ibid., p. 135). This commitment, manifested in both the social and the individual levels, is described also as 'collective asceticism' and as *sadhana*.

The notion of asceticism is prevalent in Mishra's book. He regards the ability to find water as 'asceticism of the eyes' which is 'the greatest form of asceticism' (ibid., p. 88): 'the ascetic eyes were able to perceive with fineness [the course of impermanence rivers] and thus at many chosen places it was possible to hold their waters. *Khadeens* [a sort of temporary lake; moist bed of seasonal rivers where cultivation is undertaken; pp. 89, 139] were made at such places' (ibid., pp. 88–9).

Commons values

Mishra rarely uses the *terms* 'commons' or 'commoners', but he nevertheless elaborates on the traits that characterize commons, emphasizing the collective nature of water practices and criticizing attempts to replace commons-based TEK with either private or centralized state management. Water TEK is described as inherently sustainable, since it relies on the collective commitment of communities to preserve common resources and heritage.

In the spirit of Barnes (2006) who sees the commons as a third sector of society, beyond the market and the state, Mishra (2001, p. 40) claims that 'The people of Rajasthan did not entrust the organization of such boundless [water] work to either the central or federal government, nor even to what in modern parlance is termed as the private sphere.' Mishra describes a realm that is neither private property nor the state. He argues that (ibid., pp. 50–1),

> The sacrosanct line, which divides private property from commons property, gets strangely erased when it comes to *kuin* [very narrow and deep wells]; everyone has the right to construct a *kuin* and use its water. However, the *kuin* is constructed on land which is the collective property of the village. . . . Constructing a *kuin* . . . means sharing the humidity present there and that is why, though the *kuin* is a private property, it falls under the control of the village society. It is only in case of dire necessity that permission is granted to build a new *kuin*.

Likewise, *kundis* can also be private or collective. 'Private *kundis* are made in front of houses, in the front courtyards (*angan*), in the enclosure facing the house (*ahates*) and in backyards (*pichwadas*). The collective *kundis* are made on common *panchayati* ground or usually between two villages' (ibid., pp. 56–7). The collective *kundis* are described as gifts by the well-off to society, and as based on intergenerational obligations. There are numerous places in which Mishra regards water traditions as inherently intergenerational. For instance, regarding maintenance and cleanliness of water ponds, Mishra (2001, p. 56). says that '. . . such precautions are taken at the *agor* [catchment area of a pond] that it is only necessary to remove sand . . . every 10–20 years. Such is the care with which one generation looks after its *kundi* that only the next generation needs descend inside to clean it.'

Regarding the mixing of private and collective, Mishra also speaks of *svamitva visarjan*: *svamitva* means ownership and *visarjan* refers to the immersion of an icon after a ceremonial prayer. 'Here the expression means an offering of one's ownership which, in a way, gets immersed into society' (ibid., p. 142). An example is *tankas*, which are sometimes built in uninhabited places: 'Those who undertake the construction of such *tankas* do not do it for themselves but for society. There could hardly be a better example of *svamitva bisarjan*'[8] (ibid., pp. 62–3).

Respect for nature

In Mishra's work, as well as in Montout's commentary on his work (Montout 2001), the traditional way of life is described in non-anthropocentric terms. For instance, animals (both domestic and wild) were allowed to drink from open ponds. The ponds were maintained using aquatic life: fish, crabs, sometimes crocodiles or turtles pierced with golden rings (for good luck). Vegetation was used to purify water; banyan and neem trees were often planted on banks of ponds, conceived as the abode of gods and goddesses and therefore sacred (Mishra 1993, pp. 36–9).

Non-anthropocentric values have often been ascribed to India, which was hailed as a counter-example to the anthropocentric Western culture, especially the Judeo-Christian monotheistic tradition (White 1967; Callicott & Ames 1989; Callicott 1994). Mishra (and Montout) take non-anthropocentrism for granted, primarily when they describe local values that are antagonistic to the Western 'dominant' anthropocentric paradigm (Capra 1987). Shiva (1996, pp. 105–06) claims that Western industrialism and colonialism is responsible for the transformation of nature from a living and self-organizing system to dead materials for human exploitation, or a chaos that needs to be organized and controlled by man. She argues that the Western concept of human domination is a consequence of a fear from nature, which motivates the urge to control it. Similarly, Mishra depicts nature as sacred, as a source of knowledge and inspiration, and as an inherent part of the complex social system of life in the desert. He describes traditions, which are non-violent towards nature and show respect for other beings and the environment at large.

Describing Nature as a giver and not as crude resource is manifested when Mishra (2001, p. 87) discusses Jaseri Lake: 'The livestock of many villages is totally dependent upon the prosperity of Jaseri. People describe it as *ajalapoorna*, the one who fulfils the needs of water in the same image as *Annapoorna*, the one who fulfils the food needs.' Moreover, following the teaching of 'Mother Nature' assures enjoying her abundance and wealth:

> There can be water scarcity in a high rainfall region like Cherrapunji and hilly regions like Uttarakhand and there can be sufficient water available in low rainfall regions like Rajasthan – it all depends on how we follow the directions provided by nature . . . Mother Nature has always provided us enough water to survive.
>
> (Mishra, quoted in Bajpayee n.d.)

Mishra as a virtue ethicist

According to Aristotle, there were two types of virtues: the moral and the intellectual, including *phronesis*, practical wisdom or intelligence, manifested in crafts, arts and every human undertaking which requires expertise and excellence. Mishra speaks of both types of virtues, and in a way which seems to blend the distinction together. For example, the *savai* (intended perfectionism, doing something wholeheartedly, with more than 100%): does it stand for moral virtue or for practical reason? The same goes for *tapasya* (ascetic dedication), by which Mishra describe the tedious techniques of collecting drops of water, which requires dedication and patience. And what about *swayamsiddha* (self-achieved empowerment or accomplishment) and *samaysiddha* (having achieved realization or perfection which defies time)? Are these moral or for practical virtues? What Mishra offers us is a new set of virtues, embedded in spiritual, religious and social conceptions, which are moral, spiritual and practical. For instance, when he speaks of a well (Mishra 2001, p. 97):

> . . . [the Rajasthani word] *sir* has two other meanings, one of them being sweet and the other source of regular income. In a way these two meanings are very apt to describe the water of the well. The well, too, is a regular source of water; however *without a proper ethic of management, in other words without frugality and sobriety*, this income does not fructify.
>
> (Italics added.)

Mishra does not directly discuss vices. However, he is very critical towards the blind preference for the New over the Old, and of forgetting past achievement or of failing to preserve the built heritage of water ponds and infrastructure as a living practice. Hence, some new environmental vices could be defined, including disregard or not being respectful for TEK.

Finally, key components in virtue ethics are the exemplary figures, or models of righteous conduct. Mishra gives several such examples, all of which pertain to experts of well digging and builders. Thus, 'Whosever made a pond was revered by the people as *Maharaj* or *Mahatma*, i.e., a pious soul' (ibid., p. 5). Moreover, generations of experience led to remarkable proficiency:

> A good *gazdhar* (builder) was defined as one who did not touch the tools. He decided [where to locate the pond] by just checking the site and also what has to be done. The *gazdhars* would sit at one place and everything was carried out at their oral instructions only.
>
> (Ibid., p. 14)

In addition, that eventually led to the fact that these people 'elaborated a whole science to translate their experience into practicality' (Mishra 2001, p. 37). Likewise, 'At one time the difficult work of the *bariyo* (well-builder) was greatly respected. When a marriage procession would enter the village, the first to be served with a lot of respect would be the *bariyo*' (ibid., p. 100).

It is interesting to note that Mishra often speaks of virtues and achievements in the plural, *collective* form, and not as achieved by individuals. He says that, 'The people of Rajasthan scaled the peaks of trade, culture, art and standard of living because of the depth of their philosophy of life. This philosophy gave a special space to water' (ibid., p. 24). Water expertise and knowledge exists in the society as a whole: 'To access this ambrosia [water] the people of the desert . . . elaborated a whole science to translate their experience into practicality' (ibid., p. 44). 'This culture which knows how to propitiate earth, water and heat through its asceticism, also knows how to keep flowing water as well as still water pure' (ibid., p. 104). According to Mishra (p. 105):

> In Rajasthan, more particularly in the desert, water work was never considered as work but came to be viewed as a moral duty; that is why it was able to rise far above what is today called community project, to take the beautiful shape of a *samagra jal darshan*, a perfect water philosophy.

Therefore, 'Every village, as a whole knows how to build a pond' (Mishra 1993, p. 16). Knowledge and skills are also collective, not merely individual achievements and perfection. In Mishra's own words:

> It is . . . not at all possible to have a centralised arrangement to distribute water to such widely dispersed villages. That is why society kept the water work totally decentralised, distributing the responsibility in this matter, drop by drop, between themselves. Thus *instead of being reduced to being a dry technique and a mechanical device, this practice was transformed into an ethos.*
>
> (Ibid., p. 61, italics added)

Decentralized practice and collective ethos are accompanied by yet another virtue, of which Mishra speaks with a sense of wonder, of aesthetics and beauty:

'To know how beautiful these *kundis* can be, one can go to the villages of Jaisalmer' (ibid., p. 61).

Conclusions

Mishra is a unique virtue ethicist. He is unique because he discusses special virtues, which evolved in the particular context of the Rajasthani desert and culture, and he is unique because of the special mixture, which he makes of TEK and virtue ethics, of spirituality and of individual and collective virtues. He is also unique since he not only writes about water virtue ethics, he himself plays a special role in promoting water TEK in modern India. For this reason, Mishra himself should be regarded as an ecological hero, a model and an inspiration for ecologists and social and environmental activists. In this spirit Vandana Shiva (2001, p. 1) says that 'Anupam's work on the indigenous water systems of Rajasthan is a work of poetry as well as a work of science. It is this work that has inspired the water conservation movement of Tarun Bharat Sangh.'[9] His profound philosophy, his deep moral and spiritual intuitions and rhetoric also testify as to his achievements in the area of character-building and perfection of invaluable environmental (and other) virtues. Only a truly virtuous person would say that 'The water project is a project of love and virtue' (Mishra 1993, p. 10).

Mishra's water philosophy enable us to consider ethics in a wider sense: as a worldview, as a spiritual commitment, as both collective and individual virtues, as an arena for exercising perfection in character and in craftsmanship. He also teaches us humility (and a sense of wonder and awe) in the face of past accomplishments, and thus he serves as the best advocate for water TEK and for TEK in general. As such, Mishra presents an intriguing applied environmental virtue ethics in its purest form: setting an example and an inspiration, teaching both skill and practices (TEK) while practising and instructing moral and spiritual virtues in one inseparable and convincing form.

Notes

1 This chapter is dedicated to the memory of the late Anupam Mishra.
2 Authors' names in alphabetical order.
3 For a discussion on the biases of patent theory against reinventions and rediscoveries, see Sariel et al. (2015).
4 The National Inquiry Commission on Water Management in Israel, Haifa, March 2010 (in Hebrew).
5 www.unesco.org/new/en/social-and-human-sciences/themes/comest/, accessed 22 July 2016.
6 For a discussion of commons rights and enclosure, see Mishori (2010, 2014).
7 For a short discussion on the debate regarding the effects of British 'colonial hydrology', see Levi and Mishori (2015).
8 The word is spelled *svamitva bisarjan* on page 63 and *svamitva visarjan* on page 142.
9 Rajendra Singh, who ran 'Tarun Bharat Sangh' (TBS), received the Stockholm Water Prize (known as 'the Nobel Prize for water') in 2015, for his work in India, especially in Alwar district, Rajasthan.

References

Agarwal, A 2000, 'Can Hindu beliefs and values help India meet its ecological crisis?', in C Chapple & M Tucker (eds), *Hinduism and Ecology: The Intersection of Earth, Sky and Water*, Harvard University Press, Cambridge, MA.

Andersen, N 2010, 'Exemplars in environmental ethics: taking seriously the lives of Thoreau, Leopold, Dillard and Abbey', *Ethics, Place and Environment*, vol. 13, no. 1, pp. 43–55.

Bajpayee, N n.d., '"Mother Nature has always provided us enough water", an interview with Anupam Mishra', *Harmony Magazine*, accessed 22 July 2016, www.harmonyindia.org/hportal/VirtualPageView.jsp?page_id=11291&index1=7.

Barnes, P 2006, *Capitalism 3.0: A Guide to Reclaiming the Commons*, Berrett-Koehler, San Francisco, CA.

Becker, C & Ghimire, K 2003, 'Synergy between traditional ecological knowledge and conservation science supports forest preservation in Ecuador', *Ecology and Society*, vol. 8, no. 1 (online).

Berkes, F 1993, 'Traditional ecological knowledge in perspective', in J Inglis (ed.), *Traditional Ecological Knowledge: Concepts and Cases*, IDRC, Ottawa, Canada.

Berkes, F, Colding, J & Folke, C 2000, 'Rediscovery of traditional ecological knowledge as adaptive management', *Ecological Applications*, vol. 10, no. 5, pp. 1251–62.

Berkes, F, Folke, C & Gadgil, M 1995, *Traditional Ecological Knowledge, Biodiversity, Resilience and Sustainability*, Springer, Dordrecht, The Netherlands.

Brown, P & Schmidt, J 2010, *Water Ethics: Foundational Readings for Students and Professionals*, Island Press, Washington, DC.

Cafaro, P 2000, 'Thoreau's virtue ethics in Walden', *The Concord Saunterer*, vol. 8, pp. 23–47.

Cafaro, P 2001, 'Thoreau, Leopold and Carson: toward an environmental virtue ethics', *Environmental Ethics*, vol. 23, no. 1, pp. 3–17.

Cafaro, P 2004, *Thoreau's Living Ethics: Walden and the Pursuit of Virtue*, University of Georgia Press, Athens.

Callicott, JB 1994, *Earth's Insights: A Survey of Ecological Ethics from the Mediterranean Basin to the Australian Outback*, University of California Press, Berkeley.

Callicott, JB & Ames, RT 1989, *Nature in Asian Traditions of Thought*, State University of New York Press, Albany.

Capra, F 1987, 'Deep ecology: a new paradigm', in G Sessions (ed.), *Deep Ecology for the 21st Century*, Shambhala, Boston, MA and London.

Devall, B & Sessions, G 1985, *Deep Ecology: Living as if Natured Mattered*, Gibbs Smith, Publisher, Salt Lake City, UT.

Dobson, A 2009, *Green Political Thought*, 4th edn, Routledge, London and New York.

Gagnon, CA & Berteaux, D 2009, 'Integrating traditional ecological knowledge and ecological science: a question of scale', *Ecology and Society*, vol. 14, no. 2 (online).

Gómez-Baggethun, E, Reyes-García, V, Olsson, P & Montes, C 2012, 'Traditional ecological knowledge and community resilience to environmental extremes: a case study in Doñana, SW Spain', *Global Environmental Change*, vol. 22, no. 3, pp. 640–50.

Groenfeldt, D, 2013, *Water Ethics: A Values Approach to Solving the Water Crisis*, Routledge, London.

Hamad, F 2016, 'Attitudes, perceptions and traditional practices of sustainability and the commons in Sur Baher (East Jerusalem): the past as an inspiration for the present', The Kibbutzim College of Education, Technology and the Arts, Tel-Aviv, Israel.

Hopwood, B, Mellor M & O'Brien G 2005, 'Sustainable development: mapping different approaches', *Sustainable Development*, vol. 13, pp. 38–52.

Hursthouse, R 2007, 'Environmental virtue ethics', in RL Walker & PJ Ivanhoe (eds), *Working Virtue: Virtue, Ethics and Contemporary Moral Problems*, Oxford University Press, London.

Inglis, J 1993, Traditional ecological knowledge: concepts and cases', IDRC.awaOtt.

Kuhlman, T & Farrington, J 2010, 'What is sustainability?', *Sustainability*, vol. 2, pp. 3436–48.

Levi, R & Mishori, D 2015, 'Water, the sacred and the commons of Rajasthan: a review of Anupam Mishra's philosophy of water', *Transcience: A Journal of Global Studies*, vol. 6, issue 2.

Macintyre, A 1981, *After Virtue: A Study in Moral Theory*, Duckworth & Co. Ltd, London.

Martin, B, Bright, A, Cafaro, P, Mittelstaedt, R & Bruyere, B 2009, 'Assessing the development of environmental virtue in 7th and 8th grade students in an expeditionary learning Outward Bound school', *Journal of Experiential Education*, vol. 31, no. 3, pp. 341–58.

Martin, JF, Roy, ED, Diemont, SA & Ferguson, BG 2010, 'Traditional ecological knowledge (TEK): ideas, inspiration, and designs for ecological engineering', *Ecological Engineering*, vol. 36, no. 7, pp. 839–49.

Mishori, D 2010, 'Conceptualizing the commons: reflections on the rhetoric of environmental rights and public ownership', in OF von Feigenblatt (ed.), *Development and Conflict in the 21st Century*, JAPSS Press, Bangkok, Thailand.

Mishori, D 2014, 'Reclaiming commons rights: public ownership, human rights and future generations', *Journal of Law and Ethics of Human Rights*, vol. 8, no. 2, pp. 335–66.

Mishra, A 1993, *Aaj Bhi Khare Hain Talaab* (in Hindi), Gandhi Shanti Pratishthan, New Delhi (special English edition 2012; translated by Parvesh Sharma).

Mishra, AL 1995, *Fundamentals of Gandhism*, Mittal Publications, New Delhi.

Mishra, A 2001, *The Radiant Raindrops of Rajasthan*, Research Foundation for Science, Technology and Ecology, New Delhi, India, www.arvindguptatoys.com/arvindgupta/anupam.pdf

Montaut, A 2001, 'Preface', in A Mishra, *The Radiant Raindrops of Rajasthan*, Research Foundation for Science, Technology and Ecology, New Delhi, India.

Narain, S 2006, 'Growth with equity: understanding ecological poverty', in M Malhoutra (ed.), *India: The Next Decade*, Routledge, London.

Passmore, J 1974, *Man's Responsibility for Nature*, Scribner, New York.

Pierotti R & Wildcat D, 2000, 'Traditional ecological knowledge: the third alternative', *Ecological Applications*, vol. 10, no. 5, pp. 1333–40.

Redclif, M 2005, 'Sustainable development (1987–2005): an oxymoron comes of age', *Sustainable Development*, vol. 13, pp. 212–27.

Risse, M 2013, 'The human right to water and common ownership of the Earth', Harvard Kennedy School Faculty Research Working Paper Series RWP13-003, accessed 5 May 2016, https://research.hks.harvard.edu/publications/citation.aspx?PubId=8814&type=FN&PersonId=170

Rist, L, Shaanker, RU, Milner-Gulland, EJ & Ghazoul, J 2010, 'The use of traditional ecological knowledge in forest management: an example from India', *Ecology and Society*, vol. 15, no. 1 (online).

Sandler, RL 2009, *Character and Environment: A Virtue-oriented Approach to Environmental Ethics*, Columbia University Press, New York.

Sandler, RL 2013, *Environmental Virtue Ethics*, Blackwell, Oxford.

Sandler R & Cafaro P (eds) 2005, *Environmental Virtue Ethics*, Rowman and Littlefield, Lanham, MD.

Sariel, A, Mishori, D & Agassi, J 2015, 'The re-inventor's dilemma: a tragedy of the public domain', *Journal of Intellectual Property Law and Practice*, vol. 10, no. 10, pp. 759–66.

Shaw, B 1997, 'A *virtue* ethics approach to Aldo Leopold's land ethic', *Environmental Ethics*, vol. 19, no. 1, pp. 53–67.

Shiva, V 1993, *Monocultures of the Mind: Perspectives on Biodiversity and Biotechnology*, Zed Books and Third World Network, Penang.

Shiva, V 1996, *Biopiracy: The Plunder of Nature and Knowledge*, South End Press, Cambridge, MA.

Shiva, V 2001, 'Introduction', in A Mishra, *The Radiant Raindrops of Rajasthan*, Research Foundation for Science, Technology and Ecology, New Delhi, India.

Shiva, V 2012, *Biopiracy: The Plunder of Nature and Knowledge*, Natraj Opposite Publishers, New Delhi.

Statman, D 1997, 'Introduction to virtue ethics', in D Statman (ed.), *Virtue Ethics*, Edinburgh University Press, Edinburgh.

Taylor, PW 1981, 'The ethics of respect for nature', *Environmental Ethics*, vol. 3, no. 3, pp. 197–218.

Taylor, PW 1986, *Respect for Nature: A Theory of Environmental Ethics*, Princeton University Press, Princeton, NJ.

Vinyeta, K & Lynn, K 2013, 'Exploring the Role of Traditional Ecological Knowledge in Climate Change Initiatives', United Sates Department of Agriculture, General Technical Report.

van Wensveen, L 2000, *Dirty Virtues: The Emergence of Ecological Virtue Ethics*, Prometheus Books, Amherst, NY.

van Wensveen, L 2005, 'The emergence of ecological virtue language', in R Sandler & P Cafaro (eds), *Environmental Virtue Ethics*, Rowman and Littlefield, Lanham, MD.

Wenzel, GW 1999, 'Traditional ecological knowledge and Inuit: reflections on TEK research and ethics', *Arctic*, vol. 52, no. 2, pp. 113–24.

White, L Jr 1967, 'The historical roots of our ecological crisis', *Science*, vol. 155, pp. 1203–07.

Part III

Water ethics charters and charting water

13 *I yá.a<u>x</u>ch'age?* (Can you hear it?), or *Héen Aawashaayi Shaawat* (marrying the water)[1]

A Tlingit and Tagish approach towards an ethical relationship with water

Eleanor Hayman with Colleen James, Mark Wedge and David Katzeek[2]

> When we say *yáa át wooné*, that means to learn about it [water]. You are going to meditate on it; you are going to think on it; you are going to develop a relationship with it.
>
> (David Katzeek, personal communication, 6 September 2014)

Introduction

Conversations with coastal Tlingit and inland Tlingit/Tagish Elders and intellectuals illuminate a clear and resilient perception of water as a relative, facilitator, connector, educator, healer and transformer. This perception of water works within a Tlingit and Tagish cosmology in a number of ways that challenges the dominant hydrological imaginary. Water provides a living normative framework for the circulation of virtues, in particular, respect that establishes a mode of morality for the continuation of human existence (Langdon 2013). Water functions metaphorically within many traditional Tlingit and Tagish oral narratives offering fluid, transformative imaginaries that act as guides for human relationships with more-than-human worlds. Water inscribes a place-based cultural identity through its agency, shaping Tlingit and Tagish clan names and clan origin stories, and almost all Tlingit and Tagish place names (toponyms). Lastly, as the primordial element in Tlingit and Tagish creation myths, it is water that gives definition, hydrological character, life and sacred qualities to the land.

In Tlingit, respect is *yáa át wooné* and is considered one of the most powerful words in the Tlingit language (Katzeek, personnel communication/interview at the Tlingit Clan Conference: Sharing Our Knowledge, Juneau, Alaska, 6–10 November 2013).[3] *Yáa át wooné* is the cornerstone of Tlingit beliefs and values, and a prerequisite for Tlingit identity and even existence.[4] Implicit within *yáa át wooné* is a call to listen; to pay attention. Not just to other human voices, as in the ancient Tlingit and Tagish oral tradition of storytelling, songs and ceremonial oratory, but to the greater non-human ecological voices.

Modern Western philosophy rooted in the Aristotelian tradition, is slowly recognizing that respect, not just for humans and other animals, but for ecosystems and planetary life more generally is a prerequisite for the so-called 'leap' or 'great transformation' needed for industrial societies to evolve towards holistic greener societies. However, it is not simply about changing anthropocentric practices, but challenging a now normalized and ingrained post-Enlightenment shift that privileges anthropocentric thinking, otherwise known as the monocultural imaginary.[5] In this sense, many indigenous cultures precede this 'new' form of thinking. For example, the Tlingit and Tagish cultures have extended respect (and intrinsic value) to all of the more-than-human world for many thousands of years. Not only this, but the Tlingit and Tagish cultures have been pursuing holistic ecological practices and what is touted to be the 'new' ecologically minded virtues of the twenty-first century,[6] since time immemorial.[7]

Using examples from Tlingit oratory, Tlingit place-based education, Tlingit and Tagish storytelling and Tlingit and Tagish place names, we explore the critical importance of listening within a Tlingit and Tagish cosmology. Building on our previous collaborative water work (Hayman, with James & Wedge 2015), we examine how a Tlingit pedagogy that emphasizes listening, meditating and keen observational skills is critical for developing relationships with more-than-human worlds that in turn sustains *yáa át wooné* embedded within storytelling. As storytelling and *yáa át wooné* mimic the circulatory, replenishing and sustaining power of water, we show how water, respect and identity intersect within the Tlingit and Tagish culture.

Tlingit and Tagish peoples co-evolved and co-constructed their social, moral and linguistic structures with and within southern Yukon and Alaska – subarctic lands formed and shaped by water and ice, salty and fresh. The character and unique aquatic personality of this region has been imagined for thousands of years in the minds and oral narratives of the Tlingit and Tagish, something we call a *hydrological* imaginary. However, in contrast to the dominant hydrological imaginary where 'modern water' is perceived as an economic resource with only instrumental value, the Tlingit and Tagish ontology[8] differs radically. Glaciers, rivers and lakes, for example, are understood 'to apprehend the world from distinct vantage points . . . to have points of view, exhibit agency, and engage in reciprocal responses' (Cruikshank 2012).

The Tlingit language has a precise and indeed sacred phrase to describe the spirit and agency of all things: *Yakg wahéiyagu*. *Yakg wahéiyagu* is described by Tlingit language scholar Lance Twitchell (2016) in his new Tlingit online dictionary as 'the ability of everything to comprehend language and intentions'. This concept forms the basis of respect, which David Katzeek has referred to as 'the fruit of education'. *Yakg wahéiyagu* and *yáa át wooné* are therefore intimately linked in the Tlingit worldview, and provide a profound point of departure for our thinking about knowledge production in this chapter.

Anthropologist Julie Cruikshank frames the recognition or oppression of different knowledges very elegantly with what she calls two key ontological questions: 'What is not recognized as knowledge in dominant regimes? What is lost when local knowledge in Canada is trimmed and transformed to fit the requirements of science, policy and governance?' (Cruikshank 2012). We find these questions highly pertinent to this volume on water ethics, and colonial modes of knowledge production more broadly. A further hope for our chapter in this water volume is to destabilize the 'binaries inherited from European philosophy in which Indigenous peoples appear as either victims of colonization or heroically resistant' (Cameron et al. 2014). Being cautious not to perpetuate what we have called hydrological violence (Hayman, with James & Wedge 2015) – that is, reproducing colonial hydrological ways of knowing and being – we present Tlingit and Tagish hydrological knowledge in the spirit of ontological pluralism and as a critical step towards actively decolonizing and hearing water.

Carcross/Tagish[9] First Nation (CTFN)

Since February 2012, collaborative water research with Carcross/Tagish First Nation has followed the participatory action research path, informed by post-colonial and (eco) feminist/feminist political ecology studies which overlap with critical indigenous theory. CTFN community consultants Colleen James and Elder Mark Wedge are actively involved in much of the fieldwork among and on the headwaters of the Yukon River and continue to revise and advise on all written work, publications and the mapping of Tlingit and Tagish aquacentric place names.[10] Equally, the inspiration for the title and much of this chapter is due to coastal Tlingit Elder, educator and fluent Tlingit speaker David Katzeek's guidance. This chapter therefore reflects both coastal and inland Tlingit practices and philosophies, tightly woven with the Athabaskan linguistic group, the Tagish peoples, with whom the coastal Tlingit inter-married some 500 years ago.

CTFNs citizens are of Tagish and coastal/inland Tlingit ancestry. In October 2005 the CTFN government signed the final agreement with both the Yukon and Canadian governments to become one of 11 self-governing (out of 14) First Nations in the Yukon Territory. In 2012 the CTFN government asked that this collaborative water research be the basis for a CTFN legislative Water Act, rooted in Tlingit and Tagish cosmologies, but being able to speak with Canadian environmental law. As such, this collaborative water research explores critical *hydrologies* of indigeneity through an analysis of hydrological imaginaries embedded in traditional oral narratives, aquacentric toponyms and cultural practices. These are deeply geographical themes as 'knowledge is deeply embedded within spatial environments' (Warf 2015, p. 134). Deep Mapping (or as we have called it on account of its acutely aquatic perspective, deep charting) is one way to represent such complex, empirical, sensuous hydrological Tlingit and Tagish knowledges, while at the same time supporting the shift towards democratizing and decolonizing mapping practices. The exciting performance of such critical hydrologies of indigeneity is shown not least by the forthcoming CTFN water legislation, but the fact that this Tlingit/Tagish water legislation might well be a first,[11] not just in the Yukon Territory, but in Canadian legal history since colonization.

Narrative, environment (water) and virtue ethics

It is increasingly well-documented and recognized at the national policy level[12] that the global water sector is in a critical shape. Driven by utilitarian ideals, technological advances and a narrow idea of water simply as a resource, water management strategies and goals are more often than not unsustainable. In an attempt to reimagine human relationships with water, there have been calls for a global water ethic from diverse groups including civic society, indigenous organizations and movements[13] and policy-makers.

From an academic perspective, the field of environmental philosophy is gaining increasing prominence as it challenges anthropocentric and destructive ways of valuing of nature (water). One of the most exciting recent contributions to biocentric, ecocentric and anthropo-relational[14] thinking is from environmental philosopher Brian Treanor in *Emplotting Virtue: A Narrative Approach to Environmental Virtue Ethics* (2014). Treanor's critical analysis of virtue ethics within an ecological context cross-pollinates with narrative theory and provides this chapter with a profound point of departure. In Treanor's words, 'what we need is a *narrative, environmental virtue ethics* . . . that indicates the significance of both narrative and environmental virtue in a complete understanding of what it means to live, act, and fare well in the twenty-first century' (2014, p. 187). Indeed, Treanor argues convincingly that a 'fully developed environmental ethic should address not only what we should *do*, but also who we should *be* . . .' (2014, p. 160, italics in original). One of Treanor's key themes is the critical role played by narrative in not just the formation and cultivation of a virtuous character, but its transformative role too.

In *Poetic Justice* (1995), philosopher Martha Nussbaum addresses in great depth the role that narrative plays in human ethical education and ethical judgement. Nussbaum argues that reading '. . . can be a bridge both to a vision of justice and to the social enactment of that vision' (1995, p. 12). Nussbaum refers specifically to novel-reading from a Western historical perspective and not storytelling/listening within far older oral cultures, but the outcome we would like to suggest might very well be the same, if not more pronounced.[15] The stories different cultures tell themselves mirror a particular set of virtues and values that can inform local policy-making and public life. One example where this approach is used in a living legislative way is CTFN's Family Act. Drawn up in 2008, the Family Act follows the CTFN (Tlingit and Tagish) clan-based governance model (see Visions North 2009 report[16]) which draws on the two CTFN statutes of government books on virtues and values. Book One is titled *Traditional Beliefs and Practices: Our Place, Our Responsibilities* and Book Two, *Government of Carcross and Tagish Traditional Family Beliefs and Practices*.[17] The Family Act is guided by virtues and values distilled from pertinent Tlingit and Tagish traditional oral narratives sketched out in Book Two. The CTFN Water Act that this collaborative water work lays the framework for will follow a similar formula. Tlingit and Tagish fluent speaker and Elder Angela Sidney is much quoted when she made clear the value and inter-generational critical importance of storytelling. Sidney's famous quotation "my stories are my wealth", has profoundly influenced how indigenous and modern ontologies are both intersected and understood in the Canadian North.

As mentioned at the beginning of this chapter, the role and agency of water within Tlingit and Tagish culture provides a living model for linguistic, technological, metaphorical, corporeal, navigational and storytelling practices and performances at a profound cultural level. We argue that the unique Tlingit and Tagish embodiment of water as model within their cosmology offers a particular vision of water, or a water ethic that can inform applied philosophical and legal debates concerning water governance and management. It is both a source for water justice, but also furthers discussion within international environmental law on the definition and nature of both personhood and inter-generational justice (see Hayman, with James & Wedge, forthcoming 2017). The evolution of this particular sophisticated water consciousness has been developed through the art of listening that is at the heart of the Tlingit virtue of *yáa át wooné*.

A daa tutan i yux'atangee, yáa át wooné and the art of listening

Most obviously the Tlingit and Tagish cultures are rooted in a 9000-year-old *oral* tradition. Storytelling is a core cultural practice drawing on and living out rich and encyclopaedic oral histories. Listening well is a critical virtue developed from a young age. Tlingit pedagogy in fact begins before a child is born. In coastal Tlingit elder David Katzeek's words:

> I want to describe the approach and methodology of our people, when we are beginning to learn. At the very beginning the mother is bearing the child in her womb. *Hayn too ye ati*. She has water in and around that child. They [Tlingit family] would speak to that child before it was even born. They would say [to the child in the womb] when the mother came into the tribal house . . . 'For you desiring to come to our world, we are very grateful and we are very happy.' *Gunalscheesh A!* They would say thank you without question or debate. 'You are a precious child of the Earth'. . . Then they would say . . . 'You are a human being, and being human means that you will listen to those who instruct you, to those who teach you, to those who love you, and care about you.' 'Most precious child of the Earth'. They are talking to that little baby that's not even born! . . . They are telling that child already the first cornerstone of the Tlingit language and the Tlingit educational system: this is to learn, to listen, to pay attention, to be still, to be quiet; to become focused. That is the essence of being Tlingit. That is the essence of being a human being.

These qualities, particularly listening skills, that are developed within Tlingit pedagogy are foundational for developing relationships with the more-than-human. Many Tlingit oral narratives describe marriages between humans and other animals, humans and trees, humans and fire sparks, and humans and fish scales. When it comes to learning about water, David Katzeek explains with an example of 'the woman who married the water' as an analogy to these ancient traditional Tlingit oral narratives for this research. 'Marrying' in the Tlingit sense of the word means developing a powerful intimate and spiritual relationship. In terms of 'marrying the water' it is about hearing the cry of the water and answering it:

> When we say is *yáa át wooné* that means to learn about it. . . . You are going to meditate on it, you're going to think on it, you're going to develop a relationship with it . . . They tell the story of the woman who married the water. And they [some of the audience] are going to say 'how can a woman marry the water?' You see in Western languages, in European languages, marrying means the man and woman coming together and having a real intimate relationship . . . No! . . . When you develop a relationship . . . you are going to learn about that individual. You are going to begin to study that individual in such a way that you appreciate them . . . We [Tlingit people] say in the English translation 'the woman who married the water'. However in reality you are developing a relationship with that resource called water, so that you can really appreciate it. When things come that might be detrimental to that relationship, you are able to raise a yellow flag and be able to think carefully. You see education *is* developing a relationship.

Another critical Tlingit concept that follows from *yáa át wooné* is *Woochéen*. As David Katzeek again explains, *Woochéen* is to be in harmony with mind, body, soul, spirit and the environment, especially water:

> A lot of people translate *Woochéen* as 'we are all supposed to work together as human beings'. That is part of it but not all of it. We are to work *with* it, we are to be in harmony with it. We are not to destroy it, we are to work *with* it. . . . You are going to work with the water – you are not going to work against it. Big words '*Woochéen*'. You are to work together. And what it means is that the mind, the body, the soul, and the spirit – all of that being is put together and focused on one particular subject. That is how you begin to develop a relationship.

Learning to listen. Actively listening, are the core Tlingit teachings here. However, the other side of listening well in Tlingit pedagogy is to speak well. Tlingit teachings emphasize the concept of *a daa tutan i yux'atangee* which means 'to weigh your words with care', and reflects the power of spoken words themselves.

Coastal Tlingit elder Nora Marks Dauenhauer has worked extensively on Tlingit oratory and narratives. In *Haa Tuwunaagu Yis*: *For Healing Our Spirit – Tlingit Oratory* (1990), she and her husband, anthropologist Richard Dauenhauer, comment that 'ceremonial oratory is a gift and an invitation to respond appropriately, either verbally, if words are called for, or emotionally, through listening and becoming involved' (1990, p. xx). Precise protocols exist so that words are respectfully acknowledged and also shown to be received. Coastal Tlingit elder George Jim explained the importance of this protocol in an interview conducted by the Dauenhauers in 1989:

We don't like to see words go to waste.
We lay our *at.oow*[18]
For the words to fall on.
This is when the opposite clan
Knows that its words
Have been received.
(George Jim, interview, 1989, in
Dauenhauer & Dauenhauer 1990, p. 263)

A daa tutan i yux'atangee is not only reserved for humans, but is highly relevant for all other more-than-human interactions as well, often with survival implications:

They [brown bears] listen to us
When we say to them,
I'm over here for my food, too
This is when they turn away from us
We call them
Big ears,
Big ears
They are listening to us right now
From a while back
They are listening
To the way we have been speaking of them.
(Coastal Tlingit elder Austin Hammond,
Fairbanks, Alaska, 1988, in ibid., p. 227)

A daa tutan i yux'atangee is critical at all times, even when not in the direct presence of another animal, glacier or river.[19] Other animals, and particularly glaciers[20] are always listening, and by example, encouraging humans to do the same. The whole geographical region is conceived and imagined in fluid, animate and sensory terms where listening is privileged above the many other senses such as sight and smell.

Another example is the older Tlingit tradition of speaking respectfully to salmon as they return from the sea to spawn in fresh, cold, oxygen-rich water upstream, known as *ish* in Tlingit. The precise Tlingit expression is *Hoi Hoi* and respectfully acknowledges the salmon for their journey, and potential sacrifice as food (anthropologist Richard Langdon, Sealaska lecture, Juneau, 2013). A last example taught by inland Tlingit and Tagish Elders is never to throw stones into the water, as this is met with meteorological consequences such as strong winds whipping up lake waves (James, personal communication, 21 June 2016). The lakes listen and react.

Inland Tlingit and Tagish oral narratives further emphasize the role and function of listening through two fundamental stories of the region. The first is the story of Game Mother/Animal Mother and her role as creator of all animals over the Bennett Lake and Nares Lake systems which lie to either side of the village of Carcross.

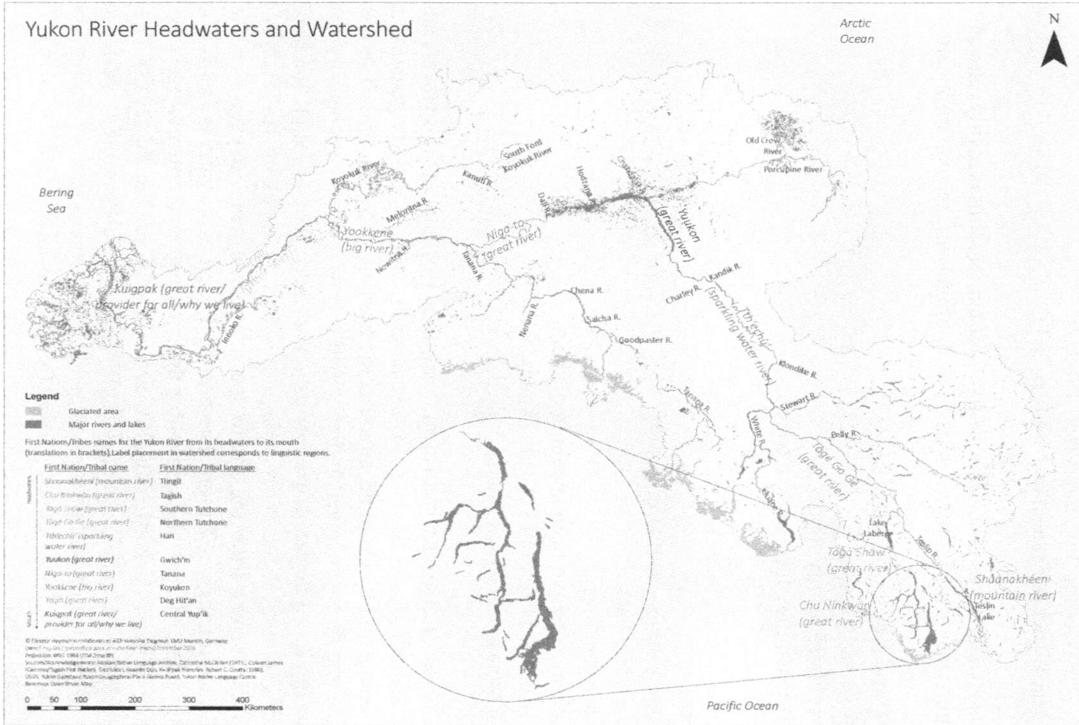

Figure 13.1 'Thinking like a watershed'. Yukon River watershed and headwaters (Carcross/Tagish First Nation's traditional territory) showcasing First Nation and Tribal indigenous language names for the Yukon River.

Source: Eleanor Hayman in collaboration with Veronika Degmayr, LMU Munich, Germany.

By listening patiently to all the songs the animals sung as they were created and modified by Game Mother, the Tagish people came to know each animal's unique song, and understand the worldview through that animal's song expression:

> A Tagish man of the *Dakleiweidi* moiety reputedly saw Animal Mother give birth to all animals in the interior. Following the event she hung a 'swing' up between four mountains near Carcross. . . . After each kind of animal had danced on the swing and sung a song, Animal Mother then gave them their characteristic attributes of today, often exchanging the teeth, horns or other feature of one animal for something from another. She told them what to eat and how to behave in general. The Tagish man sat for two summers and one winter, during which time he did not even eat. He also learned the song sung by each animal so that the Tagish *Dakleiweidi* can sing them today.
>
> (McClellan 1975, p. 90)

Colleen James remembers the Pike song taught to her and many others by CTFN Elder Angela Sidney, and specifically sung at *Taaseliyi aayi*, literally in Tlingit 'Pike Lake', but today officially known as Nares Lake after the British naval officer and Arctic explorer George Nares.[21] This points to an inland Tlingit and Tagish understanding of many songs concerning relationships with aquatic life as well as mammalian life. However, we should be clear that separating out the senses into what might be called a Eurocentric bias and analysis is not our intention. It is clear from Tlingit and Tagish oral traditions that senses blend and interconnect. The narrative of Game Mother blends kinaesthetic movement with an acoustic pulse that provides the memory framework within oral tradition. Many Tlingit and Tagish oral narratives are about movement and journeys, and this movement is reflected within both place names and the verb-based Tlingit language itself. The majority of Tlingit and Tagish place names are aquacentric and are concerned with movement, either literally, corporeally or metaphorically (see Figure 13.2). Tlingit and Tagish place names reveal complex, sophisticated cultural, ecological and even acoustic knowledges. Over 75% of Tlingit and Tagish place names are water related – they are 'aquacentric', also see Hayman, with James & Wedge, forthcoming 2017).

At the same time, there is the realization by the Tlingit and Tagish that the art of listening, so well-practised by the ancestors, is being lost. Coastal Tlingit Elder George Davis lamented this loss, articulating it in poetic Tlingit, in Sitka, Alaska in 1980:

> Yes, now
> When the sound,
> Yes,
> Kept fading,
> Of what we heard faintly
> Of the way our grandparents used to say things,
> Yes
>
> (George Davis (3), Sitka, 1980, in
> Dauenhauer & Dauenhauer 1990, p. 227)

Figure 13.2 Haa saaxú, haa latseení, hà Kus Teyea 'Our names, our strength, our way'. Counter-mapping with Carcross/Tagish First Nation.

Source: Eleanor Hayman in collaboration with Veronika Degmayr, Colleen James and Mark Wedge.

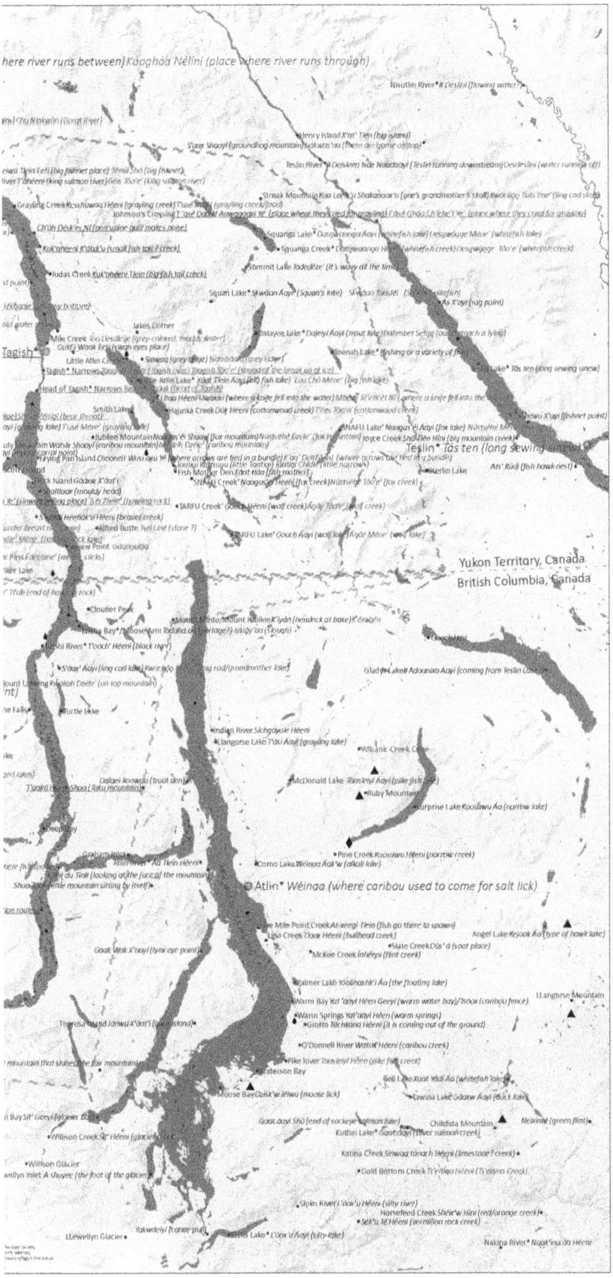

Haa saaxú, haa latseením[1], hà Kus Teyea[2]

(our names, our strength, our way)

Tlingit and Tagish place names[3] of the Yukon River headwaters, Canada (Counter-mapping: Charting Tlingit and Tagish "aqua-centric"[4] empirical scientific knowledge)

Legend

- Carcross/Tagish First Nation traditional territory in the Yukon agreed on 22nd Oct 2005 with CTFN, Gov. of Canada & Gov. of Yukon[6]
- Carcross/Tagish First Nation outstanding treaty settlement land in British Columbia (treaty in negotiation since 10th June 1997)[6]
- Yukon River *(Shaanakhéeni)* watershed[7]
- Alsek River *(Aalseix)* watershed
- Northern coastal waters of B.C. drainage area *(Kluhini, S'e Shuyee, Jilkàat, Jilkoot etc.)*
- Nakina *(Naak'ina.áa Héeni)* River/Taku *(T'aaku)* River Watershed
- Ice *(t'éex')* fields and glaciers *(sít')*
- Salt water *(Éil')* /Pacific Ocean with bathymetry in meters
- Fresh water *(Héen)* [8] with flow direction
- Lake bathymetry[9] in meters
- Settlements *(aan)* (current human population over 100)
- Ice *(t'éex')* patches/archeological sites (up to 9000 BP - Before the Present)[10]
- Volcanic centers (last eruptions: uncertain)
- Hot/thermal springs *(Yat'aayi Héen)*
- Lewes Dam (first installed 1899), rebuilt several times, most recently in 1976
- Whitehorse hydroelectric dam - 40 MW capacity (1958) with fish ladder[11] (1959) and hatchery (1984)
- Goat Lake Hydro facility - 4 MW capacity (since 1997)[12]
- Yukon Energy's potential hydro-electricity projects (as of February 2016)[13]
- name — Official place name[14] (Geographic Board of Canada 1897 onwards)
- *name* — Tlingit language place name with translation
- *name* — Tagish language place name with translation
- *name* — Tagish and Tlingit place name composition
- name — Place name related to sound[15]
- * — Anglicized Tlingit or Tagish place name
- # — Aquatic directional in Tlingit or Tagish place name[16]
- ° — US & UK military acronym (from World War II)[17]
- Klondike[18] and Alaska Highways, and other roads
- White Pass & Yukon Route: narrow gauge railroad (since 1898)
- Borders (in evolution since 1867)

In spite of coastal Tlingit Elder George David's lament, there are many and various examples from our collaborative water research and in the literature, of the continued importance of listening carefully, speaking and acting[22] respectfully. Coastal Tlingit Elder David Katzeek has offered crucial insights into listening to the Tree People and to the Salmon People[23] through his work with Tlingit scholar Lance Twitchell, as well as interviews recorded in and around his home Tlingit village of Klukwan, Alaska in 2014. David Katzeek opened his first interview with a question that has framed not only this chapter, but much of our collaborative water research since. David Katzeek asked: '*I yá.a̲xch'age?* [Can you hear it?].' David was referring to the competency of this research to actually listen to and understand the voice of water.

> I am up here working with Eleanor Hayman who is from Germany. We are doing a study on human relationships with the water, and why water is so important to human beings. Particularly how we [Tlingit peoples] understood the scientific kind of information that we have regarding the water, and all the stories that relate to water.
>
> It is so important. And I will just say this. . . . This is what the elders would say . . . Can you hear it? Can you hear it? *I yá.a̲xch'age? I yá.a̲xch'age?* It is important that this type of work be called out today, for it is so easy to walk through life not hearing the Earth calling out to the people regarding the way in which we are interacting with it. Can you hear it? *I yá.a̲xch'age*? What kind of words are speaking to you and to me? Are we listening? Are we paying attention?
>
> (David Katzeek, 6 September 2014, address to youth at the Klukwan Tribal Longhouse, Klukwan, Alaska)

David Katzeek further illustrated the power of listening to water's voice by composing a song in Tlingit to the *Klehini* river as we sat by it, but also emphasizing the profound connection that the Tlingit language has with water:

> Can you see it? Can you see it? Can you see it? Can you see how the Chilkat River is flowing? Can you feel the river flowing? . . . It is flowing like joy, like happiness, like celebration! Chilkat River is flowing like the river that it is!
>
> I can compose a song just by looking and feeling the spirit of that water. I sang to the water. I sang about the water. The joy, the celebration . . . The connection that our [Tlingit] language has with the resource. And it comes – the spirit of love for the resource. Not taking it for granted. . . . The water gave me the words. It is like the water saying, 'can you see the Earth celebrating?'
>
> (David Katzeek, 6 September 2014, Chilkat River, Alaska)

Sensory geographies and the ethnography of sound

Theories of the intersections and interactions of space and place with humans and other animals have long interested geographers and philosophers (see Tuan 1972; Malpas 1999, Casey 2013), anthropologists (Cruikshank 1998; Ingold 2000; Thornton 2010) and ecopsychologists (Gibson 1977; Abram 1996). Most recently, human ecologists and environmental philosophers have contributed to this broad field inspired by particular shifts in thinking, be it spatial, ecological or sensory. Theories of perception initially tended to focus on vision, driven by the Eurocentric 'visual' history of 'landscape'. Shifting this visual bias sociologist John Urry writes, 'Each sense contributes [in its own way] to people's orientation in space; to their awareness of spatial relationships; and to the appreciation of the qualities of particular micro- and macro-spatial environments' (Urry 2011, p. 388). Anthropologist Tim Ingold in *The Perception of the Environment* (2000), and phenomenologist and geographer Edward Casey (2013), have further successfully shown how space is converted to place through a process of engagement; a combined multi-sensory experience that is shot through with the cultural values assigned them.

Only in the last 15 years has sound become established as a serious focus for many human–ecology projects inspired particularly by musician and ecologist R. Murray Schafer, and anthropologist Steven Feld's various urban and rural anthropological work (Feld 1990, 1993). Although linguistically the Tlingit language does not use waterway terms for music intervals, rhythms or melodies, many Tlingit linguistic concepts are rooted in water imagery, movement and orientation, particularly the orientation of the human body towards and/or away from salt and fresh water bodies; and upstream or downstream. This is the kernel of Tlingit understandings of the human condition in relation to water. From an eco- (or as we prefer *aqua*) linguistic perspective, this is very powerful. For example, *yán* means the shore, but also a positioning linguistically such as shoreward and landward; and even in its third interpretation/meaning in the Tlingit language refers to 'work' being done or completed (Twitchell personal communication 2016; and dictionary 2016). The shore as a liminal, in-between zone is also a reference for estranged or problematic Tlingit kinships such as in *neechkaḵáawu*[24] which is a person without a clan or a person nobody wants; literally in Tlingit and as an extremely insulting term, it means a 'person on the beach' (Twitchell 2016).

Further descriptions of other Tlingit relatives – the more-than-human world – often have a water directional/reference or directional description embedded within them. From an eco-linguistic (Stibbe 2014) analytical perspective a Tlingit worldview can be understood to present a relational, respectful discourse not just between the human and other animals but between many animals and water itself. Activities and types of movement are in relation to water or the hydrology and geography of water. For example, the fish that in the English language is called mackerel is *dákdesax'aak*, literally translated in Tlingit as 'swims underwater out to sea' (Twitchell 2016). The polar bear, *héen-táak-xóots-Ÿí*, is literally translated in Tlingit as 'in the bottom of the water bear' (Crippen 2013). Three excellent bird examples are *eek̲ lukak̲ées'i* which translates as snipe, literally 'flood on the point of the beach'; *hinyikl'eix̲i* is the dipper or water ouzel and is 'dancer in the water'; and *hinkag áax̲i*, the red-throated loon or arctic loon, is 'cries on the water' (Twitchell 2016). These few examples in themselves suggest a particular, indeed profound understanding and relationship with water based on 'a scientific understanding of how organisms (including humans) depend on interactions with other organisms and a physical environment to survive and flourish, and also an ethical framework to decide why survival and flourishing matters and whose survival and flourishing matters' (Stibbe 2014).

Locational formative metaphors in Tlingit thought are shown in some aquacentric place names. For example, *Shtax' héen* (Sitkine River) literally means 'water biting itself' (James, 9 Jan 2016, personal communication) while *Daa kahéeni* or amniotic fluid, translates literally as water from beginning to end, all encompassing. It also means body of water. Looking at water as a 'body' (Katzeek, 16 June 2014, personal communication). Some idiomatic expressions couple water with intelligence. For example, 'water in the head' refers to a smart person (Twitchell 2016; James, personal communication 2015). Lastly, water animates almost all Tlingit and Tagish place names (toponyms) around the headwaters of the Yukon River. Unique to toponym research in the southern Yukon region is the recognition that many of these key place names mimic or reflect the soundscape (or soundmark) at that location; and almost all those soundscapes reflect the agency of water and ice, or the interaction of wind/fish/geese/earth with water, an example of Bull and Back's concept of 'deep listening'.

In 2003, Michael Bull and Les Back published the first *Auditory Culture Reader*. By privileging listening over seeing, they argued that listening, or what they term *deep listening*, reveals four aspects of sound as a core way of knowing and being in the world:

- Sound makes us re-think the meaning, nature and significance of our social experience.
- Sound makes us re-think our relationship to community.
- Sound makes us re-think our relational experiences, how we relate to others, ourselves and the spaces and places we inhabit.
- Sound makes us re-think our relationship to power (Bull & Back 2003, p. 4).

We would like to suggest that such deep listening is in fact a key pedagogic teaching element and virtue instilled in the Tlingit educational system, facilitated by traditional oral narratives and a particular relational relationship with water and water bodies.

Crow's critical role in shaping the Tlingit/Tagish hydrological imaginary and Tlingit/Tagish identity with water

CTFN have adopted the Tlingit social structure; that is that Wolf and Crow head the two moieties. This is rooted in the inland Tlingit and Tagish understanding that Wolf was the last to dance and sing on Game Mother's swing/trampoline above

Carcross, and that Crow was the creator of the world. The wolf and crow moieties that structure the Tlingit and Tagish socially, showcase another critical Tlingit philosophical principle – that of Wooch ya<u>x</u>h datí or balance.

> It should be noted that Wooch ya<u>x</u>h datí (balance) is the most important aspect and the essence of Tlingit beliefs. This means that it is important to maintain the balance. It is as simple as that. If you lose your balance you will become dizzy lose your equilibrium and actually become dysfunctional.

The Tlingit use of metaphors to accentuate water-related phenomena or abstract ideals is comprehensive, if not unique in the circumpolar north. These combined narrative traditions have not only provided a sustainable social structure for thousands of years, but have been the genesis of the complex Tlingit and Tagish oral tradition of songs, greeting songs, potlatch songs, love songs and honouring the dead oratory (Swanton, 1901; Dauenhauer & Dauenhauer 1990; Wedge, personal communication 2014; James, personal communication 2014). Many traditional stories, not just from the Tlingit and Tagish, originate in, on or around lakes, rivers and the sea. This is not limited to the Tlingit and Tagish, but seen throughout many other indigenous peoples from the Inuit in north Alaska and the Yukon, to Greenland and northern Siberia; many have similar aquacentric oral narratives, something common perhaps for the circumpolar north (for example, Kiana Oman 1995).

More broadly water is fundamentally tied to the Tlingit migration history as David Katzeek notes (Katzeek, personal communication 2016):

> According to our migration history it was a great eagle that led us into the land that we now call Southeast Alaska, Lingít Aani. The great eagle led us from spring water to spring water until we got to the Chilkat Valley, Klukwan. Goon héen (ground spring water) is the heart and life of Tlingit people and the resources of the Chilkat Valley. Ishka Héeni is spring water within the water system vital to the millions of salmon that spawn in the Chilkat River.

Water again is a key actor in the cultural and geographical making of place for the Tlingit people.

Critical themes that flow through much of indigenous oral literature are those of right relationship or right kinship or the right (good/appropriate) way of dealing with situations. Exemplified by the Tlingit *yáa át wooné*, many of these virtues can be said to be spiritual at base because as Allen emphasizes, 'harmony, relationship, balance and dignity are [its] informing principles because they are principles that inform our spiritual lives' (1989, p. 11).

Tlingit scholar Nora Marks Dauenhauer identifies some specific themes running through Tlingit oral narratives and Tlingit ceremonial oratory. These include the conflicts between virtues and vices such as loyalty, alienation, self-concept, pride, arrogance, revenge, and of course *yáa át wooné*. One of the most unusual aspects of Tlingit and Tagish storytelling is the theme of transformation. At the beginning of a Tlingit and Tagish understanding of time, humans and other animals could all talk with and understand each other. However, in more recent storytime, one of the key ways in which humans understand the worldview and ways of other animals is by 'marriage' as David Katzeek has already alluded to earlier in this chapter. Many Tlingit oral narratives speak of humans marrying the spirits of spruce trees, sparks from fires, stars, fish scales, bears and salmon. Other ways of understanding more-than-human worlds include humans transforming or being transformed into otters and salmon to inform and educate their fellow humans about alternative worldviews (water animals). Dramatically, in the Raven cycle of stories, *Yeil* or Raven/Crow transforms himself first into a hemlock needle, then into a human baby to enable him to release the sun, moon and stars imprisoned in three boxes for the benefit of the whole Earth. In this narrative, water provides the rhetorical device for enabling these transformations through its symbols of flow, transformation and source of life.

In another foundational Tlingit narrative, *Yeil* steals water not simply to distribute it over the earth, but to form and shape all the lakes and rivers in the Yukon and Alaska. *Yeil*'s role in these two narratives shapes not only the physical tangible world, but establishes the agency, function, relationship and idea of water in the Tlingit and Tagish imagination, something we call a 'hydrological imaginary'. Crow lays down crucial ethical and relational water management (a counter to Eurocentric water management) concepts through these narratives: sources of life should never be owned, tamed or imprisoned. Crow also establishes a baseline water ethic, founded in *yáa át wooné.* Other ancient or age-old stories in Tlingit and Tagish oral narratives in the southern lakes region reference a flood, confirmed in the paleo-hydrological record (Kennedy & Bond 2003; Smith 1994), and outline a crucial aquacentric phase in the cultural memory of the Tlingit and Tagish. Survival relied on reading the late-glacial advances and retreats roughly 10,000 years ago, and predicting potential floods. Simple but effective teachings for our contemporary water crisis point to shifts in conservative and damaging valuing of water that will require a number of qualities. These include intelligence, the willingness to transform and a profound *yáa át wooné* for all life.

These oral narratives, including other Tlingit and Tagish songs and ceremonial oratory are remarkable, not least for the messages embedded within them, but also the fact that they have been kept in a stable form over centuries, and even millennia. These are oral traditions that rely entirely on the human memory for storage and transmission. Recent research that connects oral traditions with cognitive psychology, neuroscience, linguistics, folklore and rhetoric shows how complex and sophisticated oral storytelling cultures endure through a practice of what we introduced above as 'deep listening' (Bull & Back 2003).

As psychologist David C. Rubin argues (1995), 'oral traditions are systems of multiple constraints. These constraints include the organisation of meaning, imagery, and patterns of sound . . . these constraints lead to stability without fixing the text; they limit the choices of what can be sung in a given context, and they cue memory' (p. 10). Rubin also suggests that sensory cues within oral narratives are key to being retained in the memory (1995, p. 78f). We want to expand the notion of deep listening here to include a type of cultural memory practice that allows for oral narratives' biological survival. Indigenous scholars again confirm much of Rubin's findings, for example, that there is a powerful spatial element coupled with movement and specific locations in many (Tlingit and Tagish) oral narratives[25] (Rubin 1995; Kiana Oman 1995; Dauenhauer & Daunenhauer 1990). Within Tlingit and Tagish oral narratives, it is water again that provides the medium and model for almost all spatial narrative exploration. Similarly, water provides the model for the mnemonic device of looping. Storytelling often involves a looping back on itself – which as indigenous scholars remark, is a mnemonic device so listeners can pick up the thread of a long tale. Our previous research has also shown how oral narratives themselves have social lives that are circulatory (Hayman, with James & Wedge 2015).

Shuka, and Rosalind Hursthouse's new virtue of 'respect for nature' or 'being rightly oriented towards nature'

Philosopher Martha Nussbaum argues that political justice can be developed through a virtue-based, ethical outlook (Nussbaum 1993). Nussbaum further contends that a virtue-based ethical outlook can help answer core sustainability questions such as what do we owe future generations? And how can we specify what we/they need? Environmental philosophers Rosalind Hursthouse (2007) and Brian Treanor (2014) might well argue that such an approach needs a fundamental revision to include the environment as a core value in this thinking.

We would like to suggest that the Tlingit notion of *Shuka* addresses both issues, and has been a powerful Tlingit term for many thousands of years (Katzeek, personal communication 2016). According to the Dauenhauers in *Haa Tuwunaagu Yis: For Healing Our Spirit – Tlingit Oratory* (1990), *Shuka* as a concept in Tlingit culture is:

> pivotal because it is ambiguous and faces two directions. It means, most literally, 'ahead', or 'before'. It refers to that which is before us or has gone before us in time – predecessors, 'one before', 'one who has gone before', those born ahead of us who are now behind us, as well as those unborn who wait ahead of us. Thus the term refers to the past and also to the future – to that which lies ahead.
>
> (Dauenhauer & Dauenhauer 1990, p. 19)

This reflects a critical element within a Tlingit and Tagish aquacentric cosmology. For example, the Dauenhauers emphasize that the 'narratives in *Haa Shuka* recall the fundamental covenants between humans and [other] animals, the physical and the spiritual and the symbolic representation of these covenants through *at.oow*' (1990, p. xxiii). These covenants are practised through the Tlingit/Tagish principle of 'all my relations'. Indeed CTFN Elders make it very clear within our collaborative water research, that all life on Earth is related through water, salty and fresh. They confirm that in their belief system they are 'part of the earth, part of the water' (after Virginia Smarch, Teslin inland Tlingit elder, in McClellan 1975). Water therefore offers a literal temporal model where water is the primordial agent that materially connects present generations with both the deep past and a possible deep future (Hayman, with James & Wedge 2017).

Water is not only a model for the circulation of the Tlingit and Tagish virtue of *yaa at* woone,[26] but also the model for the pivotal Tlingit concept of *Shuka*. It is most importantly a form of what Eurocentric practices and international law call inter-generational justice. From an eco-linguistic mandate, these are the story-tellings, the sustainable discourses that can inform global water ethical debates.

Aquacentrism

Aquacentrism or aquacentric positioning/thinking is a term we use to counter and expand the notion of terracentrism. We understand terracentrism[27] to be the tendency to privilege land-based/biased histories and narratives in Western thought. Although they do not use the term terracentric, environmental historian John Gillis (2011) and eco-critic Dan Brayton (2012) have both noted how environmental history and literary studies have been viewed through a terracentric ecological lens that has typically been 'green' and not 'blue'.

We employ the term 'aquacentrism' to underline an aquatic orientation, cultural evolution and fluid thinking within Tagish and Tlingit cosmologies. Environmental historian Rainer Buschmann has previously used the term 'aqua-centric' to draw attention to cultures displaying an 'oceanic aqua-centric consciousness' (2005, p. 2), but Buschmann's angle on this term is more ideological highlighting how 'Western historians are quick to point out that the Pacific is in fact a European artefact, and they generally allow for little indigenous involvement' (2005, p. 2). We understand the Tlingit and Tagish culture to be specifically aquacentric based on a critical Tlingit and Tagish relationship with water, evidenced by both material culture and a place-based hydrological imagination referenced through oral narratives, local place names and the Tlingit and Tagish languages themselves.

The coastal and inland Tlingit and Tagish peoples have co-evolved with the significant bodies of water around them (the southern Yukon lakes, Yukon River and Pacific Ocean) for over 9000 years. This aquacentric reciprocal unfolding is reflected tangibly by Tlingit clan names, Tlingit and Tagish place names over 75% of which are directly water-related, and many significant traditional oral narratives that still have high currency and agency today. Examples include 'Salmon Boy', 'The Boys who drifted down the Yukon River' and 'Wealth Woman'.

Intangibly, this aquacentric evolution is witnessed within the Tlingit verb-based language where water and blood use interchangeable verb forms (Twitchell, personal communication 2014). Core directionals are in relation to the direction of water flow (upstream, downstream) as well as to the relative position of the sea (inland from the sea, towards the sea). The agency of water has provided a model for defining Tlingit and Tagish social structure and identities through the power of names and naming. This is clearly seen in Tlingit clan names as well as clan/house origin stories. For example, *Ishkeetaan*, a CTFN clan, means 'People of Deep Pool in the River House'; *Goon Hít* in Alaska means 'Spring (of water) House'; the *Yaakwdáat K̲wáan* means 'People of the Mouth of Body of Salt Water' that includes communities such as Yakutat, Icy Bay, etc. *Lk̲óot K̲wáan* or the People of the Chilkoot Area, literally 'People of the Lake that Puked', includes communities such as Haines, Chilkoot, Skagway, Dyea, Chilkoot Trail and Lynn. In fact,

Shgagwéi or Skagway literally means in Tlingit 'bunched up water' or 'whitecaps on the water' (all references in Twitchell 2016).

One meaning of the word Tlingit is 'people of the tides' which reflects why the Tlingit and Tagish have followed the lunar (aquacentric) calendar. Water (and ice) have further provided geographical organizational reference points as well as dictating the seasonal movement for hunting, trapping and fishing for the Tlingit and Tagish on, in and among the southern Yukon lakes. The agency of water (including glacial flows and surges, hot springs and melting patterns) have impeded and/or facilitated the exchange of ideas, goods, canoe design and technology,[28] as well as marriages between the coastal Tlingit and inland Tlingit and Tagish peoples.

The construction of hydrological space, but also linguistic and imaginative space has long been dominated by Western constructs and their parcelling out of the world, whether it be cartographically, imaginatively or literally. Understanding a Tlingit and Tagish cosmology through an aquacentric, or 'thinking *with* water', perspective reimagines post-colonial space. Tlingit and Tagish place names of this region are highly sophisticated and rich aquacentric understandings of place that are remapping or reimagining the current, dominant and imported hydrological imaginary of the region through what we call 'deep charting'.

We further argue that such an aquacentric approach facilitates a reconsideration of what it means to be human within the context of contemporary international legal and philosophical scholarship. Water begs greater attention from the international legal community as a medium to think with for a possible inter- and intra-species future (Hayman, with James & Wedge forthcoming 2017). A potential aquacentric path to international law and environmental ethics requires serious dialogue with far older aquacentric cultures and philosophies, many of which have a deeply intimate hydrological knowledge that has provided an ethical model for behaviour. Our collaborative water research provides a framework for Tlingit and Tagish water legislation, but also acts within broader philosophical and legal systems.

Deep Charting and deep listening

Thinking *with* water through and with Tlingit and Tagish oral traditions opens up ethical spaces where 'alternative ways of storying and mapping waters can give voice to inclusive and evolving vocabularies of water places, thereby transforming collective ways of thinking' (Chen 2013, p. 9). As a post-colonial and critical indigenous theoretical approach, we introduce our collaborative counter-charting project that showcases Tlingit and Tagish place names. Many of these place names are unique for global toponym research as they either personify a specific location, or act as a soundmark for a seasonal and/or hydrological site.

We are currently producing both 2D and 3D Deep Charts. The latter utilizes the free Google Earth Pro platform which through its multi-media, multi-layered GIS technologies is able to privilege the profound and unique acoustic (listening) or audio/oral element foundational to Tlingit and Tagish cosmologies (see Hayman, with James & Wedge forthcoming 2017). Such shifts in how geographic knowledge and in particular cartographic and GIS applications and knowledge are (re)produced has only recently become possible through such digital online applications which include Neatline, Geolive as well as Google Earth (Pro) platforms. Sensory studies scholar and critic David Howes suggests that:

> The practice of cartography has metamorphosed from the production of two-dimensional scalar projections into cybercartography or 'multisensory mapping'. This development is partly due to advances in technology. But it is also inspired by a growing awareness of what the study of indigenous knowledge systems, which tend to be nonpictorial, such as Inuit wayfinding, can contribute to our understanding of human spatial orientation. At the Geomatics and Cartographic Research Centre at Carleton University, there are many innovative cybercartographic forms being developed, which take their inspiration from indigenous practice (see Lauriault & Fraser Taylor 2013).

Our Deep Charting responds to the recognition that 'paradigmatic knowledge' (that here specifically includes geo-spatial technologies such as remote sensing and Geographical Information Science (GIS)) – based in scientific thought, has been privileged to the detriment of 'narrative' knowing – located in the construction of experience through narrative (Cleaver 2009). Deep Charting is a form of counter-mapping or reclaiming the map and the practice of mapping itself. Deep Charting critically utilizes new geo-spatial technologies and associated media to weave together multiple narratives in an attempt to capture experiential, emotional, sensual, acoustic, spiritual and metaphorical space as well as more quantifiable signatures of a specific region. As such Deep Charting practices with the inland Tlingit and Tagish peoples can be seen as a move towards decolonizing both 'water' and 'mapping practices' themselves (Hayman with James & Wedge forthcoming 2017).

Specifically, our 'Deep Chart' *counters* all current maps and narratives of this region in specific ways. Tlingit and Tagish deep charting counters the:

- privileging of land over water (maps by definition do not reveal bathymetric (water) depth, water temperature, qualities, currents or navigational hazards);
- privileging of modern water (valued as commodity) over Tlingit/Tagish water (valued as relationship, reciprocity and respect);
- privileging of written (visual) 'text' over oral (acoustic) storytelling and place name 'texts' – an acoustic ecology;
- privileging of explorer/entrepreneur/military (male) place names (glory/narcissism) over Tagish and Tlingit place names (which detail precise hydrologies and ecologies of the region from water ecology to navigational hazards, to geomorphological change over time, as well as references and mnemonics for sites of oral storytelling narratives);
- privileging of egocentric over eco- (or aqua)-centric directionals (Eurocentric philosophical and corporeal constructs privilege an egocentric relational concept – left and right directionals are in relation to the body. Tlingit and Tagish directionals are in relation to the flow of water – upstream/downstream – and in relation to the sea – inland from the sea/towards the sea);
- privileging of recently imposed colonial borders over First Nations traditional territorial borders;

- privileging of modern navigational practices over ancient indigenous salmon cultural harvesting practices. For example, the introduction of the Lewes Dam (see Deep Chart – Figure 13.2) to 'flush out'[29] ice in the Yukon River for the Tutshi paddle-steamer to pursue its tourist goals. This effectively stopping the salmon run to the southern Yukon lakes and fundamentally altered many First Nations' reliance on salmon as core protein resource. More than that, salmon provided a literal and metaphorical value system as keystone species in the Tlingit and Tagish culture. Salmon shaped and still shapes key attitudes and values. The Tlingit and Tagish are a 'salmon culture' (James personal communication 2015; Wedge personal communication 2014).

Deep charting intentionally creates space for ontological difference, specifically, in this case, for different hydrological imaginaries that are either marginalized or have been silenced. Our deep charting gives voice to aquacentric wisdom embedded within a unique hydrologically rooted indigenous ontology.

Conclusion

The Western terracentric cultural construction of space and place, along with Western perspectives on virtue ethics continue to influence (water) policy-making and attitudes towards a societal transformation to a greener (we argue it should be a 'bluer') ethos at a global level. Tlingit and Tagish cultures have withstood the assimilation strategies associated with aggressive colonial thinking for over a hundred years. However, both Tlingit and Tagish cultures continue to remain under threat from state-led systematic marginalization of their values and virtues. Many of the core emancipatory principles that the Tlingit and Tagish cultures promote based on their aquacentric cosmology may well be a blueprint for informing a global water ethic, for example, the principles of *yáa át wooné* and *Shuka.* These principles have shaped a Tlingit and Tagish hydrological imaginary and water ethic, and offer remarkable potential to inform water ethics dialogues and are increasingly democratically showcased via modern technological GIS spatial platforms. Only via a collaboration on these terms – a complete revaluing of how water is listened to (Can you hear it?), thought *with* and understood as a relative/relation (marrying the water) – can the human species evolve into *homo-**hydro** sapiens*. Embracing an aquacentric ontology – a deep water consciousness – and actively working to decolonize not only water but knowledge itself, may very well be a deciding factor for planetary survival on this, our shared blue Earth.

Acknowledgements

I would especially like to thank coastal Tlingit elder David Katzeek/Kingeisti, for inspiring this water ethics chapter, and to Colleen James and Mark Wedge as CTFN community consultants for providing such guidance and clarity. This is based on much preparatory work already conducted with the CTFN community and government – *Gunalcheesh A*! I am also extremely grateful to all funding from the Heinrich Böll Stiftung, Berlin, Germany; the LMU Geography Department, Munich, Germany, and the Northern Research Endowment Fund, Yukon, Canada.

Notes

1 'To take care of, protect and provide for it with unconditional love' (David Katzeek, personal communication, 31 May 2016).
2 This chapter results from the collaborative research of Eleanor Hayman with Carcross/Tagish First Nation inland Tlingit and Tagish community consultants Colleen James/Gooch Tláa and Mark Wedge/Aan Gooshú in the Yukon Territory, Canada, and coastal Tlingit Elder David Katzeek/Kingeisti in Alaska, United States.
3 Elder David Katzeek, abstract for Sealaska Heritage Institute lectures, 2013.
4 Elder David Katzeek's spoke to me of 'the woman who married the water' in reference to this collaborative water research (September 2014).
5 Coined by feminist philosopher Lorraine Code (2006).
6 See especially Rosalind Hursthouse (2007) and the Roman Catholic Pope's 2015 encyclical.
7 See, too, the emerging academic field of 'new materialisms' that again, reinvents indigenous holistic thinking.
8 Other Indigenous ontologies are also open to perspectivism, but we focus only on the Tlingit and Tagish cosmology here.
9 Carcross and Tagish are the two main settlements within CTFN traditional territory. Carcross is a compilation of 'Caribou Crossing'. Tagish, on the other hand, refers to the village, but also to the southern Athabaskan peoples who first named this region. Tagish also refers to the language that the Tagish people spoke. Although there are no fluent Tagish speakers alive, there are efforts to revitalize the Tagish language.
10 Of the 130 Tlingit and Tagish toponyms recorded in CTFN traditional territory, over 75% are directly or indirectly related to water; for example, flow direction, turbidity, colour, temperature, mineral and oxygen content, etc. (see Hayman forthcoming 2017). This is one out of many Tlingit and Tagish cultural indications that privilege detailed empirical hydrological knowledge. We use the term 'aquacentric' to describe both the orientation of inland Tlingit and Tagish toponyms, but along with other aquatic-laden indications, the Tlingit and Tagish cosmology.
11 Champagne and Aishihik First Nation, Syilx First Nation, and the Anishinabee first peoples are also in the process of developing indigenous water legislation in Canada. See http://news.umanitoba.ca/elders-to-release-preliminary-report-on-anishinaabe-water-laws/ accessed 27 December 2016.
12 Interview with Professor Martin Grambow, civil engineer and Bavarian ministry for the environment, interview, accessed May 2015, www.professoren.tum.de/en/honorary-professors/g/grambow-martin-g/.
13 For example, *Idle No More*, the *Mother Earth Water Walk*, and the *Indigenous Environmental Network*.
14 For more information, see www.drze.de/in-focus/biodiversity/modules/anthroporelational, accessed 21 June 2016.
15 Cognitive psychology studies conclude huge differences in memory within oral tradition: active engagement/involvement by listening; passive, detached by reading.
16 www.aadnc-aandc.gc.ca/DAM/DAM-INTER-YT/STAGING/texte-text/vnss09_1100100028411_eng.pdf, accessed 18 July 2016.
17 See www.ctfn.ca/documents/legislation, accessed 18 July 2016, for pdf copies of CTFN Statutes of Government showcasing Tlingit and Tagish virtues and values through traditional oral narratives.

18 *At.oow* is a key and complex concept in Tlingit culture and is closely translated as 'property'. Ice bergs and glaciers can be *at.oow*. For a more detailed explanation, see Dauenhauer and Dauenhauer (1990).
19 See inland Tlingit elder Elizabeth Nyman talking about the Taku River with Tlingit linguist Jeff Leer (1993).
20 For more information, see anthropologist Julie Cruikshank's ethnography of the region (1998).
21 A classic and very common example in this region. Almost all indigenous place names have been replaced by obscure European and American male military officers and explorers, often called a phallocentric and systematic oppression of First Nations' philosophy of place and experiential connection with the land and waters (see Hayman, with James & Wedge forthcoming 2017).
22 CTFN community consultant Colleen James reminds us that even throwing stones into the water is disrespectful (James, personal communication 2014).
23 The Tlingit language reflects the Tlingit worldview. There exist *Aas K̲wáani* Tree People, Salmon People and Water People that have their own agency in the world (Katzeek, personal communication 2014). For example, in the Tlingit language *Aas K̲wáani* refers to the Tree People, literally 'People of the Trees'. As Tlingit scholar Lance Twitchell states in his new Tlingit dictionary, 'this term is used ceremonially, often when speaking to the spirits of the trees' (2016).
24 *neéhkkàawu*: "This is a term used, to describe someone new and who knew nothing about living in the north, it is still used, however it's 'Cheechacko'. It was used by people around me to describe a person, or a person was just called Cheechacko when I grew up! It's Tlingit and is turned into a recognizable word to describe a 'greenhorn' in English. 'Person on the beach', nobody wants to be responsible for this person!" (James, 28 December 2016, personal communication).
25 For example, the oral narratives of Smart Beaver Man, Crow, The Two Boys who Drifted Down the [Yukon] River, Wealth Woman, Skookum Jim and the Frog and the long-ago Flood stories.
26 Anthropologist Steven Langdon writes that 'it is through the continuous circulation of respect in thought and deed exhibited in connections and fulfilment of obligations in various socially and ritually prescribed ways that the Tlingit pursue a morality to ensure the continuation of existence' (Sealaska Heritage Institute lectures, October 2013).
27 Terracentrism as a concept was introduced by Marcus Rediker in his chapter 'Hydrarchy and Terracentrism', in *Aquatopia*, a book that accompanies a major exhibition at Nottingham Contemporary and Tate St Ives in 2013. Other references can be found in Bozovic and Matthew (2016).
28 See Twitchell's (2016) new Tlingit dictionary for the relationship between different types of water and different Tlingit canoe designs.
29 See Yukon Energy's report on the history of the dams in relation to the current project to keep the southern Yukon lakes raised in the winter months to provide extra hydro-electric power, accessed 18 July 2016, www.yukonenergy.ca/media/site_documents/Southern_Lakes/FACT%20SHEET%20Southern%20Lakes%20Enhanced%20Storage%20Concept.pdf.

References

Abram, D 1996, *The Spell of the Sensuous: Perception and Language in a More-than-human World*, Vintage/Random House, New York.

Allen, P 1989, *Spider Woman's Granddaughters: Traditional Tales and Contemporary Writing by Native American Women*, Beacon Press, Boston, MA.

Bozovic, M & Miller, D 2016, *Watersheds: The Poetics and Politics of the Danube River*, Academic Studies Press, Boston, MA.

Brayton, D 2012, *Shakespeare's Ocean: An Ecocritical Exploration*, University of Virginia Press, Charlottesville.

Bull, M & Back, L (eds) 2003, *The Auditory Culture Reader*, Berg, Oxford.

Buschmann, R 2005, 'Oceans of world history: delineating aquacentric notions in the global past', *History Compass*, vol. 2, no. 1, doi: 10.1111/j.1478-0542.00068.

Casey, E 2013, *The Fate of Place: A Philosophical History*, University of California Press, Berkeley.

Chen, C 2013, 'Mapping waters: thinking with watery places', in C Chen, J MacLeod & A Neimanis (eds), *Thinking with Water*, McGill-Queen's University Press, Montreal & Kingston, pp. 274–98.

Cleaver, D 2009, 'Polarising narrative and paradigmatic ways of knowing: exploring the spaces through narrative, stories and reflections of personal transition', AARE 2009: Australian Association for Research in Education Annual Conference 2009: Inspiring Innovative Research in Education, 29 November to 3 December 2009, Canberra, Australia, accessed August 2014, http://eprints.usq.edu.au/7715/.

Code, L 2006, *Ecological Thinking*, Oxford University Press, Oxford.

Crippen, J 2013, Tlingitology seminar, http://tlingitlanguage.com/wp-content/uploads/2015/01/Tlingitology-Seminar.pdf.

Cruikshank, J 1998, *The Social Life of Stories: Narrative and Knowledge in the Yukon Territory*, University of Nebraska Press, Lincoln.

Cruikshank, J 2012, 'Are glaciers "good to think with"? Recognising Indigenous environmental knowledge', *Anthropological Forum*, vol. 22, pp. 239–50.

Dauenhauer, N & Dauenhauer, R 1990, *Haa tuwunaagu yis: For Healing Our Spirit – Tlingit Oratory*, University of Washington Press, Seattle.

Feld, S 1990, *Sound and Sentiment: Birds, Weeping, Poetics and Song in Kaluli Expression*, 2nd edn, University of Pennsylvania Press, Philadelphia.

Feld, S 1993, 'Reading R. Murray Schafer in the Papua New Guinea Rainforest', in *Essays from Ethnomusicology to Echo-muse-ecology*, accessed 18 July 2016, www.acousticecology.org/writings/echomuseecology.html.

Gibson, JJ 1977, 'The theory of affordances', in R Shaw & J Bransford (eds), *Perceiving, Acting, and Knowing: Toward an Ecological Psychology*, Erlbaum, Hillsdale, NJ, pp. 67–82.

Gillis, J 2011, 'Filling the blue hole in environmental history', unpublished paper and presentation at the Rachel Carson Center, accessed 1 February 2015, http://johnrgillis.com/works/bluehole2.pdf.

Hayman, E with James, C & Wedge, M 2015, 'Héen Kas'él'ti X_oo (among the ragged lakes) – storytelling water, north of the future: collaborative water research with Carcross/Tagish First Nation, Yukon Territory, Canada', in S Buechler & A Hanson (eds), *A Political Ecology of Women, Water and Global Environmental Change*, Routledge, London, pp. 185–205.

Hayman, E with James, C & Wedge, M forthcoming Spring 2017, *A Deep Chart: The Aquaface of Deep Mapping*, International Journal of Humanities and Arts Computing, Edinburgh University Press.

Hursthouse, R 2007, 'Environmental virtue ethics', in R Walker & PJ Ivanhoe (eds), *Environmental Ethics*, Oxford University Press, Oxford, pp. 155–72.

Ingold, T. 2000, *The Perception of the Environment: Essays on Livelihood, Dwelling and Skill*, Routledge, London.

Kennedy, K & Bond, JD 2003, 'Evidence for a late-McConnell readvance of the Cassiar Lobe in Seagull Creek, Pelly Mountains, Central Yukon', in DS Emond & LL Lewis (eds), *Yukon Exploration and Geology*, pp. 121–8, accessed 18 July 2016, http://emrlibrary.gov.yk.ca/ygs/yeg/2003/2003_p121-128.pdf.

Kiana Oman, L 1995, 'The epic of Qayaq: the longest story ever told by my people', Carleton University Art Gallery.

Langdon, S 2013, Lecture at the Sealaska Heritage Institute, 'Spiritual obligations and connections: the foundation of Tlingit existence', https://vimeo.com/78779584.

Lauriault, TP & Fraser Taylor, D 2013, *The Map as a Fundamental Source in the Memory of the World*, Geomatics and Cartographic Research Centre Carleton University, Canada, accessed 18 July 2016, http://gcrc.carleton.ca.

Malpas, JE 1999, *Place and Experience: A Philosophical Topography*, Cambridge University Press, Cambridge.

McClellan, C 1975, 'My old people say: an ethnographic survey of Southern Yukon Territory', part I and part II, *Ethnology*, vol. 6, nos 1 & 2, National Museums of Canada, Ottawa.

Nussbaum, M 1993, 'Non-relative virtues: an Aristotelian approach', in M Nussbaum & A Sen (eds), *The Quality of Life*, Clarendon Press, Oxford.

Nussbaum, M 1995, *Poetic Justice: the Literary Imagination and Public Life*, Beacon Press, Boston, MA.

Nyman, E & Leer, J 1993 '*Gágiwduł.àt*: Brought Forth to Reconfirm: the Legacy of a Taku River Tlingit Clan', Yukon Native Language Center.

Rubin, DC 1995, *Memory in Oral Traditions: The Cognitive Psychology of Epic, Ballads, and Counting-Out Rhymes*, Oxford University Press, Oxford.

Smith, D 1994, 'Glacial Lake McConnell: paleogeography, age, duration, and associated river deltas, Mackenzie River Basin, Western Canada', *Quaternary Science Reviews*, vol. 13, pp. 829–43.

Stibbe, A 2014, An ecolinguistic approach to critical discourse studies, *Critical Discourse Studies*, vol. 11, no. 1, pp. 117–28.

Swanton J 1901, *Tlingit Myths and Texts*, Smithsonian Institution, Bureau of American Ethnology, Washington, DC.

Thornton, T 2010, *Being and Place among the Tlingit*, University of Washington Press, Seattle.

Treanor, B 2014, *Emplotting Virtue: A Narrative Approach to Environmental Virtue Ethics*, State University of New York Press, New York.

Tuan, YF 1972, *Topophilia: A Study of Environmental Perception*, Columbia University Press, New York.

Twitchell, L 2016, Online Tlingit language dictionary (updated August 2016), http://tlingitlanguage.com/wp-content/uploads/2015/01/dictionary-4.pdf (accessed 19 December 2016).

Urry, J. 2011, 'City life and the senses', in G Bridge & S Watson (eds), *The New Blackwell Companion to the City*, Blackwell, Oxford.

Warf, B 2015, 'Deep mapping and neogeography', in D Bodenhamer, J Corrigan & TM Harris (eds), *Deep Maps and Spatial Narratives*, Indiana University Press, Bloomington and Indianapolis.

14 Developing an ecumenical framework for water justice

Susan Lea Smith

The Water Justice Framework discussed in this chapter has been developed by the Ecumenical Water Network (EWN), a policy initiative of the World Council of Churches (WCC). Since its conception in the ecumenical movement during the 1920s and 1930s and its establishment in 1948, WCC has grown to a fellowship of 346 regional and national Christian churches in more than 110 countries and territories representing some 500 million Christians around the world. Members include most of the world's Orthodox, Anglican, Baptist, Lutheran, Methodist, Reformed, United and Independent churches. WCC works in partnership with the Roman Catholic Church on many issues as well as with the World Evangelical Alliance and other ecumenical and interfaith organizations.

The Ecumenical Water Network emerged in 2006 from the 9th WCC Global Assembly held in Porto Alegre, Brazil, which proclaimed in its *Statement on Water for Life* (2006), 'Because water is truly the source of life, it requires responsible action from us as human beings: action to preserve and share water for the benefit of all creatures.' EWN became the means for World Council of Churches to encourage that responsible action.

EWN has become a strong advocate for universal access to water and sanitation, promoting community-based solutions to the water crisis, and advocating for water justice at the local, regional and international level. WCC member churches work in a variety of ways at all levels, from providing poor communities with financial and technical assistance for clean water and sanitation, to engaging in advocacy at the international, national and local levels for just and sustainable water resources policies. EWN itself focuses its efforts at the global level, seeking to serve as a Christian voice on behalf of water justice and to educate WCC member churches on the challenges and opportunities for water justice.

The concept of 'water justice' is part of our proper relationship to nature (God's creation) and to each other, given our utter and complete dependence upon the ecological integrity of the Earth. Protecting water and assuring just distribution of this life-giving resource is among the most critical eco-justice issues.

History and process of creating the Water Justice Framework

The EWN-WCC Water Justice Framework developed out of both the organizational needs of EWN-WCC as well as its vision about the WCC's ability to help transform the international water policy discourse, which had become dominated by a neo-liberal economic paradigm contrary to Christian principles. Details of the origins and motivations of the Water Justice Framework are contained in *The Pilgrimage towards Water Justice: A Theological Foundation for Christian Action* (Smith et al. forthcoming). What follows is a summary version of that longer work.

As EWN moved beyond the initial issues that gave birth to the network into new and more challenging territory, EWN and its policy advisory group felt a need to articulate the faith basis that grounds EWN's program, priorities and policy positions. Second, having confidence that a solid, convincing faith-based conception of water justice could be articulated, EWN entertained a more expansive vision that global religious organizations, from both Christian and other faith traditions, might develop a consensus statement of water justice ethics for people of faith throughout the world.

Two EWN statements in particular articulated the faith basis for EWN's water justice efforts. The 2013 Berlin Statement, *God, Lead Us to Abundant Life and Peace through Water Justice* (EWN 2013) gave recommendations for the Sustainable Development Goals (SDGs) on water and sanitation and also addressed water injustice and conflict, and the role of the church in achieving water justice. The 2014 Palestine Statement, *Water Crisis in Palestine* (EWN 2014) focused on the water injustice faced by the Palestinian people and called for interfaith dialogue within Israel and Palestine on measures to remedy water injustice.

During 2014, EWN recruited theologians and water justice advocates from around the world to prepare papers articulating the theological foundation for water justice. This theological reflection on water justice, to be published as *The Pilgrimage towards Water Justice: A Theological Foundation for Christian Action* (Smith et al. forthcoming), served as the basis for developing the text of the EWN Water Justice Framework in 2015. In July 2015, the EWN's policy advisory group approved the Water Justice Framework in principle. Once ratified by WCC member churches, EWN members and partners, EWN anticipates launching an effort to develop a consensus statement of water justice ethics with the Pontifical Council for Justice and Peace (PCJP), the World Evangelical Alliance (WEA), and other interested parties, and mounting an effort to develop interfaith consensus shortly thereafter. Such a consensus in the articulation of water justice ethics among the six billion people who adhere to one of the major faiths should substantially assist governments in managing water resources on an ethical basis and peaceably resolving conflicts over water resources.

The substance of the Water Justice Framework

The Water Justice Framework outlines 10 fundamental ethical principles that water justice requires along with commentary to explain the principles, and suggestions on a way forward to foster water justice. Among the preliminary issues facing the drafters of the Water Justice Framework were: (1) could we develop a global, universal ethical framework to articulate water justice principles applicable across societies with distinctive cultural and political characteristics; and (2) how could we identify common ethical principles amidst the diverse set of Christian traditions represented by the WCC, in the face of sometimes conflicting theological concepts, approaches to authority and hermeneutics of scriptural interpretation.

The universality concern proved to be less significant than it might superficially appear. The significance of water and the deep realities of water injustice resonate across cultures and even across the political spectrum. At least at the high level of generality required to create a global Water Justice Framework, there has been a strong international consensus on the need for water justice. This consensus found expression in the 2010 UN actions recognizing the human right to water and sanitation. Not a single nation, including the United States, was willing to vote against the General Assembly resolution recognizing the human right to water and sanitation – and when the role of well-regulated and controlled private service providers was acknowledged, the UN Human Rights Council resolution was agreed to unanimously (Barlow 2012). The universality of the EWN-WCC Water Justice Framework has been assured through the involvement of theologians and water justice advocates from around the world, bringing to bear their knowledge, faith and practical experience.

The Water Justice Framework successfully identified common ethical principles despite conflicting theologies, differing senses of what constitutes legitimate religious authority, and disparate approaches to interpreting biblical texts. Three factors may have contributed to this success. First, EWN has long enjoyed the active involvement of people from all major Christian traditions, including Roman Catholic theologians and water activists, as well as secular partners. Second, EWN has also drawn upon the extensive experiences of WCC in fostering ecumenical and interfaith dialogue. Third, given the extensive involvement of EWN members in the practical work of providing water and sanitation to communities in need, EWN has remained grounded in the basic tenets of our faith, rather than being drawn into doctrinal subtleties irrelevant to the work at hand.

Substantive judgments embodied in the Framework

The proper role of economics in water issues

One key area of discussion was the tension between economic and human rights approaches to water and sanitation services and protection of the environment. Despite a constructive dialogue, that tension remains. Purists in the economic camp say 'Put a price on it' and purists in the human rights camp say 'Give it away for free.' Together purists in both camps force a false choice. The real question is whether we will address water resources issues primarily from an ethical perspective, with a hesitant nod to the appropriate use of economic instruments and incentives or whether we will address water resources issues primarily from an economic perspective, with an even more hesitant nod to ethical principles, such as equitable sharing of water, water conservation and protection of water from pollution. After extended reflection, the World Council of Churches, has chosen to take an ethical approach to water resources and other resource issues, relegating economics to an appropriate secondary status. Indeed, faced with conflict between the ethical precepts of their faith and the prescriptions of neo-liberal economic schemes, we believe all people of faith are apt to choose an ethical approach to water.

The Water Justice Framework rejects the neoclassical economic paradigm as the basis for human decision-making about water resources. Specifically, the Framework relegates a key principle of neoclassical economics – allocative efficiency – to secondary importance as a goal of resource management, and puts use of economic incentives in its proper place as an implementation device that must be scrutinized for its effect on core ethical values.

Using core biblical values to address water issues

The Framework recognizes that ethical precepts drawn from two core biblical values must be given primary importance in deciding water issues. The first core biblical value is that all human beings are children of God entitled to equitable treatment and to live with full human dignity. The second core value is that care for God's Creation is a human responsibility demanding conscientious environmental stewardship. Interestingly enough, these were the two values that formed the basis of Luna Leopold's pioneering discussion of water ethics (Smith, Chapter 2 in this volume).

While the EWN-WCC theological discussion accompanying the Framework's ethical principles discusses uniquely Christian theological concepts such as the Trinity, the Framework carefully confines itself to water justice principles drawn from these two core biblical values. Thus, the Framework seeks to avoid unproductive and unnecessary ecumenical and interfaith disputes about the ethical principles that should govern water management as a whole.

The core value of equitable treatment of human beings is reflected in several water justice principles:

- Principle II prescribes intergenerational equity in water management to benefit both current and future generations.
- Principle IV calls for universal access to clean drinking water and adequate sanitation and priority given to providing the poor and marginalized with access.
- Principle V gives priority to access by small-holder farmers, herders and fishers.
- Principle VI suggests that all three priority uses (ecosystem, drinking water and sanitation, and small-holder) can be met simultaneously in most cases, but prescribes an ordering of priority uses in exceptional cases.
- Principle VII affirms the right of all those affected to be involved in decisions about water management.

The core value of care for Creation is also reflected in several principles.

- Principle III provides for preservation of biodiversity and protection of the integrity and resilience of aquatic ecosystems.
- Principle VI specifies that water for ecosystem purposes is a priority use of water.
- Principle X address care for Creation and specifically human responsibilities for water stewardship.

Providing strong, unequivocal ethical guidance

The Framework forthrightly addresses highly controversial contemporary issues that have arisen with respect to management of water resources, ranging from privatization of water and sewage services and private water rights, to the ethical issues implicated in using new technologies for water conservation such as desalinization. The contribution of the Framework to the discussion of these issues lies in its application of consistent water justice principles derived from core biblical values, rather than relying on ad hoc decisions prone to inconsistency and manipulation. For that reason, the principles have been deliberately phrased to limit discretion and avoid manipulation. This approach reflects the conviction of the drafters that the crises caused by current water management practices around the globe can only be remedied by strong and unequivocal ethical guidance. Reform of those practices is desperately required to put us in right relationship with all of humanity and the rest of Creation.

The EWN Water Justice Framework

Except as noted, the text that follows is verbatim from the Water Justice Framework and is reprinted by permission of the Ecumenical Water Network-World Council of Churches. The 10 ethical principles presented in the Water Justice Framework are the ones that EWN considers most fundamental in ensuring water justice. They might indeed be termed 'the Ten Commandments' of water, for they set forth ethical imperatives from the viewpoint of Christian faith. Read together, they provide a consistent and coherent ethic for dealing in a just manner with God's gift of water.

Principle I: ethical management

Water justice requires ethical management of water.

Allocation and management of water must be based on ethical principles grounded in compassion and justice, rather than on economic interests, wealth, profit, or supposed economic efficiency.

Principle II: intergenerational equity

Water justice requires that we manage water as a gift from God to be available for life-sustaining purposes to the current and all future generations.

We must not deplete groundwater reserves, we must protect the sustained availability of surface water, and we must not destroy the ability to use ground or surface water by polluting it.

The hydrological cycle has always assured that water would continue to be available for all creation. Humanity has disrupted that cycle through deforestation and greenhouse gas emissions that are causing climate change, melting glaciers and reducing snowpack. We must halt this disruption and reverse it to the extent possible for many reasons, including maintenance of water supplies for all creation.

In addition to the profound effects on water resources caused by climate change, current water management practices compound the threats to availability of clean water for current and future generations.[1] Humanity has begun to use water in unsustainable ways by polluting both surface and ground water, by withdrawing groundwater at rates in excess of recharge, and by mining ancient groundwater. All of these practices are unethical and should be prohibited.

Principle III: protect biodiversity and aquatic ecosystems

Water justice requires responsible human action to preserve biodiversity and to maintain ecological integrity and resilience of aquatic ecosystems.

All creation has intrinsic value – and is critically important to human and other life dependent upon it. We must take whatever actions are necessary to preserve biodiversity as well as maintain the ecological integrity and resilience of aquatic ecosystems. Ultimately, we should seek to restore their integrity and resilience when they have been degraded by human action.

Principle IV: assure universal access to water and sanitation

Water justice requires universal access to safe water and adequate sanitation. Efforts to provide such access should give preference to the poor and marginalized.

All human beings enjoy a human right to water and sanitation. This means that they are entitled to 'sufficient, safe, accessible, culturally acceptable and affordable water and sanitation services for personal and domestic uses . . . delivered in a participatory, accountable and non-discriminatory manner'. Under international law, governments are required to progressively realize this human right and all governments have committed to achieve universal access to water and sanitation by 2030. Global efforts to realize universal access, whether by churches, faith-based organizations, non-governmental organizations or governments, should focus first and give priority to providing access to the poor and marginalized in each society.

Principle V: assure access to water by smallholders

Water justice requires that water be available for use by subsistence and smallholder farmers, herders and fishers.

Availability of water for the use of smallholders is particularly important because their activities provide meaningful livelihoods for an enormous number of people and lead to efficient production of, and sovereignty over, life-sustaining food. The set-aside is necessary because, despite the importance of this work, smallholders tend to be poor, marginalized by their societies, and lack the economic and political power of corporate agriculture and urban consumers.

Principle VI: assure availability of water for priority uses

Water justice requires that three uses of water take precedence: household uses; agricultural use by smallholder farmers, herders and fishers; and use to maintain aquatic ecosystem integrity and resilience.

Human societies may differ somewhat on priorities set for water use, but three priority uses are universal: (a) sufficient safe and affordable water for household uses, (b) sufficient clean and affordable water in and out of stream for smallholder uses, and (c) sufficient clean water freely available in stream to assure the integrity and reliance of aquatic ecosystems and the lives of creatures that are dependent on them. By meeting these priority uses, indigenous peoples will ordinarily also have the water required for ceremonial and other traditional uses.

In the unusual case where there is not enough water to meet priority uses, water managers should seek practicable alternatives to reduce the amount required for each priority use and adjust all three priority uses to best protect the most vital of uses. In the extraordinary case where no adjustment can protect the most vital uses in all three categories, urgent human needs for water and food take priority over protection of aquatic ecosystems. It is not ethical or consistent with human dignity to require a human being to starve or die of thirst so that a fish or snail might live.

Principle VII: democratic water governance

Water justice requires democratic governance of water.

Water should be allocated and managed based on sustainable ecosystem management through a process that assures democratic governance of water resources. Decisions about water management should be made at a level appropriate to the scope of the decision, but decisions should be made as close to the community level as possible, in accordance with the subsidiarity principle. Water governance must be transparent and accountable, and involve effective and equitable participation of all interested parties, especially the poor and the marginalized.

Principle VIII: water is a common good

Water justice requires that water be considered a common good. No compensable private rights in water should be created.

Water is a gift from God and must be regarded as a common good whose use may be allocated, subject always to limitations imposed for the commonweal. Perpetual ownership of land was not allowed in biblical times. Even more than land, exercising or granting rights to perpetual use or control of water, intended to be a gift of God available to all, is unethical. Governments should not allow water to be permanently privatized. Where governments confer use rights on private actors, those rights should be conditioned to allow them to be revoked and the water reallocated if necessary to meet priority uses of water so that water is available for basic food and water needs of human beings, other creatures and protect the integrity of aquatic ecosystems upon which they depend.

Principle IX: prevent economic exploitation of water

Water justice requires that use of water for commercial purposes be strictly regulated. Economic exploitation of water is unethical.

Using water to produce food and other goods is natural and beneficial. However, because of the value of water, some unscrupulously seek to profit by controlling more water than they can use for legitimate purposes. Seeking to profit by the commodification or control of excess water is unethical. Use of water for commercial purposes, such as corporate agriculture and industrial production, must be strictly regulated to avoid economic exploitation of water. A few examples illustrate how economic exploitation of water can be avoided.

Water marketing

Water allocated to priority uses should not be marketed or transferred. If the water is not needed for a priority use, the priority allocation should be reduced. The community can then allocate that water in accordance with its normal allocation process.

The democratic governance process should best reflect social priorities about water allocation. Ordinarily no water marketing should be necessary if a sound allocation process has been used. However, if a community decides to use water auctions to allocate water for non-priority uses, it should retain authority to approve allocations and impose appropriate limitations and conditions. If it decides to use water marketing to reallocate water available for non-priority water uses, it should retain the authority to approve transfers. For example, approval of transfers could be limited to situations where (a) all priority uses of water have been satisfied; (b) the water to be transferred has been conserved from the original amount allocated through changes in practice and technology – i.e. not just initial overestimates or 'paper rights'; (c) the conserved water being reallocated is being allocated to uses considered more socially beneficial; and (d) considering all impacts of transfer based on adequate information, the transfer provides net social benefits.

Private provision of water and sanitation services

Because of the extraordinary significance of water and sanitation to health, human dignity and life itself, water and sanitation services should ordinarily be provided by a democratic, accountable government body. Private provision of such services is rarely acceptable. The international experience suggests that government generally cannot effectively regulate private providers, government dealings with private providers are seldom done in a transparent and accountable manner, and private providers rarely provide service that is affordable, reliable, universally available and non-discriminatory and that meets health and environmental standards. While private providers are especially apt to exploit the weakness of regulatory systems in developing countries, recent experience in places like Jakarta indicates that the human right to water and sanitation may be violated by government reliance on private providers even in middle-income countries.

Commercial water bottling

Where safe drinking water is accessible through a community water supply, it is unethical to produce or consume bottled water. Bottled water wastes water and energy and creates large amounts of solid waste. Due to the waste of water, materials, and energy involved in commercial water bottling, production of this product ordinarily should not be considered a beneficial use of water.

Water pricing/user fees

Governments frequently use water pricing or water user fees as a means to raise revenue to support water infrastructure, ensure water conservation and encourage the most productive use of water. Use of tiered water rates is an appropriate means of harnessing economic incentives to avoid waste. However, sufficient water for priority human uses must be made freely available or available at affordable rates to realize the human right to water and sanitation and the human right to food. Water prices should be set to implement sound policy based on ethical principles rather than allowing economics to drive policy and management.

Water speculation

Economically exploiting the value of water by speculation in water rights is unethical and should be prevented by effective regulation. Similarly, acquiring an interest in a company involved in speculation of water rights is also unethical. Speculation involves purchasing land with water rights or purchasing water rights outright, with an intent to profit from fluctuations or increased in value of water rights and without an intent to put the water to full beneficial use.

Principle X: water stewardship responsibilities

Water justice requires that water users fulfil their profound stewardship responsibilities towards water.

Because humanity has special gifts that allow us to reason, plan and transform our environment as no other creature does, we also have special responsibilities to care for other human beings and other life on the Earth, and to limit our activities in a manner consistent with preserving the tapestry of life on Earth. We have the responsibility not to waste or pollute water, not to significantly disturb its natural flow and not to significantly alter its fundamental composition. Human water use, and development of water resources for human use, must be in accordance with these limitations.

Waste

We should not waste water and should uniformly employ sound water conservation, recycling and reuse practices. Priority water users should receive subsidies as necessary to allow them to employ water conservation, recycling and reuse practices.

Pollution

We should not pollute water, utilizing water treatment to the extent necessary to restore the natural condition of the water before returning it to the environment. Priority water users should receive subsidies as necessary to allow them to avoid polluting water or to treat polluted wastewater.

Natural flow

We should hesitate to alter the natural flow of water and should preserve its natural flow unimpaired, except where necessary to meet priority uses of water. Small disruptions such as construction of sand dams to capture intermittent streams in arid areas and construction of small reservoirs to store drinking water or irrigation water are appropriate. Massive disruption of mighty rivers to construct hydroelectric dams causing widespread destruction of riverine ecosystems and displacement of local communities is unlikely to be ethically justified. Such projects are presumptively ethically inappropriate. Such projects should not be undertaken without full consideration of their impacts and unless the projects meet stringent ethical criteria established through the water governance process. For example, the ethical criteria applied to such projects might be: (a) the project is the only means available to realize fundamental human rights, (b) the benefits of the project will primarily be directed toward realization of those rights, especially by affected communities and by the poor and marginalized in the broader society, (c) affected indigenous peoples have consented to the project, (d) those displaced by the project will be fully compensated before the project is constructed, (e) the project has been designed to minimize to the greatest extent possible ecological damage and social damage to affected communities, and (f) both social and ecological damage will be fully mitigated by mitigation measures completed prior to or simultaneously with the construction project.

Fundamental composition

We should hesitate to alter the fundamental composition of water provided in nature. In particular, we should refrain from acts that unnecessarily alter its salinity, such as poor irrigation practices, over-drafting coastal aquifers and desalinization of water for other than priority uses.

Note

1 Rephrased from original.

References

Barlow, M 2012, 'Foreword', in F Sultana & A Loftus (eds), *The Right to Water: Politics, Governance and Social Struggles*, Earthscan/Routledge, London.

EWN 2013, 'God, lead us to abundant life and peace through water justice', Statement of the Ecumenical Water Network, World Council of Churches, Berlin, accessed 21 May 2016, www.oikoumene.org/en/resources/documents/wcc-programmes/justice-diakonia-and-responsibility-for-creation/climate-change-water/ewn-statement-from-berlin-consultation.

EWN 2014, *Statement of the Ecumenical Water Network*, World Council of Churches, Jerusalem, accessed 8 July 2016, www.oikoumene.org/en/resources/documents/wcc-programmes/justice-diakonia-and-responsibility-for-creation/climate-change-water/ewn-jerusalem-statement?set_language=en.

Smith, S, Zachariah, G & Suna, D forthcoming, *The Pilgrimage Towards Water Justice: A Theological Foundation for Christian Action*, World Council of Churches, Geneva.

World Council of Churches 2006, *Statement on Water for Life*, Statement of the Ninth Assembly, World Council of Churches, Porto Alegre, Brazil, accessed 16 November 2016, www.oikoumene.org/en/resources/documents/assembly/2006-porto-alegre/1-statements-documents-adopted/international-affairs/report-from-the-public-issues-committee/water-for-life

15 Developing a global Water Ethics Charter

David Groenfeldt

Creating charter statements enumerating aspirational principles that would improve conditions for humanity has become a widespread practice in many fields, and water governance is no exception. There is the 2007 Baltimore Charter,[1] a Berlin Water Charter (Härlin in this volume), and the 2009 Istanbul Water Consensus (Acuto 2013). But charters about the ethics of water are particularly rare. The most comprehensive effort to instil environmental ethics at a global level has been the Earth Charter, which grew out of the Rio World Summit on Sustainable Development held in 1992 (Sanders 2014). The Earth Charter, however, does not give specific attention to water.

The specific idea for a global charter on water ethics was a recommendation from the 6th World Water Forum held in Marseille, France in 2012. The concept emerged from the ad hoc 'Working Group on Ethics, Culture and Spiritualities' which organized a session to identify commonly shared value principles across religious, cultural and philosophical traditions. The group noted that even well-known ethical principles about water, environment and social justice, whether derived from major religions, Indigenous cultures or secular philosophy have had little influence on water policies. What was needed was the 'recognition of spiritual and ethical values and principles and their consideration in decision-making process in the water sector' (6th World Water Forum 2012). The group recommended that a set of universal ethical values about water should be developed and formulated into a universal charter.

The basic principles that the group agreed should inform the process of developing the charter should be the following (according to an informal position statement issued by the group):

- Honour and promote justice, equity, the rights of nature, care and compassion in decision-making.
- Learn from and engage all disciplines and schools of thought, including science, law, philosophy and religion, as well as from the traditional knowledge of indigenous cultures.
- Learn from and engage the private and the public sectors, at the local, regional and global levels.

By the following year (2013) a core group of three organizations – the French Water Academy, UNESCO's Division of Water Sciences and the Water-Culture

Institute (based in Santa Fe, New Mexico) – agreed to work together to develop a 'Water Ethics Charter'. Along with this core team, six other organizations and individuals joined the Steering Committee to get the process underway: Alliance for Water Stewardship, Botin Foundation, Club of Rome, Indigenous Environmental Network, Water Youth Network and Amb. Magdy Hefny from Egypt.

Nearly two years after the Marseille Water Forum, the Steering Committee met at UNESCO-Paris in March 2014 to establish the broad framework for a water ethics charter. The outcomes of this meeting were a provisional outline of an Ethics Charter, agreement on broad aims and expectations, and an agreed short-term course of action. The development of a charter document, envisaged as a concise statement of just a few pages, was seen as the leading edge of an integrated set of outreach and awareness-raising activities to promote ethical principles in water policies. Once finalized, the charter document could be endorsed by private companies, public agencies, local governments or any decision-making entity. The campaign to seek endorsements would contribute further to public awareness and commitment to the charter's principles.

The process of developing the content of the charter relied on a list of some 80 experts, representing broad, but admittedly opportunistic geographic, cultural and technical diversity, compiled from the personal contacts of the Steering Committee members. These experts were sent an annotated outline of the prospective Charter (Draft 1.0) with an invitation to provide feedback about issues the charter should address and any other guidance. Some 30 experts submitted substantive comments, which were then compiled into a spreadsheet and then incorporated into a new version (Draft 2.0) which was the first comprehensive draft of the charter. Subsequent revisions will follow a similar procedure of evolution through comments solicited from an ever-widening circle of experts, reflecting steadily greater diversity. Successive versions will be posted on the Water Ethics Network website (waterethics.org). The present discussion of the Charter's contents is based on Draft 2.0, the current version at the time of this writing (May 2016).

Content of the Water Ethics Charter

The Charter (Draft 2.0) is divided into five substantive sections corresponding to five 'domains' of water ethics: (1) Environmental, (2) Economic, (3) Social, (4) Cultural and (5) Governance. Within each of these five domains, general (value) principles as well as operational (ethical) principles are specified. But first, the Introduction, lays out the rationale and scope of the Charter, and highlights several general principles common to all five domains of water ethics.

Introduction to the Charter

The broad context for the Charter is given in the opening sentence:

> This Charter establishes the moral and ethical foundations to guide decision-making around the use of water and the protection/stewardship of water

> basins and water-reliant ecosystems. The intent of this Charter is to engender water policies and practices that are environmentally sustainable, economically responsible, socially just, respectful of cultural and spiritual diversity and which will help safeguard the welfare of future generations.

The Charter is aimed at both water policy-makers and the general public. Its purpose is:

> (1) to educate water policy makers, water users, and the public at large about their moral responsibilities in making choices which involve water directly or indirectly, (2) to foster an ethical attitude towards water bodies, and in so doing, (3) to improve water management and governance.

The earlier focus on bridging the perspectives of culture, religion and philosophy, which framed the discussions at the 2012 World Water Forum in Marseille, has been revised to a very much applied perspective. The new focus is promoting water ethics to policy-makers and the members of civil society and the private sector who influence the policy-makers.

The Charter distinguishes between general principles of value and belief and more specific 'operational principles' which can be more directly incorporated into policies. The introductory section of the Charter addresses four general principles: (1) Precaution ('We should approach this interconnectedness between humans and nature with an attitude of humility and adopt the fundamental principle of precaution to guide our management interventions.'), (2) Water as a commons ('We all depend on water and we all have a shared responsibility for its management'; see Kallhoff in this volume), (3) Intergenerational justice ('We have a responsibility to all future generations to be good stewards of their water today') and (4) Knowledge and education ('We have a moral obligation to generate knowledge about water in all its aspects and attend to the governance of that water knowledge').

Environmental ethics

The general principle of 'rights of nature' is a core feature of the Water Ethics Charter, explicitly expressed as an environmental value: 'Water ecosystems have inherent rights, and intrinsic value over and above their utilitarian value to people.... It is our moral responsibility to adjust human demands for water to accommodate healthy ecological functions.' The value that water ecosystems provide to people (and what is generally meant by the term 'ecosystem services') constitute economic or social values but are not, strictly speaking, environmental values. Rather it is the inherent value we accord to healthy natural ecosystems which can rightly be considered as true environmental values.

Given these general principles about rivers and other water ecosystems having inherent value, the operational value is to respect and protect their healthy function (see Ziegler et al. in this volume). Here the precautionary principle becomes an important guideline, since 'The complex physical inter-linkages between water-ecosystems and the rest of nature, as well as the complex inter-linkages with socio-cultural systems render interventions inherently unpredictable. The principle of precaution should be applied when taking decisions which will, or which reasonably could, have severe and long-lasting negative impacts.'

Other guiding operational principles, which have been suggested for the Charter, but without unanimity of views, include the following:

- 'No net loss from current conditions or (where local impacts are unavoidable) offsetting environmental destruction with (nearby and reasonably equivalent) environmental restoration.'
- A commitment to advancing 'regenerative watershed sustainability, which would imply not only respecting the needs of nature, but also contributing to maintaining and improving these. Examples of regenerative activities here could be day-lighting of urban streams (to provide fish habitat), natural water filtration through wetlands, etc.'

The health of water ecosystems is inevitably affected by human water use, whether for agriculture, manufacturing and extractive industries, or urban and domestic water systems. While some environmental impacts cannot be avoided, others can (and therefore 'should') be avoided. This logic alludes to the reasonableness or legitimacy of water use. Is water being used for 'best purpose'? Wasting water violates an economic ethic (see discussion below), but using water efficiently to produce useless things violates an environmental ethic of 'reasonable use'.

For example, the biofuel ethanol, can be manufactured from corn that has been efficiently cultivated using careful water management techniques, but the net environmental impact of the production process (irrigating and cultivating the corn and processing the ethanol) is negative (Yang et al. 2012). Its production entails an unreasonable environmental impact, since there are more environmentally friendly alternatives. In this example, we are considering only the economic and environmental ethics. A full ethical assessment would also need to consider additional domains of value, for example, social, cultural and governance ethics.

Economic ethics

The current draft of the Water Ethics Charter echoes the 1992 Dublin Principles in stating that 'Water has an inherent economic dimension in all its uses, and economic principles are essential for comparing impacts and benefits from proposed water investments or interventions.' The notion of water as an economic good is foundational to both international and national water policies, yet remains controversial among civil society organizations focusing on human rights, health and environment. These controversies stem, in my view, from an overly narrow concept of 'economic principles'.

Economic thinking need not be limited to questions of monetary value, but applies equally (though with far less precision) in considering trade-offs and opportunities related to non-economic values (e.g. social and environmental values). Economic analysis is most robust when it is applied within a value category (e.g. comparing one type of social outcome with another type of social outcome). For example, the economic costs of installing a community well might be cheaper if it is done as a turn-key operation by a large engineering contractor, rather than through a community development process where the villagers are coached in how to construct, finance and repair their own well. But the potential benefits of the community approach would include not only the new water supply, but also the enhanced institutional capacity of the village to solve their own problems.

The Charter recognizes four general economic principles about water use:

1 Water use should be reasonable and frugal, using only as much as needed for a given purpose.
2 The re-use of water should be favoured over extracting fresh water from nature.
3 Existing water stocks should be maintained and their resilience and sustainability protected (e.g. in the management of aquifers and lakes).
4 Given the inherent nature of water as a commons (both globally and locally), private ownership of water must be balanced with accountability to the larger society. (This is also a governance principle.)

This last principle about water as a commons, implying a moral 'accountability to the larger society' has a number of important corollaries, which are included in the Charter as operational principles:

- Water for basic human needs (e.g. the right to water and sanitation) should be effectively free, whereas water used in economic activities should have a market cost (see Feitelson in this volume).
- Water markets can be important tools for good management but must be subject to social justice principles.
- Ecosystem services are inherently common goods to be shared equitably with people (right to healthy ecosystems), nature (rights of nature) and future generations (intergenerational equity).

This principle about ecosystem services is both a social principle (discussed in the next section), as well as an economic principle. Additional principles guiding the use of water, for example, for economic activities, also include the following:

- User-pay and polluter-pay principles.
- Principle of cost-recovery for water services.
- Principle of fair, transparent, and efficient financing arrangements for water-related investments.
- Principle of monitoring and sharing information about the status of water stocks (feeding into the principles of education and knowledge governance).

Social ethics

Water is intimately connected to society and is part of a socio-ecological system that addresses social needs and provides social opportunities. Water is a human right, but it also carries individual and community responsibilities for engaging in water decisions and partic ipating in common management tasks. Everyone has a right to safe water and sanitation to meet basic needs, protect water quality and promote good health.

General principles:

- Access to safe water and sanitation is a basic human right. (This principle was formally endorsed by the UN General Assembly, in 2010.)
- Water is a common good which belongs to everyone, under the principles of fairness, equity, solidarity and social justice.
- There is a moral 'right' to a healthy water environment (not only clean water, but ecologically healthy water ecosystems) for purposes of both practical (economic) and aesthetic (cultural) enjoyment.

The operational principles implied by these general values include the following:

- Promote universal access to safe water and sanitation.
- Ensure water security, and water justice for all segments of society, including youth, women, minorities and most especially future generations.
- Protect and restore water ecosystems and ensure access for all segments of society.

Cultural ethics

Water and water ecosystems provide cultural and spiritual meaning of fundamental importance to individual and social identity. Since these values are intangible (though often represented in architectural monuments and other material and artistic expressions) they are often overlooked. Water ecosystems (e.g. rivers) and water-related infrastructure (e.g. dams, bridges, canals) have both existing and potential cultural value which can change over time.

The rights of Indigenous Peoples to practise their customary cultural traditions have been recognized by the United Nations in the Declaration on the Rights of Indigenous Peoples (DRIP). A general ethical principle which derives from this Declaration is that: 'The cultural meanings of water traditionally recognized by a society constitute an aspect of cultural sovereignty. Beliefs about water and water ecosystems, for example, that a river is a sacred, living being, enjoy a fundamental protection as a cultural right.'[2] A second and related general principle is that both natural water ecosystems (e.g. springs, rivers, lakes) and built infrastructure (e.g. canals and fountains) can convey important cultural meaning and identity which contribute to well-being (Volker & Kristemann 2011; Nichols 2014). Just as water has an inherent 'economic dimension' in all its uses, so too it has an inherent cultural dimension.

The operational principles implied by the concept of cultural water ethics include the following:

- Water infrastructure development (e.g. dams, levees, river diversions, etc.) should accommodate customary cultural uses as a matter of priority. Rather than pursuing a strategy of compensation for cultural impacts which are incommensurable with monetized settlements, the preferred strategy would be to adjust the infrastructure design to meet cultural parameters.
- Proposals for water development within the territory of Indigenous Peoples, should be subject to the 'free prior and informed consent' of the local stakeholders.

The current draft of the Water Ethics Charter does not address the potential for creating *new* cultural meanings for water, but in subsequent versions, this issue will be addressed. For example 'creative place-making' enhances the cultural significance of water and offers a promising approach not only to derive more total benefits from water partly by fostering new water values within the community (Gadwa Nicodemus 2013). Furthermore, through strengthening the cultural values of water, the other categories of water values (environmental, social, economic and governance) might also be strengthened.

Governance ethics

The arrangements by which water is managed and decisions taken about water policies, practices and investments reflect governance values. In a sense, the realization of the other water ethics principles hinge on the form of governance, which in turn is linked to the societal values about what constitutes 'good' water governance. Do water stakeholders have a right to information that will affect them (e.g. plans for a dam on their river, or reliable data about water quality)? Even the principle of 'free, prior and informed consent', cited above as a cultural principle for safeguarding the cultural rights of Indigenous Peoples, can also be viewed as a governance principle. It could also apply to non-Indigenous communities whose homes, livelihoods and cultural traditions would be threatened from a dam or other form of water development.

The role of stakeholders in day-to-day management of water, whether irrigation canals, aquifers or urban water supplies, presents a complex range of ethical issues involving governance principles. Irrigation canal networks, for example, could, in theory, be managed by farmer organizations, or through local government bodies, or by centralized state agencies. Choosing one option over another depends not only on local physical conditions (smaller, simple canal networks will be more amenable to farmers' management); the social and institutional context (e.g. is there a strong local tradition of cooperation?) but also on the values which the stakeholders wish to promote. From the perspective of climate resilience, strong local community management systems have important benefits, but those same local institutions might make centralized national governments a bit nervous.

There is also the question of governance scale and boundaries. The river basin has become the core reference for 'integrated water resources management' (IWRM) but there are subjective values involved in setting governance boundaries. Should the river basin be defined in terms of surface hydrology only? Or should we conceptualize a virtual water 'shed' that encompasses everyone having an interest in the value chain of the water, for example, food produced in the physical water basin which feeds urban consumers in a different physical basin, perhaps in a different country?

The current draft of the Water Ethics Charter notes that 'Water systems are closely integrated socio-ecological systems . . .' Two general ethical principles of water governance are implied by this fact:

- The boundaries of governance should incorporate the whole water ecosystem and a broad ecological frame to incorporate whole watersheds, aquifers, and interactions between freshwater and marine ecosystems.
- Water governance should reflect the interests of all stakeholders, with particular emphasis on those groups who have the least political power.

Two additional general governance principles are:

- The principle of subsidiarity (management at the lowest practical level) should be favoured for specific water management functions (e.g. irrigation system management).
- Water governance should also include 'knowledge governance' and the guidance of investments in research and development.

At the level of operational principles of governance, it is helpful to distinguish between the internal processes of governance (how it operates) and the intended outcomes, that is, what the governance system tries to accomplish. On the process side of water governance, the concept of 'integrity' has emerged as a broadly accepted standard of good governance. The operational principle can be framed as follows:

- The principles of 'integrity', namely transparency, accountability and participation/engagement of stakeholder groups should be applied to water governance.[3] (This implies the need for well-organized civil society stakeholders that can hold both government and the private sector accountable for complying with the standards of both formal laws and informal ethics.)
- Informal water systems that may co-exist with public systems also need to be accommodated.
- Water conflict mechanisms need to be designed into the governance arrangements.

A further ethical responsibility of water governance, though not addressed in the Water Ethics Charter, is to avoid distractions and focus on issues of real significance. Setting water governance goals is a values-based task. Which of the many potential concerns about the river should be accorded the highest priority? The governance of an impaired river, for example, should directly take on the challenge of restoring ecological function to benefit both nature and poor river-dependent communities. Those who are in positions of governance responsibility bear a particular moral responsibility to steer towards ethical priorities.

Applying the Charter

There are three intended purposes of the global Water Ethics Charter, and perhaps additional purposes will also be discovered. First and perhaps most importantly, the Charter is intended to raise awareness and provide education about water ethics, for example, that water ethics provides a valuable guide towards truly sustainable, just, and in a word, ethical management of water and water ecosystems. Of course, no ethics charter will have impact unless people know about it, so awareness raising campaigns, conference presentations, articles and other forms of getting the word out will be needed to fulfil the Charter's potential role in awareness raising.[4]

A second intended purpose of the Charter is to be used as its name implies, as a 'charter' which stipulates the ethical principles which have been agreed upon. The individuals and organizations involved in drafting the Charter will be the direct authors of the Charter, but we are writing the document with the intention and hope that it will serve as a practical guide for setting policies and making decisions about the use of water, and the management of water ecosystems. Our original intent in developing a water ethics charter was to seek endorsements from cities, companies, countries and perhaps eventually from the United Nations General Assembly. But we soon adjusted our expectations as we confronted the paradox of crafting a meaningful charter that could also be broadly acceptable to vested interests. Rather than weakening the charter to make it palatable to the water status quo, we opted for developing a strong charter reflecting a diversity of perspectives. We are no longer seeking broad approvals or any specific endorsements, but we will vet each successive draft of the charter through a diverse set of expert reviewers.

The third, and in my view the most important, purpose of the global Water Ethics Charter is to serve as an inspiration and template for facilitating community-led local-level water ethics charters. Any organizational entity, whether a village, city, corporation, NGO, river basin group, etc. can create their own water ethics charter in their own way. The global charter provides a template that can be replicated, modified or ignored. Any group would be welcome to label their statement a 'water ethics charter' or to use a different term if they wish. The distinguishing feature would be written documentation of value principles about local water that can serve as an aid to making good decisions about water-related policies, practices and investments. There is no external quality control over the term 'water ethics charter'. The quality of each charter, and the monitoring of the charter's implementation, will be in the hands of the community of stakeholders who develop the charter in the first place.

The loose structure of the Water Ethics Charter development process reflects the loose organizational structure of the initiative itself. Several of the original sponsors of the charter initiative, including the French Water Academy, UNESCO-IHP and Club of Rome have withdrawn, and the initiative today is being implemented by a new coalition coordinated through the Water Ethics Network (waterethics.org), and its parent organization, Water-Culture Institute. New coalition partners include Green Cross International and the Ecumenical Water Network of the World Council of Churches. Continuing partners include Water Youth Network, Indigenous Environment Network, Botin Foundation and Alliance for Water Stewardship.

We anticipate that our coalition of partner organizations will continue to expand in both geographic and sectoral (e.g. agriculture and mining) representation. We also expect that the focus on developing the global water ethics charter will continue, and in addition, the coalition partners will also engage in local applications of the charter principles. As discussed in the chapter on ethics in decision-making (Groenfeldt, Chapter 5 in this volume), there is a natural synergy in working at both the local and global levels simultaneously.

Conclusions

Developing a global water ethics charter sounds like a daunting enterprise, but much of the work of generating global consensus on value principles related to water has already been done. Countless international conferences, workshops and meetings convened by UN agencies, special projects, research institutes, NGOs, water organizations (and particularly the World Water Council's World Water Forums and SIWI's Stockholm World Water Week) have forged a remarkably consistent set of value principles about responsible water management. Evidence of the emerging global norms for water management can be found in the literature of integrated water resources management (IWRM) (Aldaya et al., Chapter 8 in this volume), and in the 2010 UN General Assembly resolution declaring access to safe water and sanitation to constitute a fundamental human right.

The project to formulate water ethics principles into a systematic global water ethics charter is consistent with the evolutionary trend of seeking global consensus on core issues of water governance. Ethics, however, offers a distinct kind of influence, simultaneously motivating both internally (the ethical sensibilities of an individual or organization) and externally (peer pressure and reputational risk). An ethics approach complements the influences of soft law, established best practice and the multitude of conference declarations. The Water Ethics Charter packages water ethics into a systematic framework in an easily accessible written form.

The Water Ethics Charter can be seen as a way of highlighting already established norms by repackaging them into a coherent framework that makes it easier to identify interactions and potential synergies. In this sense, the Charter provides a logical structure to water decisions. But the Charter is not only about thinking clearly; the Charter can also empower civil society to take a more active role in water decisions, implying a responsibility to employ 'ethical imagination' in anticipating diverse impacts and opportunities lurking within water decisions. The Charter's multidisciplinary framework reminds us that water decisions are just as much about environmental, social and cultural values as they are about economics and technology. The ethics charter can become a reminder that water decisions are everybody's business.

Notes

1 http://sustainablewaterforum.org/baltimore.html, accessed 22 July 2016.
2 For a discussion of the water implications of DRIP, see Groenfeldt (2013, pp. 135–54).
3 The best resource on water integrity is the Water Integrity Network (www.WaterIntegrityNetwork.net, accessed 22 July 2016).
4 Education and awareness-raising is precisely the mission of the Water Ethics Network (http://waterethics.org, accessed 22 July 2016).

References

6th World Water Forum 2012, 'Post-forum highlights', World Water Council, Marseille, France, accessed 1 July 2016, www.worldwaterforum6.org/fileadmin/user_upload/pdf/publications_elem/Highlights_web_BD_en.pdf.

Acuto, M 2013, 'City leadership in global governance', *Global Governance: A Review of Multilateralism and International Organizations*, vol. 19, no. 3, pp. 481–98.

Gadwa Nicodemus, A 2013, 'Fuzzy vibrancy: creative placemaking as ascendant US cultural policy', *Cultural Trends*, vol. 22, nos 3–4, pp. 213–22.

Groenfeldt, D 2013, *Water Ethics: A Values Approach to Solving the Water Crisis*, Routledge, London.

Nichols, WJ 2014, *Blue Mind*, Little, Brown, New York.

Sanders, SR 2014, 'The dawning of an earth ethic', *Ethics & International Affairs*, vol. 28, no. 3, pp. 317–24.

UNESCO 2011, *Water Ethics and Water Resource Management*, Ethics and Climate Change in Asia and the Pacific (ECCAP) Project Working Group 14 Report, UNESCO, Bangkok.

Völker, S & Kistemann, T 2011, 'The impact of blue space on human health and well-being – salutogenetic health effects of inland surface waters: a review', *International Journal of Hygiene and Environmental Health*, vol. 214, no. 6, pp. 449–60.

Yang, Y, Bae, J, Kim, J & Suh, S, 2012, 'Replacing gasoline with corn ethanol results in significant environmental problem-shifting', *Environmental Science & Technology*, vol. 46, no. 7, pp. 3671–8.

16 The Berlin Water Charter

Water ethics from an activist's viewpoint

Dorothea Härlin

Without water, there would be no life on this planet. No animals, no plants – indeed over 70% of the human body consists of water. So, it stands to reason that we should all be concerned about this elixir of life. But that is just not the case. Here in the industrialized countries of the Global North, most people take it for granted that an unlimited amount of clean water will flow from the tap. We don't need to think any more about it than that. We know that elsewhere, in the Global South, water is a matter of life or death. But it seemingly has no bearing on our everyday life. The problem seems to be a long way away, even though lifestyles in industrialized countries increase the water shortage in other countries.[1] However, rather than further discussing these issues, which fall under the headings of 'water footprint', 'virtual water', environmental refugees or the growing problems associated with our heavy consumption of meat, this contribution aims to discuss, with reference to the specific example of Berlin's water supply, the extent to which a people's water charter can create a legal framework that is able to bring about a rethinking of the politics and corporate management of a water company – away from its previous profit-oriented water policy to a focus on the common good. This charter rests on a consensus across the whole of society that water is an inalienable commons and not a commodity; a consciousness based on an 'ordinary ethic', which is discussed below. Although local in many of its details, it is part of a worldwide movement of water activists and thus also expresses an alternative or new water ethic (Barlow 2013).

Position of the researcher

This contribution is based on the viewpoint of a water activist who is part of a successful movement against water privatization in Berlin. I see myself in the tradition of critical theory. In contrast to the critical rationalism of Karl Popper and others, which aims to depart from philosophical reflection and practical interests and move towards a search for objectivity based on the methodologies used in the natural sciences (Hillmann 1994, p. 678ff), the Frankfurt School of Jürgen Habermas, Max Horkheimer and others attempt to find the connection between theory and practice, thus leading to self-awareness and a continuous reflection on one's own subjectivity and social position. My contribution here is thus based on both my knowledge as a political scientist and my own practical experience as an activist in the local and global water movement (Hillmann 1994, p. 235ff).

As a high-school teacher, I never had the opportunity to address the scientific aspects of water ethics. I draw my knowledge primarily from over ten years spent fighting the privatization of the water in Berlin. This was a collective learning process that I have helped to shape as a member of the Berliner Wassertisch – the Berlin Water Table[2] – from the time it was established in 2006. This continues to be my local mainstay to this day.

My second mainstay, as a member of Attac, is rooted in the anti-globalization movement. It was at the World Social Forum (WSF) in Porto Alegre in 2002 that I first learned about the global movement for water as a commons at a workshop run by Maude Barlow, the world renowned water activist.[3] Since then, the topic of water has held me in its grip.

Since the first successful protests against water privatization in South America (including the water war in Bolivia, to mention just one), this worldwide movement has grown rapidly, and there are no signs of a slowdown. On the contrary, the global trend towards remunicipalization is constantly growing. Activists against water privatization have recorded some significant successes around the world. Over the last 15 years, at least 180 cities in 35 countries have returned their water utilities to the public sector. This is a rising trend (Kishimoto et al. 2014).

Against this background, my contribution seeks to articulate the ethic of the anti-water privatization actors worldwide. This is an ethic that insists that water must never be used for profit and must always serve the common good. It is enshrined locally in the Berlin Water Charter (see Box 16.1).

Some people would dismiss these thoughts as unscientific, but I regard myself as a bridge-builder between science and the movement. There is often a gaping void here, which occasionally leads to mistrust and lack of respect towards the other viewpoint. The arrogance of some academics is met by some activists' accusation of academics' general aloofness and ineffectiveness.

Do we need a new global water ethic?

The global water corporations hold world water forums every three years.[4] At these events, the global water movement has organized its own alternative forums in order to discuss ideas for what to do about water and how to counteract the neoliberal trend towards the increased commercialization of water.

The global grassroots movement for water as commons has for decades been engaged in the formulation of a comprehensive water ethic from below. For decades, this movement has struggled against the privatization of water and for the conception of water as a common good for all. These movements have often also insisted that nature is endowed with an independent right to be preserved. This can be summarized briefly by the following motto: 'We do not inherit the Earth from our ancestors; we borrow it from our children.'[5] Much of this ethic has most comprehensively been set out by Maude Barlow in her books and many lectures and talks given around the globe (see, for example, Barlow & Clarke 2002;

Barlow 2007, 2013). Most recently, in her speech to COP21, the climate summit in Paris 2015, Maude Barlow not only criticized the ever-intensifying overexploitation and contamination of water, but also the squandering of water, land and forests. She also presented a water ethic that must be implemented if we are to ensure that life continues on our planet:

> We can start with a new water ethic. Rather than seeing water as a resource for profit, we need to understand that it is the essential element in all living ecosystems. All policies and practices must be planned with the preservation of water at their core. Not only do we have to reject the market model for our water future, we must put ourselves at the service of undoing what we have done to the natural world and hope it is not too late. [. . .]
>
> Imagine a city where all who are there want to be there and are not dispossessed of their rural lands and livelihood. Imagine caring for our water as a fiercely managed public trust based on the principles of justice and sustainability. Imagine a world in which water becomes nature's gift to teach us how to live in peace with one another and dwell more lightly on this lovely planet.[6]

I thus argue that we do not need a new water ethic, but need to draw on and learn from the water ethic that people have for decades been formulating in their local struggles for water as a commons and right. This also means that I question a water ethic that is formulated abstractly from 'above'. The battle for water in Berlin, which I discuss in more detail below, is embedded in this ethic and movement that is active on every continent on the planet.

A key debate for this movement is that of the commons and how it should be managed in the interest of a common good that needs to be defined more precisely. But one thing is clear. There is a clear contradiction between common good and private ownership, between a good that belongs to everyone and one that lies in the hands of a few. Private ownership always implies the exclusion of others. What started in England in the eighteenth century with the enclosure of common land continues today with the privatization of water and other aspects of public services. This has resulted in the economization of areas for which the state used to be responsible, for which we pay taxes in order to guarantee the basic needs of all people living in this society, regardless of their financial situation. (The author understands that this welfare state-oriented approach has never existed in many countries of the Southern hemisphere and thus cannot be expected in the context of neoliberal policy.)

I therefore feel that it is impossible to hold a debate about water ethics that ignores the question of ownership. Private property and commons are irreconcilable and a contradiction in terms. Profit-orientation and an orientation around the commons are irreconcilable. By definition, the commons belong to all. Private property and profits generated from it always flow into the pockets of the few who, by virtue of their ownership, demand for themselves the power of control over an asset that belongs to all. Attempts by water corporations to address water ethics and to thus legitimize their water policy should therefore be treated with considerable suspicion; these are groups such as Suez and Veolia which together hold two-thirds of all privatized water worldwide (excluding China, although they are gaining a foothold there as well). They have started to use our own language; these corporations publicly acknowledge the human right to water, but then conveniently forget that they naturally set high prices for the provisioning and disposal of water and thus make profits. In contrast, an approach oriented towards the common good will rely on raising charges for a publicly owned water supply that are based on the actual costs of provision and disposal of the water, including reserves for necessary investment and careful maintenance. Transparency is essential for the management of a commons as this is the only way to allow participation by the owners, that is, the citizens. Civil law, however, expressly excludes the possibility of transparency.

Social aspects reduce the profit from the shareholders' viewpoint, whether it be the working conditions of employees, or people who are unable to pay their water bill and who risk having their water supply cut off. The well-being of future generations also reduces profit, for example, expensive improvement, encouragement of organic farming as a form of water pollution control, support for sustainable measures to improve the groundwater or the maintenance of wetland areas to prevent them from being drained and used as lucrative building land. The new term 'grandchild-friendly' (in German, *Enkeltauglichkeit*) does not appear to have reached the vocabulary of private companies.

The fight against water privatization: a case study from Berlin

The impetus for the water struggles in Berlin came from Bolivia in 2004: 'If the indigenous people of Bolivia are able to drive the US Bechtel group out of their country, then we must be able to throw Veolia out of Berlin,' said some members of Attac in Berlin in their workgroup 'Arguments of Globalization Critics'. When the Berliner Wassertisch was founded in 2006, the name was inspired by a visit to the '*mesas technicas del agua*' in Venezuela. These 'technical water tables' are bodies that manage the water on a democratic basis (McMillan et al. 2014). This extremely participatory model of a water assembly that takes place at all levels and is open to all interested parties, was our model when, in 2006, we sent out invitations to an event in the Circus Cabuwazi to publicize the scandalous conditions under which Berliner Wasserbetriebe (BWB) had been part-privatized.

The BWB was subjected to a PPP (Public–Private Partnership). Such partnerships are currently the most aggressive form under private law through which public assets are currently privatized.[7] The damage done by this privatization model can be illustrated using as an example the privatization of Berliner Wasserbetriebe in 1999, the largest municipal PPP project in Europe. Only the most important features that apply to all PPPs are listed here.

The contract was secret, with a term of 30 years and a guaranteed profit for the entire term secured against Berlin's budget, which meant that profits were independent of BWB's ongoing business. Thanks to its complicated holding structure, the senior management was 100% in the hands of the private owners, even though they only held 49.9% of the shares (24.9% each for RWE and Veolia). Investment was cut back sharply. The Federal State – the majority shareholder – refrained from influencing the ongoing business. One of the main motives for the sale was claimed to be Berlin's massive debt level. Using debt as the motive for selling off public assets is not limited to Berlin. There are the many insolvent municipalities in Germany, Greece and other countries that are subject to the EU's austerity policy.

Water, which until then had been a public good in the hands of the City-State of Berlin, became subject to the logic of maximized profits as a result of the PPP. This is basically the only reason why corporations buy into public assets. In this respect, it is not really the corporations that are to blame, as it is well known that the maximization of returns is their fundamental goal. The criticism is directed more at our political representatives who sell off our shared public property, often against the will of the population. They do not administer water as a common good on the behalf of the people.

For this reason, water movements are often also democracy movements. Water should not be administered secretly under private and profit-oriented conditions and without the consent of the population. On the contrary, water should be administered as a commons in the public sector and under public law with the participation of the population. The Italian water movement worded this as follows: '*Si scrive acqua, si legge democrazia!*' ('We write water and read democracy!'). The Berlin Water Charter states, 'We wish to take this further and democratize both the Berliner Wasserbetriebe and water policy as a whole, and so achieve transparent, socially just, and environmentally sustainable water management in Berlin.'[8]

Resistance arose against the sale of Berlin's water. The protest was organized by the Berliner Wassertisch. However, it took six years – until 2011 – before the disclosure of the previously secret contracts was achieved in the successful 'Our Water' referendum. The vote was overwhelming: 98.2% of more than 666,000 Berliners voted in favour of the disclosure act (Wiesenau 2015). It was in fact the first successful referendum in Berlin. The success of this direct democratic instrument has since inspired many other citizens' initiatives to dare to do the same.

For legal reasons, in 2011 'Our Water' referendum was not able to directly demand remunicipalization. And yet remunicipalization was still the declared aim of the Wassertisch and Berlin's population.[9] The fight for Berlin's water thus continued after the successful referendum. In the end, public pressure was so great that privatization had to be reversed. The Senate finally, and begrudgingly, bought back the shares from RWE in 2012, and from Veolia in 2013. Today the Federal State is once more the 100% owner of Berliner Wasserbetriebe. In the opinion of the Berliner Wassertisch, the buy-back price was far too high. Water customers will be paying off the loan for the buy-back of what was once their own water utility. They will do so every month with their water charges for another 30 years. This is one result of the long terms agreed in the PPP contracts; the corporations will be owed their promised profit right up to the end of the contract – or until 2028 in Berlin's case.

The change of ownership from private to public is just the start

Nevertheless, this first step towards remunicipalization is a success that was received very positively in many other countries as encouragement to do the same. The recognized international organizations Public Services International Research Unit (PSIRU), Transnational Institute (TNI) and Multinationals Observatory have been studying a growing trend towards remunicipalization for years. The introduction to the German edition published by the AöW (Allianz öffentliche Wasserwirtschaft – Public Water Management Alliance) reads as follows: 'In the last 15 years, 180 water remunicipalizations have taken place in 35 countries, in both the global North and South, with spectacular cases in Europe, America, Asia and Africa. These include capital cities such as Accra (Ghana), Berlin (Germany), Buenos Aires (Argentina), Budapest (Hungary), Kuala Lumpur (Malaysia), La Paz (Bolivia), Maputo (Mozambique) and Paris (France).'[10]

And yet we are far from having reached our goal. The declared aim of the Berlin Senate is now to rake in the profits (which to date they have had to share with RWE and Veolia) and to use them solely to aid the financial recovery of Berlin's budget. The City-State of Berlin has thus retained its profit-oriented management and continues the business logic started under the management of the private corporations RWE and Veolia.

The Berliner Wasserbetriebe will thus continue to be used as a source for the maximization of returns. The Berliner Wasserbetriebe is thus still far from being a public company oriented towards the common good. The situation is similar in many other cities as well. The change of ownership from private to public is nowhere near from having been moved away from market-oriented water management with an interest in high returns. The management of our water for the common good requires us to break away from this logic.

For this reason, the Berlin Wassertisch took the following steps after the shares were bought back from RWE and Veolia. First, it developed the Berlin Water Charter as a recommendation for a common good-oriented water policy in Berlin. It contains the significant features as sketched out generally above and specifically related to Berlin's situation. The Berliner Wassertisch regards this Charter as the starting point for a debate in Berlin on how water can be managed as a commons in the future. It hopes to create a legal framework for future legislation relating to Berlin's water policy. This demands a broad-based public debate in the city, however. The Berliner Wassertisch is using the Water Charter to start this discussion.

This is not an easy proposition in a city in which good drinking water flows out of the taps with very few problems, day and night. Most people in Berlin have more pressing problems. We should therefore not disguise the fact that the public debate we had planned has barely taken off. It would be entirely fitting for a metropolis like Berlin, which has more bridges than Venice with all of its numerous rivers and canals, to make its mark as a pioneer with a water concept for the entire region that focuses on the common good. Why in fact do we not imitate the Swiss capital Bern or even Zürich and join the Blue Community project initiated by the Council of Canadians? I regard this approach as an excellent opportunity to bring the topic of water to the population in lots of tiny steps over a long period. Schools, universities, parishes, individual districts, for example, could make a start before Berlin as a whole joins this project. For this we need visionaries in a densely woven city bedevilled by corruption and where petty politics is barely able to cope with the everyday, instead lurching from one election to the next.[11]

The Berliner Wassertisch has the vision and the will to implement this utopia of Berlin as a 'Blue Community'. This will take a lot of staying power. We know this. One small first step in this direction was the founding of the 'Berliner Wasserrat' (Berlin Water Assembly) in 2014 once the Berliner Wasserbetriebe was back in public ownership.[12] The intention was for interested citizens and organizations to participate in and contribute to the Water Assembly, and to thus move towards the goal of strong citizen participation in the Berliner Wasserbetriebe. This will require transparency and participation, and will have to be implemented in small steps. The water charter was discussed in the Berliner Wasserrat, which also deliberated on how to bring the debate to the wider public. Our motto in this regard is 'First remunicipalization – then democratization!'[13]

Box 16.1 Berlin Water Charter

Preface

The private shareholdings in the Berliner Wasserbetriebe (BWB) were bought back by the City-State of Berlin in late 2013. We wish to take this further and democratize both the Berliner Wasserbetriebe and water policy as a whole, and so achieve transparent, socially just and environmentally sustainable water management in Berlin.

This demands a complete return of the formerly partly privatized company to ownership of the City-State of Berlin. To this end, the Berliner Wassertisch has drawn up a draft Water Charter for Berlin. Our intention is to develop this draft further by means of a broad-based debate within society. We wish to bring together all the different areas of expertise on the subject of water in our city, and to invite Berlin's population to actively participate. We regard the Berlin Water Charter as the basis for statutory regulations and as a guide for Berliner Wasserbetriebe.

Preamble

1 This Charter is based on the UN Resolution on the human right to water and basic sanitation of 28.07.2010 and on the initial petition of the European Citizen's Initiative, Right2Water, which was successfully concluded in September 2013.

2 These fundamental rights are inalienable and may not be restricted by national or transnational treaties.

3 All the people of Berlin should be able to participate democratically in the implementation of a socially, economically and environmentally sustainable water policy. This requires transparency at every level.

(continued)

(continued)

4 A high-quality drinking water supply and waste water treatment system is inextricably linked to the protection of nature and our natural resources.
5. The business management of the Berliner Wasserbetriebe should be oriented towards the common good. This precludes the option of a profit-oriented approach. The population's water revenue must be used solely to ensure a sustainable drinking water supply and waste water treatment. 'Water pays for water.'

The following principles form the framework for this.

1 General and political principles

1.1 The Berliner Wasserbetriebe serve the public good. Access to clean water and basic sanitation must be permanently guaranteed to all Berliners as a human right.
1.2 Water must be affordable for all Berliners. They have a right to obtain high-quality water subject to socially appropriate charges.
1.3 The Berliner Wasserbetriebe shall permanently remain a public service that is entirely owned by the City-State of Berlin and managed according to municipal law. There must be no privatization or part-privatization of the drinking water supply and sewerage system, not even in the context of so-called public–private partnerships or similar models. The provisioning of clean, safe water, which is essential to life, and its disposal, rank among the highest priority and indivisible tasks of the City-State of Berlin.
1.4 Berlin's domestic water management must be sustainably developed in a process of coordination that involves all stakeholders, particularly with the participation of the citizens of Berlin. This requires a high level of transparency.

2 Social and economic principles

2.1 The Berliner Wasserbetriebe is not run for profit. It may levy charges that cover the costs of facilities and to make provision for investment in sustainable economic and technical development.
2.2 The pricing model of Berliner Wasserbetriebe takes the burden away from small consumers and places it on large consumers.
2.3 No companies that are not associated with water may be integrated into Berliner Wasserbetriebe.
2.4 Water from Berlin's groundwater and bank filtrate should be available in at least the same high quality to the current and all subsequent generations. To this end, the technical equipment associated with the drinking water supply and sewerage system must meet the latest scientific and technical standards and incorporate an alternative water management system.
2.5 Berliner Wasserbetriebe is open to cross-body cooperation in the context of publicly run domestic water management, with the common good as its guiding principle. A profit-oriented approach to inter-regional cooperation will be rejected on principle.
2.6 The working conditions and remuneration of Berliner Wasserbetriebe personnel must satisfy the requirements of freedom, justice, safety and

human dignity defined in the guiding principles of the International Labour Organization (ILO). The salaries for all employees shall be based on the principle of 'Equivalent pay for equivalent work.'

The existing right to codetermination of employees of Berliner Wasserbetriebe will be guaranteed and further extended.

2.7 The City-State of Berlin provides, to a reasonable extent, resources to allow for democratic participation and water-related research.

3 Environmental principles

3.1 Berliner Wasserbetriebe supplies Berlin with drinking water from its own groundwater resources and the bank filtrate obtained from the Spree and Havel rivers.

3.2 The work of Berliner Wasserbetriebe and the orientation of Berlin's policy are characterized by the desire to protect resources. Berlin's environmental balance sheet must not deteriorate, and should be constantly improved.

3.3 The State implements the environmental standards of the EU Water Framework Directive (WFD) of 2000 and enshrines these provisions in statutory standards.

3.4 Organic agriculture is to be encouraged in order to protect water resources and reduce contamination of the groundwater. Berlin is in favour of a reform of European and German agriculture and biomass cultivation policy with the aim of greatly reducing the pollution of our waterways through fertilizers and pesticides.

3.5 Berlin's surface waters are to be developed in greater harmony with nature as regards riverside landscaping and environmental consistency. The City-State should not pursue any development of bodies of water that impacts negatively on nature.

3.6 Water protection areas are to be maintained and cared for. They may not be re-designated as speculative building land.

3.7 To protect groundwater resources, water extraction shall be limited to a defined, environmentally compatible extent.

3.8 The City-State of Berlin shall develop an overall concept for groundwater management in cooperation with BWB and with the agreement of the population.

3.9 The groundwater extraction charge shall be defined as a ringfenced levy and used for the protection of the groundwater and Berlin's bodies of water.

3.10 The City-State of Berlin promotes the development of its green spaces in harmony with nature with the aim of ensuring soil conservation and thus preventing water pollution. The Berliner Wasserbetriebe is jointly responsible for the urban water supply; championing the retention of Berlin's green spaces of all types thus also falls within its remit.

3.11 Soil management is oriented towards protecting the ground and surface water and watercourses. Further soil sealing is to be avoided and reversed wherever possible. The quantity, quality and structure of unsealed soils are to be retained. Wherever possible, rainwater will be allowed to seep into the ground locally to reduce the load on the sewerage system.

(continued)

(continued)

3.12 Fracking and other methods of obtaining crude oil and natural gas in and around Berlin are to be ruled out for all time. The underground storage of separated CO_2 shall also remain banned in Berlin in the future. The City-State ought to advocate for a Germany-wide ban.

3.13 The City-State of Berlin is working with the other Federal States towards the early ending of pollutant discharges into the rivers and ensuring a stable water supply for all.

4 Legal principles

4.1 The Berlin Water Charter is the basis for interpretation of existing and new laws, statutory provisions and other regulations.

Why this success specifically for water?

Just four months after the referendum in Berlin, the Italian water movement achieved a much greater success: it won a national referendum against the privatization of water. This required the agreement of 50% + 1 of the votes of all entitled Italian voters. Against all expectations, this target was met. And the story continues. In 2013, at the pan-European level, the first European citizen's initiative was won under the title Right2Water. Almost 2 million people from nine European countries voted in favour of this referendum.[14]

Why have all the referenda against water privatization been won? The explanation appears simple: water as a basic condition for life must be accessible to all equally. Unlike many other resources, water appears per se as a public good and/or commons for people across the world and across cultures (Strang 2004).[15] When we were collecting signatures for the popular initiative in Berlin, many people said 'Water just cannot be privatized.' We often didn't even have time to explain to people that Berlin's water had already been privatized since they were willing to sign immediately without asking too many questions. The idea that our politicians had already sold off the water appeared too far-fetched. In my opinion, there is an unspoken ethic that lies behind these spontaneous reactions and that exists as a moral value system for many people, even if it is not explicitly reflected. People like to call this 'common sense'. I prefer the term 'ordinary ethic' which is deeply rooted in the consciousness of most people because it is quite simply a matter of life and death (see also Lambek 2010); an ethic based on a moral framework that is deeply enshrined in society. There is practically no other commodity for which it is so clear: water belongs to everyone, so it is a paradigmatic common good. This 'ordinary ethic' exists independently of gender, skin colour or of any religious or ideological differences between people.

And it is global. In all my encounters with water activists from every continent, it is the principle behind our shared struggle, the basis for our actions. It needs no further explanation. It is so self-evident and differs only in the form in which it achieves public expression. In Italy, for example, the idea of the commons (or '*bene commune*') is much more present in public debate than it is with us in Germany, where the emphasis is more on the arguments against privatization. In Ireland, in contrast, the introduction of water meters as the first step towards water privatization led to unexpectedly fierce resistance. While Ireland had previously manifested no resistance to austerity measures, a water movement emerged incredibly quickly with a level of radicalism that overwhelmed even the organizers of the first demonstration in October 2014.[16] Indigenous peoples in Latin America, on the other hand, are much more inclined to emphasize Nature's standalone right to be preserved, independently of people.[17] For example, the well-known Bolivian water campaigner Oscar Olivera criticized the UN Resolution for the human right to water and basic sanitation (28 July 2010) because the right relates solely to people and not to Nature. Nevertheless, regardless of these local differences, there is no doubt about this 'ordinary ethic'. The Earth is our mother and no one would sell their mother!

When we talk about water, we are talking not simply about our own subsistence and Nature's right to existent independently of humans. It also relates to how we use and the way we exploit water, which must be protected against misuse by its owners, that is, humanity as a whole. This demands transparency and participation: democratic structures that cannot be found in the private economy.

Conclusion

We have a long road ahead of us. We have to continue to fight, both locally and globally against the trend towards market fundamentalism, which started in the 1970s and threatens to turn everything into commodities. This chapter aimed to show, with reference to a small local approach in Berlin, how water can be turned back into a common good after privatization and subsequent remunicipalization. The Berlin Water Charter describes one approach for marking out a legal framework with the intention of inspiring a new political consciousness leading to a water management that is oriented towards the common good. It thus illustrated one local step towards the implementation of the global water movement's insistence on a form of water management that is oriented towards a more human water economy.

The scope was limited to the topic of drinking water, so other significant aspects could not be covered. The contradiction between commons and private ownership can be demonstrated in many other fields, however, such as in the bottling and sale of local spring water, the desertification of whole tracts of land, the expulsion of people from their ancestral settlements through the construction of mega-dams, the destruction of valuable eco-topes by such dams, the displacement of small farmers by extremely water-intensive agribusinesses, or the contamination of rivers by uncontrolled waste water from industry and mining, to name just a few aspects.

Ethics, understood as a generally valid moral basis for society, cannot be restricted to drinking water alone. This leaves unresolved the question as to whether the ordinary ethic relating to drinking water can be easily transferred to the issues listed above. Since many of these measures hold up the enticing promise of progress, prosperity and modernity, I would like to open the debate with this question.

Acknowledgements

I would like to thank Judith Townsley for translating this text from the German original and Andrea Mühlebach for consultation and support with earlier drafts of this chapter.

Notes

1 The UNO reports that 3.5 million people around the world die every year because they have inadequate access to drinking water. At present, there is no data to demonstrate that the global average water quality is improving. Quite the opposite. See www.unwater.org/topics/water-quality/en/, accessed 7 June 2016.
2 Further information, see www.berliner-wassertisch.net, accessed 7 June 2016.
3 The Canadian Maude Barlow is the world's leading activist in the water field. She was instrumental in having water enshrined in the UN Charter as a human right. She chairs the Council of Canadians and Food and Water Watch. In Germany, she became known for her book *Blue Gold*, published with Tony Clark. Her latest book, *Blue Future*, was published in 2013.
4 See www.worldwaterforum7.org/main/, accessed 7 June 2016, for more information about the water corporations' World Water Forums. Information about the alternative world water forums can be found at www.fame2012.org/en/, accessed 7 June 2016.
5 One good example of this water ethic can be found at the end of the Youth Water Manifesto for Water Protection, which was presented to the European River Parliament in July 2015. The manifesto ends with the important conclusion: 'We do not inherit water from our ancestors; we borrow water from future generations.' See http://en.bigjumpchallenge.net/news-details/youth-water-manifesto.html, accessed 7 June 2016.
6 The full speech can be found at www.foodandwaterwatch.org/insight/blue-and-just-future-possible, accessed 7 June 2016.
7 See Gemeingut in BürgerInnenhand (GIB) for more information about PPP: www.gemeingut.org/, accessed 7 June 2016.
8 The Berlin Water Charter is available from https://berliner-wassertisch.net/assets/start/Wasserchartav1.pdf, accessed 7 June 2016.
9 Information on the wording of the citizens' law can be found on the Wassertisch website at https://berliner-wassertisch.net/content/docs/docs.php, accessed 7 June 2016.
10 More information about this global trend can be found in a report of the Allianz Öffentlicher Wasserwirtschaft (AÖW) at www.tni.org/files/publication-downloads/heretostay-de_0.pdf, accessed 7 June 2016.
11 See http://canadians.org/bluecommunities, accessed 7 June 2016, for the Blue Communities project of the Council of Canadians.
12 See www.berliner-wasserrat.de/, accessed 7 June 2016.
13 It is not only in Berlin that people are attempting to create a legal framework for a future water policy. A countrywide water law is also currently under discussion in Mexico. A draft law is available for discussion at 17 events in various cities around Mexico. The central demands are: water should be a common; no profit with water and a ban on privatization; transparency and social control. See http://aguaparatodos.org.mx/, accessed 7 June 2016.

14 See www.right2water.eu/, accessed 7 June 2016. Further information about water policy at the EU can be found on the website of the European Water Movement: http://europeanwater.org/, accessed 7 June 2016.

15 As the anthropologist Veronica Strang explains, the material quality of water has allowed for it to acquire similar meanings across cultures. Water's fluidity and transmutability, for example, has allowed for this substance to stand metaphorically for flow, movement and change over time. But it also has other meanings that occur cross-culturally, such as its association with life, wealth, health and spiritual and ecological regeneration (Strang 2004).

16 See Right2Water Ireland: www.right2water.ie/, accessed 7 June 2016.

17 See the recently published online article by Oscar Olivera: https://roarmag.org/magazine/oscar-olivera-water-war-commons/, accessed 7 June 2016.

References

Barlow, M 2007, *Blue Covenant: The Global Water Crisis and the Coming Battle for the Right to Water*, The New Press, New York.

Barlow, M 2013, *Blue Future: Protecting Water for People and the Planet Forever*, House of Anansi Press, Toronto.

Barlow, M & Clarke, T 2002, *Blue Gold: The Fight to Stop the Corporate Theft of the World's Water*, The New Press, New York.

Hillmann, K 1994, *Wörterbuch der Soziologie*, vol. 4, Kröner, Stuttgart, 681 pp.

Kishimoto, S, Lobina, E & Petitjean, O 2014, *Here to Stay: Water Remunicipalization as a Global Trend*, Transnational, Amsterdam.

Lambek, M 2010, *Ordinary Ethics: Anthropology, Language, Action*, Fordham University Press, New York.

McMillan, R, Spronk, S & Caswell, C 2014, 'Popular participation, equity, and co-production of water and sanitation services in Caracas, Venezuela', *Water International*, vol. 39, no. 2, pp. 201–15.

Strang, V 2004, *The Meaning of Water*, Berg, Oxford.

Wiesenau, U 2015, 'Mehr Demokratie und bürgerschaftliches Engagement', in R Kreibich & F Lietsch (eds), *Zukunft Gewinnen!*, Altop Verlag, München, pp. 98–104.

17 Water ethics and water stewardship

Personal reflections

Adrian Sym

Water stewardship is primarily concerned with patterns of behaviour in the context of managing a scarce, and shared, resource. Its focus is on the users of that resource, and the need to reconcile diverse and often competing needs. By focusing on the conduct of water users, water stewardship automatically strays into the realm of ethics, understood here to refer to the 'moral principles that govern a person's behaviour or the conducting of an activity' (*Oxford Dictionaries* 2016).

The Alliance for Water Stewardship (AWS) has been at the forefront of efforts to engage different groups to advance this inherently collaborative approach to managing shared water resources. The most notable contribution we have made is the International Water Stewardship Standard (the AWS Standard) that defines, from the perspective of a major water using site, what can be understood as global best practice in collaborative and catchment-focused water use. Water ethics were not formally codified when we began the development of the AWS Standard (and associated system components[1]) in 2010. Instead, water ethics tacitly formed the basis of our work to build an international water stewardship system, and it is this system that provides the means through which ethical understanding can be translated into context-specific actions.

Water ethics can be understood either as a starting point for engaging different groups in water stewardship, or a prerequisite for achieving the aims that water stewardship strives to accomplish, or both. With the common understanding of right and wrong that water ethics provides, a foothold can be gained to build the collaboration that forms the basis of water stewardship. From this foothold, a process can begin to translate this common understanding into more tangible and context-specific actions of water stewardship. Without this grounding, any effort to pursue the behavioural change in pursuit of shared goals that is fundamental to water stewardship will be a non-starter.

Origins of water stewardship

Over the last decade, it has become increasingly apparent that water presents a major risk factor for society, the economy and the environment. In 2009, the McKinsey & Company and the 2030 Water Resources Group noted that 'without a step change improvement in water resource management, it will be very difficult to meet related resource challenges, such as providing sufficient food or sustainably generating energy for the world's population' (2030 Water Resources Group 2009). Exacerbated by global drivers of population growth, changing consumption patterns and climate change, our freshwater resources are increasingly under stress. As the major user and consumer of freshwater resources, the private sector is now widely acknowledged as a critical partner in efforts to achieve water sustainability.

In natural resource management, the traditional role of governments has been to legislate, regulate and enforce up to a minimum compliance standard. Some government-backed market-based approaches have been set up to provide incentives to go beyond minimum requirements, but the ability of governments to drive innovation and best practice is limited. The voluntary and market-based system we have developed aims to fill this gap between the capacity of regulation and the need for improved performance. We do this by providing a voluntary framework for engaging in collaborative action (the AWS Standard) and market recognition for achieving global best practice.

Water stewardship draws on the principles outlined by Nobel Prize-winning political economist Elinor Ostrom who demonstrated that 'common pool resources' can be successfully managed through systems of transparent rules that are developed and enforced by the users of the resource. Water stewardship understands that water is best managed by those who use it, and that by engaging multiple users at the appropriate level it is possible to reconcile diverse needs and achieve outcomes of more equitable water use, protection of the resource and economic growth. Rooted in transparency, water stewardship is also non-rivalrous in nature, for example, transparent processes enable information to be shared for the benefit of all – one user accessing this information does not undermine the ability of others to access it or the benefits that can be realized through accessing information – quite the contrary.

Water stewardship also recognizes the need to have a sound understanding of the condition of the resource and how this impacts different users and their priorities, including biodiversity, environmental flows and specific sites and values that are of important to different groups. All of these demonstrate the importance of the grounding of water stewardship in ethical principles. The growing sophistication and credibility of natural capital accounting and the understanding gained through assessing the resilience of social and ecological systems can be drawn on to support water stewardship approaches.

Water use touches on a wide range of competing priorities, presenting a complex picture with multiple challenges. Traditionally, addressing these challenges has been almost exclusively in the remit of the public sector. Yet the private sector, whether in primary production (mining or agriculture), manufacturing or commercial purposes, accounts for the majority of water use worldwide. The traditional roles and responsibilities of managing water has meant that the private sector tended to limit water management systems to direct operations. We now know that in many cases in water stressed catchments, this 'within the fence-line' approach is insufficient to meet a single company's needs, let alone contribute to broader catchment sustainability. The emergence of water stewardship has helped the private sector to understand their water risk exposure within a catchment context, and through supply chains. By engaging major water users, water stewardship also serves to support public policy objectives like integrated water resources management (IWRM).

The idea of a voluntary, multi-stakeholder, standards-based approach to water management was first articulated in Australia during the country's 'millennium drought'. The Water Stewardship Initiative, subsequently renamed Water Stewardship Australia (WSA), then began the development of an Australian Water Stewardship Standard. The Nature Conservancy (TNC) had been exploring the concept of 'blue water certification' and the Pacific Institute had been combining expertise on water and research standard systems. Connected by this work and a recognition of the need for collective and catchment-focused responses, and utilizing the positive lessons from sustainability standards in areas like forestry and fisheries, these three organization joined forces in 2008 to form the international Alliance for Water Stewardship (AWS).

This group of founders recognized the need to include a diversity of perspectives and expertise within AWS in order to oversee the development of an international water stewardship standard. Over the next two years, different organizations were recruited as AWS 'board organizations'. In total, 11 organizations served on the AWS board between 2009 and 2015,[2] representing leading players in water management, conservation and corporate responsibility. This ensured a balanced governance of AWS during the standard development process and that key perspectives from business, investment, conservation, social equity and public sector were represented without risk of 'capture' by any individual organization.

AWS board organizations were not the only important component in our early days. We established regional initiatives in Australasia, Europe, North America and Latin America and the Caribbean which served to simultaneously engage regional stakeholders in the AWS processes and also to expand the range and diversity of our partners. In Australasia, regional work was led by WSA, in Europe by the European Water Partnership, while TNC supported regional initiatives in North and Latin America, through their partners The Water Council (an initiative of the city of Milwaukee) and FEMSA Foundation. In other regions AWS engaged with different business and/or environmental organizations directly to ensure input into the standard development process, including the Centre for Responsible Business in India and various country offices of WWF.

While we were successful in engaging instrumental players around the idea of a global water stewardship system, it was still not clear that a single global water stewardship standard could deliver real impact at a local level, especially in developing country contexts where infrastructure, reliable data and regulatory environments were all weak. To help us understand this potential we implemented a project to pilot test the draft Australian and European standards with horticultural exporters around Lake Naivasha, Kenya. In this project, we partnered with the German development organization GIZ[3] and British retailer Marks & Spencer which sources many products from the region.

Lake Naivasha was chosen for its environmental, social and economic significance. The lake is an important wetland ecosystem and home to more than 400 bird species, but also provides irrigation for agriculture which generates over 10% of Kenya's export revenue, contributes more than 2% of the national GDP and provides employment for 75,000 people. The project found that applying water stewardship standards delivered real benefit for producers, local authorities and for international buyers (AWS 2011, pp. 109–10). This result was instrumental in giving us the confidence we needed to push ahead with the development of the International Water Stewardship Standard.

Building an international water stewardship standard

The AWS Board decided early on that the process to develop the AWS Standard – the Global Water Roundtable that was formally launched in June 2010 – should be conducted according to the standard setting code of the ISEAL Alliance.[4] This code supports social and environmental standards in being relevant, transparent and inclusive of a balance of stakeholder interests. The AWS Board thus appointed a 15-person multi-stakeholder committee – the International Standard Development Committee (ISDC) – to develop the Standard on its behalf (see Table 17.1). ISDC members were volunteers, appointed following a transparent selection process and chosen to represent the diversity of stakeholder interests from around the world. Representation on the ISDC was based on three stakeholder groups: public sector agencies; business and water service providers; and civil society organizations. ISDC members were also appointed on a regional basis, to ensure balanced sectoral and regional perspectives in decision-making. The ISDC worked on achieving consensus to resolve important issues, with our role within AWS being to ensure they had sufficient high-quality information to guide decision-making. The standard development process was iterative, with feedback coming not just through consultation processes but also through practical experiences in diverse contexts.

The ISDC's mandate was to produce a single global water stewardship standard that would be implementable by major water users, either private sector companies or publicly held water service providers. Among the many requirements that the ISDC were set were that the standard should address impacts at a watershed/catchment level; should support rather than diminish effective water policy and regulation; and should not disadvantage small- and medium-sized enterprises. Overall, the Standard would meet the social and environmental objectives of minimizing the negative impacts of water use in relation to: human health; social and cultural well-being; ecosystems and species; and economic use.

The ISDC met a total of seven times between 2011 and 2014 in Sri Lanka, USA, Australia, Mexico, UAE, Belgium and France. In each case, local perspectives were brought into the process. Connecting with local stakeholders is a cornerstone of the AWS philosophy as it brings an appreciation of local

Table 17.1 List of original members of the International Standard Development Committee

ISDC member	*Sectoral affiliation*	*Regional affiliation*
Imane Abdel Al, Shahid Ahmad	Civil society organization	Western and Central Asia
Maureen Ballestero	Public sector	Latin America & the Caribbean
Sanjib Bezbaroa	Business & water service provider	South Asia
Peter Cookey	Civil society organization	Africa
Axel Dourojeanni	Civil society organization	Latin America & the Caribbean
Carlo Galli	Business & water service provider	
Ma Jun (withdrew participation during process)	Civil society organization	North Asia
Chaudry Riaz Khan (withdrew participation during process)	Public sector agency	South Asia
John Langford	Public sector agency	Asia Pacific
Marco Mensink	Business & water service provider	Europe
Gerphas Opondo	Public sector agency	Africa
Jiseon Matilda Park	Public sector agency	Asia Pacific
Ed Pinero	Business & water service provider	North America
Peter Ruffier	Public sector agency	North America
Lesha Witmer	Civil society organization	Europe
Hao Xin	Civil society organization	North Asia

Source: own.

application into the global processes. At each different location, members of the ISDC were able learn about different hydrological, legislative, political or other factors that influenced the group's thinking on how an internationally consistent standard could be meaningfully applied at a local level. Between meetings, members of the ISDC participated in teleconferences and reviewed documents and progress reports that informed their deliberations.

AWS and its regional initiatives provided information for the ISDC to consider, gained through stakeholder meetings, consultations and testing draft versions of the Standard. The first draft was launched at the 2012 World Water Forum in Marseilles and was then opened for public comment for a three-month period. This draft was built around four principles of water stewardship:

- Principle 1 – Water Governance: water stewards shall strive to achieve equitable and transparent water governance for all water users within the defined area of influence.

- Principle 2 – Water Balance: water stewards shall strive to achieve and maintain a sustainable water balance, and help to ensure adequate availability for all users at all times within the defined area of influence.
- Principle 3 – Water Quality: water stewards shall contribute to the maintenance of good water quality status in terms of chemical, physical and biological characteristics to maintain ecosystems and ensure adequate water quality for all users within the defined area of influence.
- Principle 4 – Important Water Areas: water stewards shall identify Important Water Areas at their sites and within their defined area of influence and shall strive to protect, manage and restore such areas as necessary.

These four principles represented the main concerns that stakeholders had voiced consistently throughout the process. Wherever stakeholder engagements had taken place we heard the same concerns about having enough water now and in the future (water balance); about the quality of the water (water quality); about whether the specific sites and values were being adequately considered or protected (important water areas); and lastly, but very importantly, about how decisions were being made about water (water governance).

Following the public comment period, the ISDC meeting in Mexico City in June 2012 considered the input received. In response to common feedback that the structure of the first draft was not 'implementable', a decision was taken to structure the next version around steps of implementation rather than the principles of water stewardship. The principles were maintained, but rather than providing the framework for implementation, they were embedded throughout the document thereby providing a clearer implementation framework while simultaneously strengthening understanding of the interconnection of the four principles. The resulting structure was six clear steps that were designed to be familiar to staff likely to be responsible for implementing the Standard at a site level. On this basis, work began on developing the Beta version of the Standard. which was launched in March 2013 alongside the UN Global Compact's CEO Water Mandate meetings in Mumbai, India.

With a revised structure and several additional improvements over the first draft, we set about organizing pilot tests of the Beta version in several sectors and locations. Working closely with regional initiatives and other partners, pilot tests were conducted in industrial and agricultural sectors in China, Australia, India, South Africa, Peru, USA and Canada. As with the previous case in Kenya, the Beta Standard proved its value in all of these locations highlighting the importance of contextual understanding, robust data and constructive engagement to understand shared water challenges. We also got a better understanding of what additional material would be required to make an international standard achieve its potential at a site level, for example, appropriate local guidance and capacity-building of key stakeholders. And of course, we understood more about the improvements that needed to be made in the Beta version. With input from such diverse settings as chemicals in China, mining in South Africa and village-based implementation in rural India (Deloitte 2013), and with the input received from another public comment period and various meetings around the world, the ISDC had its work cut out when it met for the last time in Paris in March 2014 to develop the AWS Standard v1.0.

In April 2014, the AWS Standard v1.0 was formally launched in Lima, Peru, again utilizing the meetings of the CEO Water Mandate. This marked a major milestone in the evolution of water stewardship and for AWS as an organization. For members of the ISDC the launch marked the formal end of their work that had begun three years earlier in Sri Lanka. At that first meeting of 15 individuals from different backgrounds and countries, it had been a challenge to simply imagine a pathway to the objective for a globally consistent water stewardship standard, applicable in all regions and sectors. Yet this objective enabled this diverse group to quickly become an efficient problem-solving unit. In many ways, the ability of the ISDC to work collaboratively to reach consensus on key issues is reflective of the processes that the AWS Standard is designed to support: working collaboratively to address shared water challenges in pursuit of a shared goal.

Some of the critical issues that the ISDC dealt with included ways to determine the physical scope of the Standard, for example, how far upstream and downstream of the implementing site was required to conduct analysis and interventions; level of performance and assessment criteria for actions beyond the boundaries of the implementing site, that is, defining a realistic pathway to ensure that implementing sites made meaningful contributions to the health of the catchment in which they are situated; how the Standard should deal with interactions between surface water and groundwater, for example in contexts where surface water is the primary source of the implementing site; and ensuring that the Standard is connected to and supportive of broader catchment objectives, for example, whether to adopt a top-down approach that starts from a set of catchment goals and works back from that to define the contribution of an implementing site, or whether to adopt a bottom-up approach that requires certain actions by the implementing site and then identify additional actions within a defined sphere of influence.

In working through these issues, the ISDC needed to remain cognizant that the Standard would be applied in very different regulatory contexts. For instance, in highly regulated catchments it could not be assumed that implementing sites would have a high level of discretion for the kinds of actions they could take, for example, in shaping catchment governance mechanisms. However, these contexts provide significant scope for political engagement around goals and targets for the catchment. In unregulated catchments, there is more scope for implementing sites to contribute to developing a catchment management regime and that water stewardship principles are embedded within this regime. But such processes take time and defining an acceptable minimum contribution to catchment governance within the Standard, given the diverse ranges in which it would be applied, was a challenge.

The result of the ISDC's work (the AWS Standard v1.0) provides a consistent global framework for sites to understand their own water use and impacts, and to work collaboratively towards shared catchment goals. The Standard could be described as defining processes and outcomes, while leaving sufficient space to define locally relevant actions to achieve those globally consistent outcomes. Processes include best practice in terms of transparency, contextual understanding and stakeholder engagement. The four outcomes to be achieved are: (1) good water governance, (2) sustainable water balance, (3) good water quality status and (4) healthy status of Important Water-Related Areas. The Standard also includes performance levels (AWS Gold and AWS Platinum) to encourage innovation and continual improvement.

Another key decision taken early on by the ISDC was that the Standard should be accompanied by comprehensive guidance material. By making this whole package 'open source' we are allowing a diverse range of stakeholders to engage with the Standard and the guidance, thus building a greater level of ownership and uptake (even if this uptake is not easily traceable) than would be achievable if the Standard or the guidance material or both had restricted access. Additionally, much of the guidance material points users towards other water stewardship-related initiatives or tools to support the implementation of water stewardship. Examples include water accounting mechanisms like water footprint assessments, risk assessment tools like the WWF Water Risk Filter, or water reporting initiatives like CDP or GRI.

As well as providing links to useful water-specific tools to support implementation of the Standard, the ISDC were charged with ensuring that the AWS Standard was mutually reinforcing in respect to other standards, best management practices and management system standards. Other ISEAL member standard systems, for example, commodity standards, provided a particular focus for the ISDC due to their status of having met the same credibility criteria that AWS was striving to meet.

The AWS International Water Stewardship Standard System

In the two years since its launch, the AWS Standard has made a hugely important contribution to the advancement of credible water stewardship, in particular supporting businesses all over the world. While the 'open source' approach we have taken – making the Standard and the comprehensive guidance material freely available to download – makes it difficult to track precisely who is using the Standard for what purposes, we know from independent research that many companies are using the Standard to help them understand the characteristics of their water use, and the associated risks and impacts. In addition, the Standard has helped public sector agencies to join the dots between policy and actions of major water users. But perhaps more importantly, due to its stakeholder-endorsement as a result of the robust, ISEAL-compliant development process, the AWS Standard provides a 'safe place' in which competing perspectives on water use can be discussed and consensus sought on suitable responses.

The Standard does not provide the blueprint for solving all water-related problems, but as a neutral framework that drives transparency, builds trust and understanding of shared water challenges, and motivates, recognizes and rewards best practice, the Standard is both unique and invaluable. The key to fulfilling this role is the integrity of the processes during the Standard development, as described above. The inclusiveness and neutrality of those processes provides a safeguard against actual or perceived 'capture' of the agenda. In situations where water use is highly contested and one or more stakeholders have disproportionate power, this is perhaps the Standard's most valuable asset.

Of course, certification to the Standard is a key part of its value to implementing organizations. After the launch of the Standard in 2014 we turned our attention to the necessary components of the AWS water stewardship system: a verification system through which credible claims of conformity to the Standard could be made; an accreditation program for specialist service providers to support and assess the implementation of the Standard; a capacity development program to advance understanding and implementation of water stewardship; and a multi-stakeholder governance system implemented through membership of AWS.

It is the latter element – multi-stakeholder governance – that is at the heart of the AWS approach. Balanced stakeholder participation is critical at all levels. Our members are organizations committed to responsible use of freshwater. At this stage, we have organized our membership into three sectors: private sector, public sector and civil society. Implementing the Standard at a site level requires identification of stakeholders, understanding their perspectives and developing a water stewardship plan that responds to shared water challenges. At a national or regional level, together with our partners, we strive to ensure mechanisms are in place to develop a sense of ownership of AWS among diverse groups, and to act as a conduit to ensure experiences gained from implementing or promoting water stewardship in the region inform processes at an international level. These 'local-to-international' learning pathways are critical to the success of the AWS system. Multi-stakeholder governance at an international level is tasked with making sense of these diverse perspectives to develop the system in a way that meets the expectations of our members and partners, and of other participants in the AWS system.

Examples from early implementation of the AWS Standard

While the precise benefits of implementation of a site-based standard will always be site-specific, lessons from early implementation of the Standard and of certification processes have helped us understand the value in different business and regional contexts. Due to its stakeholder-inclusive nature, AWS water stewardship delivers both private benefits and public benefits. In China, where the first AWS certificate was issued in 2015, the implementing site – a zero-discharge chemical facility of US-based corporate Ecolab – realized significant efficiency gains in water use, and as a result was able to quantify substantial financial savings. The public benefits included reducing the burden on the public supply, collection of data on the catchment condition and, crucially, the beginnings of a collective impetus to tackle the challenges in a catchment that supports the livelihoods of 140 million people.

In Australia, where an Inghams Enterprises poultry site became the first to be AWS Gold Certified, the site was also able to achieve efficiency savings and to strengthen the company's reputation with local communities. At the same time, by engaging commercial water users in the catchment goals, a more cost-effective way of implementing public policy was achieved. As with the example from China, diverse stakeholders were connected through the common understanding and shared goals that the Standard enables.

Another example of the Standard's potential to simultaneously deliver private benefits and support public water policy objectives can be seen from work in the Western Cape Province of South Africa. When UK retailer Marks & Spencer and its South African counterpart Woolworths asked conservation NGO WWF to help them examine their water risk exposure, the focus fell on suppliers of stone fruit (peaches, nectarines, etc.) in the upper Breede–Overberg catchment, with surprising results. By using the AWS Standard as a neutral framework to engage different water users, it was established that the catchment's main water quality threat was not fertilizer use as may have been assumed, but inadequate sanitation in the formal and informal settlements around neighbouring towns in which seasonal farm labourers and their families live. Having taken a transparent and inclusive approach, the potential exists to build the trust needed for a collective effort to effectively address inequality in provision of sanitation.

Likewise, in Tanzania the implementation of the Standard by Olam at its Aviv coffee plantation in the Ruvuma basin identified several ways in which direct benefits for the site could be realized, including increased productivity, ability to strategically target future investments in what is a relatively new facility, and reducing the likelihood of regulatory action in the future. More broadly, water stewardship in this context has the potential to support the government's IWRM (integrated water resources management) implementation and strengthen local catchment governance.

The intersections between water stewardship and water ethics

As the examples above make clear, implementation of the AWS Standard should not be viewed as a one-off event, but rather as a process of engagement and continual improvement around a robust framework. Implementing the AWS Standard is about making progress towards shared catchment goals, that deliver both private and public benefits. It demands transparency and supports disclosure, thereby addressing asymmetries of information that are a key factor in mistrust and inequalities.

One critical aspect that should not be overlooked is that by driving stakeholder-inclusive processes the tendency to look for single solutions to complex problems is diminished. Too often, well-intended 'solutions' fail to take account of the complexity of water resource management and lead to disappointing outcomes – one man's solution is another man's problem. The quest for a single solution inevitably undermines the importance of understanding the context in which this will be applied, and ignores the complexity of that context. Water stewardship embraces complexity, and understands that we cannot simply engineer our way to a sustainable future. It will require ethical conduct, self-organization and social learning.

The AWS Standard follows this technology-agnostic characteristic of water stewardship and provides a space for innovation, within a clear set of parameters on what is credible and responsible. Just as the dimensions of a football pitch can provide the theatre for incredible innovation, decision-making and athletic skills, the AWS Standard can provide the framework for water users to embrace innovation in pursuit of globally consistent outcomes. The pursuit of agreed outcomes can utilize all kinds of technology – whether that is 'traditional' hardware for water saving, re-use, purification or irrigation; recent innovations in areas like mapping, remote sensing or forecasting that can be applied in water-related contexts; or softer technologies around engagement and partnerships.

With major companies now aligning their water strategies to the AWS Standard, the challenge now is how we bring others on the journey. AWS water stewardship is not simply something for well-resourced (and often already high-performing) corporations, but needs to have participation that reflects the inclusivity that is fundamental to the approach. Smaller companies need to be able to join forces with the group approach we have developed. Governments, utilities and other public sector bodies need to be fully included in water stewardship processes. Broader participation from civil society will help to ensure that water stewardship is not perceived solely in environmental terms, and that its contributions to social equity, public health, sustainable consumption and production, green growth and economic prosperity are recognized.

Given this evolution and the opportunities and challenges we are facing, how can articulation of a set of water ethics support the goals of water stewardship? Returning to our understanding of ethics as moral principles provides us with a shared understanding of the fundamental rights and wrongs. A high-level statement of water ethics would speak to many different constituencies and have potential to engage diverse groups around a common theoretical and philosophical approach to water. This would support more context- and location-specific efforts to advance uptake of water stewardship. Parallels could be drawn to legal systems, which can be ethical in nature but the legal code would draw on precedents gained from interpretation of the ethical basis of the system. Similarly, articulation of a water ethics charter would provide a useful overarching framework that would complement the practical and user-focused approach of AWS water stewardship. A theoretical flow can be envisioned whereby a water ethics charter would outline the generally accepted dos and don'ts of how we interact with water; the AWS Standard would continue to define global best practice in terms of outcomes of water use and the principles of engagement, transparency and contextual understanding; and the individual user's water stewardship action plans would determine the specific approach that was adopted to achieve the outcomes defined by the AWS Standard, taking into account the local context in which they are applied. Seen in this light, water ethics and AWS water stewardship are not just complementary but also mutually supportive.

Despite several years of work, at AWS we have only just begun our journey along this long and winding road. Some major milestones are within sight, most notably strengthening multi-stakeholder governance of AWS and the first review and revision of the AWS Standard. Both of these are sure to lead us into uncharted territory. Just as we encourage major water users to look beyond their fence line, to engage with others and to understand the benefits of not being fully in control of the agenda, so must we be prepared to explore new ways of promoting and applying water stewardship. Some of the areas that could be explored include the potential for catchment-level certification; enabling municipalities to participate in and benefit from water stewardship; and building on our experiences with community-based water stewardship in India (FES 2015) to inform similar approaches elsewhere. As the AWS Standard system develops it will need to become more localized in terms of the networks and information systems to support implementation of the globally consistent AWS Standard at a local level. Above all, we must be open to learning and improving – a critical component of what is fundamentally an educational journey for all of us involved in water stewardship. The moral guidance that water ethics provides can help us to navigate our way along this long and winding road.

Notes

1 The 'AWS International Water Stewardship System' includes the AWS Standard, capacity-building, verification requirements and processes, and accreditation program for specialist service providers, and multi-stakeholder governance, implemented through membership and regional affiliates.

2 AWS Board Organizations were: The Nature Conservancy, Water Stewardship Australia, the Pacific Institute, WWF, Water Witness International, the Water Environment Federation, the European Water Partnership, the International Water Management Institute, UNEP, the UN Global Compact's CEO Water Mandate, and the Carbon Disclosure Project.

3 Deutsche Gesellschaft für Internationale Zusammenarbeit (GIZ) GmbH on behalf of the German Federal Ministry for Economic Cooperation and Development (BMZ).

4 www.isealalliance.org, accessed 22 July 2016.

References

2030 Water Resources Group 2009, *Charting Our Water Future: Economic Frameworks to Inform Decision-making*, 2030 Water Resources Group, Washington, DC, accessed 6 June 2016, www.mckinsey.com/business-functions/sustainability-and-resource-productivity/our-insights/charting-our-water-future.

AWS 2011, *Kenya Case Study Technical Report: Exploring the Value of Water Stewardship Standards in Africa*, Alliance for Water Stewardship, accessed 6 June 2016, www.allianceforwaterstewardship.org/assets/documents/AWS-Kenya-Case-Study-Technical-Report.pdf.

Deloitte 2013, *Learning Report: Implementation of Alliance for Water Stewardship (AWS) Standard (Beta Version)*, Deloitte Touche Tohmatsu India Private Limited, accessed 6 June 2016, www.allianceforwaterstewardship.org/assets/documents/Learning%20 report%20on%20AWS_DSC.pdf.

FES 2015, *Community as Local Stewards*, Foundation for Ecological Security, Anand, Gujarat, accessed 6 June 2016, www.allianceforwaterstewardship.org/assets/documents/ Community_as_Local_Stewards_FES_HUF_AWS.pdf.

Oxford Dictionaries 2016, accessed 28 May 2016, www.oxforddictionaries.com/definition/ english/ethics.

Index

For Product Safety Concerns and Information please contact our EU representative GPSR@taylorandfrancis.com
Taylor & Francis Verlag GmbH, Kaufingerstraße 24, 80331 München, Germany

www.ingramcontent.com/pod-product-compliance
Ingram Content Group UK Ltd.
Pitfield, Milton Keynes, MK11 3LW, UK
UKHW022012180926
475701UK00003B/318
* 9 7 8 0 3 6 7 2 2 6 8 1 7 *